AF412739

High Speed Optical Communications

TELECOMMUNICATION TECHNOLOGY AND APPLICATIONS SERIES

Series editor:
Stuart Sharrock, Consultant
The Barn, Sugworth Lane, Radley, Abingdon
Oxon OX14 2HX, UK

This series covers research into and the development and application of a wide range of techniques and methods used in telecommunications. The industry is undergoing fundamental change under the combined impact of new technologies, deregulation and liberalization, and the shift towards a service oriented philosop0hy. The field of communications continue to converge, encompassing al of the associated technologies of computing, networking, software, broadcasting and consumer electronics. The series presents this material in a practical and application-based manner which equips the reader with the knowledge and tools essential for an engineer working in the industry.

Titles available

1. **Coherent Lightwave Communication Technology**
   Edited by S. Shimada

2. **Network Management Concept and Tools**
   Edited by ARPEGE Group

3. **The Informatics Handbook**
   A guide to Multimedia communications and broadcasting

4. **Mobile Communications Safety**
   Edited by N. Kuster, Q. Balzano and J.C. Lin

5. **The ISDN Subscriber Loop**
   N. Burd

6. **High Speed Optical Communications**
   R. Sabella and P. Lugli

7. **Handbook of Data Communication and Networks**
   W. Buchanan

8. **Data Mining Techniques in Speech Synthesis**
   R. Damper

9. **Intellegence and Power in Global Information Systems**
   W. Whyte

10. **Broadband Access Networks**
    L.A. Ims

High Speed Optical Communications

by

R. Sabella

Ericsson Telecommunicazioni, R&D,
Italy

and

P. Lugli

Dipartimento di Elettronica
Universita di Roma
Rome, Italy

KLUWER ACADEMIC PUBLISHERS
DORDRECHT/BOSTON/LONDON

A C.I.P Catalogue record for this book is available from the Library of Congress

ISBN 0 412 80220 1

Published by Kluwer Academic Publishers,
P.O. Box 17, 3300 AA Dordrecht, The Netherlands.

Sold and distributed in North, Central and South America
by Kluwer Academic Publishers,
101 Philip Drive, Norwell, MA 02061, U.S.A.

In all other countries, sold and distributed
by Kluwer Academic Publishers Group,
P.O. Box 322, 3300 AH Dordrecht, The Netherlands

Printed and bound in Great Britain by Antony Rowe Limited

Dedication
(Roberto Sabella)

To my mother

La idea, over imaginativa, è e timone e briglia de' sensi, in però che la cosa immaginata move il senso.

Leonardo da Vinci

Translation from the ancient Italian:
The idea, that is the imagination, is the helm and the rein of the senses, since the imagined thing moves on the sense.

Contents

Acknowledgements

We wish to acknowledge many colleagues who have provided us with so many stimulating suggestions.

We would also like to express our sincere gratitude to Monica Avattaneo for her help in preparing many figures in the book and Laura Malacuti for having prepared the editorial format of the book.

Finally, we would thank all the researchers and authors of book and journals who have provided us with many figures and results. In particular, we thank Prof. K. Kikuchi, Dr. P. Spano, Dr. A. Mecozzi, Dr. O. Sahlén, Dr. M. Gustavsson, Dr. Granestrand, Dr. G.P. Agrawal, Dr. A. d'Alessandro, Dr. D.A. Smith, Dr. H. Hermann, Dr. W. Sohler, Dr. F. Wehrmann, Dr. S.J.B. Yoo, Dr. R. Ludwig, Dr. E. Iannone and Dr. A. Willner.

Preface

The use of optical technology in transmissions dates back about twenty years. A relevant figure which can express the progress of optical fiber transmissions is the capacity. During this period the transmission capacity has been increasing of about five order of magnitude: from systems working at a bit rate of a few Mb/s to system operating to the order of hundreds Gb/s. If on one hand this was made possible by the progress of high speed devices technology, either for electronic or opto/electronic components, on the other hand the maturity of the technology based on wavelength division multiplexing (WDM) has pushed significantly towards very high capacity systems.

Beyond the application of WDM technology for point-to-point links, there is an increasing interest for transparent optical networking and all-optical switch fabrics. As a matter of fact, among the different approaches toward the realisation of high-capacity, protocol transparent optical networks, WDM offers the most promise in the near term, since it not only enables significant capacity enhancements, but provide the means for new networks in which the routing path is wavelength dependent. Indeed, WDM networks offer potential advantages, including higher aggregate bandwidth per fiber, new flexibility for automated network management and control, noise immunity, transparency to different data formats and protocols, low bit-error rates, and better network configurability and survivability: all leading to more cost effective networks.

From a technological perspective, several technological breakthroughs allowed the evolution previously sketched to be made possible, with costs low enough to justify the investments for the upgrading of high speed telecommunications systems. First, the availability of high speed single mode laser sources (like the distributed feedback lasers, DFB) which allowed to reduce the limits imposed by fiber chromatic dispersion so as to significantly increase the bit rate-span product (B^2L). Second, but of a great importance, the advent of optical amplifier (precisely the Erbium doped fiber amplifiers), which provided the means for eliminating the regenerators in optical transmission links. Third, the advent of WDM technology, by means of optical filters, tuneable lasers and receivers, and WDM mux/ demux based on integrated optics, which allowed the capacity to increase up to the values reached nowadays. Finally, the advent of optical switching devices and of wavelength converters, is allows the possibility of realising optical networks

1
Semiconductor Lasers

In this chapter a brief introduction is presented of semiconductor lasers, from the principles of their operation to applications in communication systems. No thorough discussion of laser issues is intended here. Interested readers should refer to the references [1–4].

1.1 PRINCIPLE OF SEMICONDUCTOR LASERS

In an optical communication system, information is transmitted by light propagation inside an optical fiber, mainly in the form of a coded sequence of optical pulses. In fact, early stage development of semiconductor lasers and light-emitting diodes (LEDs) as light sources paved the way for optical fiber communications. These light sources are compatible with the transmission characteristics and small physical dimensions of low-loss optical fibers. The requirement of small physical dimensions was provided by sources that were fabricated from single crystals of group III–V compound semiconductors on their ternary or quaternary mixtures. The light sources can be categorised into short-wavelength (700–900 nm region) and long-wavelength sources (1300–1600 nm region). The material GaAs-AlGaAs is used for application in the former region, while the material InP-InGaAsP is used for the latter one.

Semiconductor lasers are based on coherent emission of light due to the stimulated recombination of injected carriers (electrons and holes. There are two basic types of semiconductor laser: (1) multi-mode laser and (2) single-mode lasers, which display quite different spectra from one another. The multimode lasers oscillates in several longitudinal modes simultaneously due to small gain (or loss) difference between adjacent modes, which results in several frequency components. Conversely the singlemode laser oscillates in a single longitudinal mode (dominant mode) whereas the other modes are discriminated by their higher losses, thus resulting in a single frequency component.

Semiconductor lasers offer higher power output, higher modulation bandwidth, and narrower spectral width than LEDs. In contrast to lasers, LEDs make use of the incoherent light emission from the spontaneous emission of carriers.

In its simplest form, a semiconductor laser consists of a forward biased, heavily doped p–n junction. A narrow depleted region (no mobile carriers) exists at the junction. Its width depends on the applied voltage and carrier concentration in the p and n regions. As the applied voltage is raised, the potential barrier across the junction is reduced and a net flow of carriers crosses the junction. The magnitude of the potential barrier depends on the initial positions of the Fermi levels relative to the band gap energy. If such an applied voltage is large enough, then the electrons are injected into the p-region and holes into the n-region across the transition region. This region contains a large concentration of electrons within the conduction band and holes within the valence band. If these population densities are large enough, a condition of population inversion results in this region.

Electron-hole recombination in the depleted region provides the optical gain, once a critical voltage is exceeded which guarantees population inversion conditions. Optical feedback leading to laser oscillations is provided by the formation of an optical cavity achieved by polishing the end faces of the junction diode to act as mirrors. The earliest laser prototypes, based on the structure described above, had a huge threshold current density (in the order of 10^5 A/cm^2), thus they could not operate in CW mode.

A major advance was obtained with the introduction of heterostructures. Thanks to the improvement in epitaxial growth techniques, it was possible to put together different semiconductor materials to form a single crystal structure with an artificially modulated band gap. In order for such a structure to maintain or improve the electrical and optical characteristics of the constituent materials it is necessary to have an excellent interface quality. A necessary condition is that the lattice constants of the different materials differ at most by a few percent. Typically, very good heterostruc-

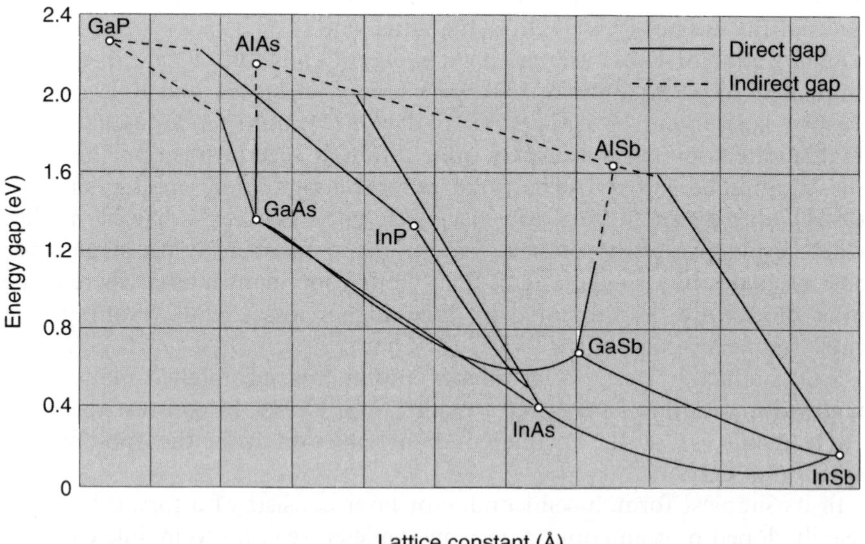

Fig. 1.1 Energy gap versus the lattice constant of III–V materials.

tures are obtained by using ternary or quaternary compounds of a binary semiconductor, as shown in Fig. 1.1. There the energy gap vs lattice constant is shown for a variety of materials. By growing for instance a ternary alloy (AlGaAs) on a GaAs substrate, a perfect lattice matching is obtained, indicated by the fact that the two materials lie on a vertical line in the figure. The same is true for the quaternary alloy InGaAsP, with a composition of 23% indium and 52% gallium, on an InP substrate. When a double heterostructure (DH) is realised, such as by sandwiching a GaAs layer between two AlGaAs regions, the higher band gap materials form a potential barrier for the electrons and holes located in the small gap layer. Thus, carrier confinement is achieved. Furthermore, the larger band gap materials have a smaller refractive index than the small gap material, which then acts as an optical waveguide. When a double heterostructure is used to fabricate a laser, whose active region coincides with the small gap layer, carrier and optical confinement lead to more efficient recombination and reduced losses hence lower threshold current densities.

A typical DH laser, called a stripe geometry injection laser, is shown in Fig. 1.2. In the x-direction, perpendicular to the heterointerfaces, carrier confinement in the active region d_p is provided by the double heterostructure which forms also the dielectric step waveguide. In the transverse y-direction, the optical field builds up only in the central part of the active region, just beneath the contact stripe, corresponding to the maximum

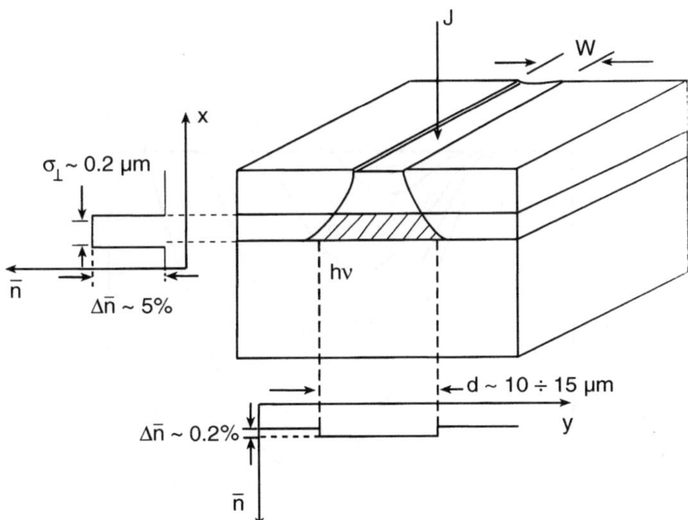

Fig. 1.2 Schematic of a stripe-geometry semiconductor laser. Guiding in x and y directions is provided by a Δn dielectric step whose order of magnitude is indicated only for illustration purposes.

current flow (i.e. to the population inversion condition). In general, because of the lateral spreading under the injection stripe, the optical field distribution in the y-direction is larger than the nominal stripe width w. Such a laser is called a gain guiding laser, compared to index guiding lasers where lateral confinement is achieved by creating a variation of the refractive index in the lateral direction, for instance by proton bombardment, isolation or regrowth.

Several physical considerations enter into the calculation of gain in a semiconductor laser. As the radiative recombination process involves interband transitions between electrons and holes (respectively in the conduction and valence band), three key elements need to be known:

- the carrier distribution function, describing the probability of finding an electron (or a hole) at a given energy;
- the density of states, describing the form of the band and defining the number of k-states available at a given energy; and

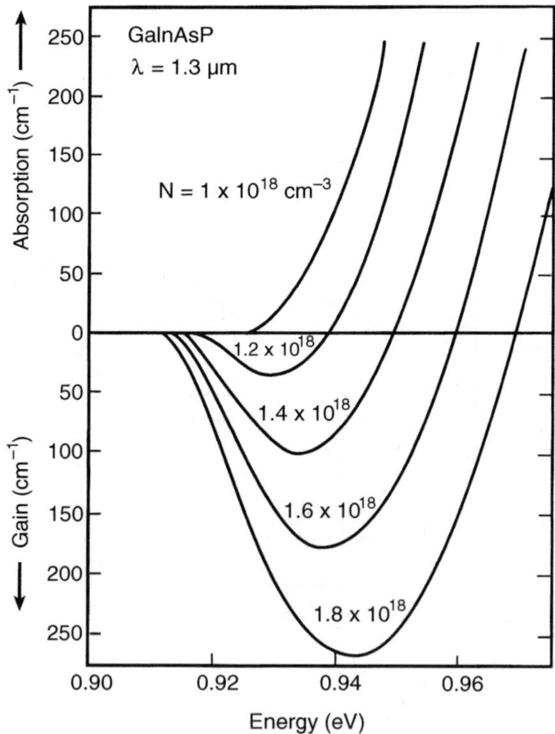

Fig. 1.3 Calculated gain as a function of photon energy for λ = 1.3 \mum InGaAsP at various injected carrier density (after [5]).

- the optical matrix element giving the oscillator strength of a given transition.

The absorption coefficient can be calculated by integrating over the energy space the product of the three elements mentioned above, with the constraint of energy conservation for the transition (that is, the photon energy $h\nu$ must be equal to $E_e - E_h - E_g$, where E_e is the electron kinetic energy, E_h the hole kinetic energy and E_g the band gap). A typical calculation for InGaAsP is shown in Fig. 1.3 [5].

As long as the difference between the quasi Fermi levels for electrons and holes is smaller than the photon energy, i.e. $\Phi_n - \Phi_p < h\nu$, no population inversion exists and the absorption coefficient is positive. As soon as that condition is reversed, for instance, in a laser structure, by the increase of the injected carrier density, gain start occurring in the energy range

$$E_g < \Phi_n - \Phi_p < h\nu \tag{1.1}$$

As the injection level increases, more and more high-energy states become inverted and the peak gain moves to higher energies, as shown in Fig. 1.3. The dependence of the maximum gain on the injected density is plotted in Fig. 1.4 for various temperatures. One can notice from the figure that a linear approximation for the differential gain can be given in the form

$$g = a_0(n - n_t), \tag{1.2}$$

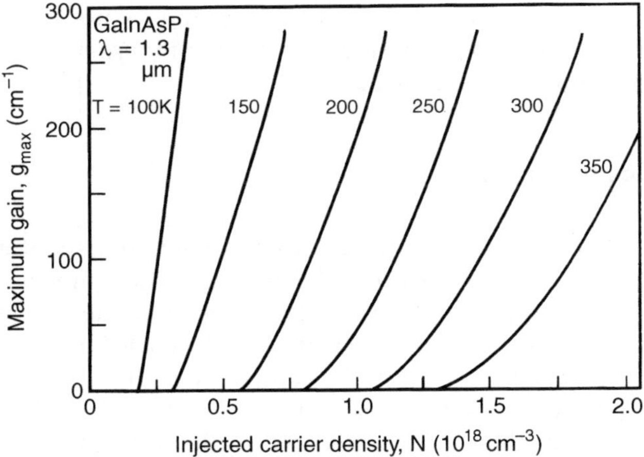

Fig. 1.4 Maximum gain versus the injected carrier density for undoped $\lambda = 1.3$ \mum InGaAsP at different temperatures.

where a_0 is the gain coefficient, n the carrier density and n_t the carrier density at transparency.

A more physical approach to the calculation of gain spectra can be achieved by using the density matrix formalism [6]. In this case, a fully quantum mechanical description is given, able to incorporate many-body effects, and therefore capable of treating non-linear gain. This approach is in any case outside the scope of a brief description.

In a DH injection laser, the optical feedback necessary to sustain lasing is provided by cleaved or etched semiconductor surfaces. If the gain g and the waveguide losses α_i are constant over the length of the laser cavity, then the optical wave grows exponentially along the length and lasing occurs if the total loop gain is unity:

$$R_1 R_2 \exp\{2(\Gamma g_{th} - \alpha_i)L\} = 1, \tag{1.3}$$

where R_1 and R_2 are the reflection coefficients at the two cavity facets. Equivalently, the gain at threshold equals the total loss:

$$\Gamma g_{th} = \alpha_i + \alpha_m \tag{1.4}$$

Where Γ is the optical confinement factor the fraction of optical power in the active layer) and α_m represents the mirror losses. The gain and internal loss have a weak wavelength dependence, with a spectral width of the order of 50 nm. The mirror loss is minimised at the Fabry–Perot modes, i.e. at the wavelengths for which the cavity length is an integral number of half-wavelengths. These resonant wavelengths are separated in wavelength by $\Delta\lambda = \lambda^2/2n_g L$. At low gain the finesse of the cavity is low and the ratio of the peaks to the troughs in the light output is not large, while at higher gain the width of a given mode can be one-thousandth of the mode spacing and the output greater than 10^4.

1.2 LASER STRUCTURES

The typical spectrum of a broad area injection laser consists of a series of wavelength peaks corresponding to different longitudinal modes within the structure. The spacing of these modes (usually a few tenths of a nanometer) depends on the optical cavity length. Each longitudinal mode is further broadened due to subpeaks caused by higher-order transversal modes. Gain-guided structures present a series of lateral modes when the stripe width is larger than a few μm's, due to the difficulty in controlling the lateral field pattern. They are simpler to fabricate, but they are characterised by high values of threshold currents, in the order of 100 mA. Furthermore, the beam pattern in the plane of the junction and the output spectrum are complex, and both may be unstable in time. The lateral extent of the laser gain distribution can be controlled with the addition of a narrow current-confining structure. Laser oscillation is then limited to one or a small number of lateral modes, and the optical source spot is confined to facilitate efficient coupling to an optical fiber. The threshold current is typically reduced by at least an order of magnitude, as the drive current is restricted to flow through a region only a few micrometers wide, rather than a few hundred micrometers wide. In general, gain-guided laser suffer anyway from 'kinks', or non-linear output characteristics, caused by jumps between sets of lateral modes. As the drive current is increased, the carrier density at the centre of the laser stripe grows more slowly than at the edges, because of the higher optical field and subsequent higher carrier recombination. The resulting gain distribution then favours the next highest lateral mode and the laser power tends to dip on the mode change.

The problems just mentioned are greatly reduced by introducing some refractive index variation into the lateral structure of the laser. In some cases the active waveguide thickness is varied by growing it over a ridge or a channel in the substrate. In other cases, loading of a uniformly thick, planar active waveguide can be obtained by the lateral variation in the confinement layer thickness or refractive index. The best solution is offered by buried heterostructure (BH) lasers, in which the active volume is completely buried in a material of wider band gap and lower refractive index. The optical field is well confined not only in the transverse direction but also in the lateral direction, providing strong index guiding of the optical mode, in addition to excellent carrier confinement. Typically BH structures are obtained starting with a standard DH system, by forming a narrow mesa stripe by etching, and regrowing the large band gap material around the mesa. Excellent performance has been demonstrated for both AlGaAs/GaAs and for 1.3 μm and 1.5 μm GaInAsP/InP BH lasers. For 1.3 μm operation of a BH laser with an active region of 1–2 μm, a room temperature threshold current of 20 mA can be obtained, with CW output power of 30 mW per facet and operation up to 100 °C [2].

The structures just described provide control of the lateral modes. As pointed out earlier, several longitudinal modes generally have sufficient gain to reach threshold and oscillate simultaneously. Although some lasers show a tendency towards single longitudinal mode operation, under CW excitation, most FP lasers are prone to multimode behavior under high-speed current modulation. This leads to pulse spreading during propagation through dispersive fibers and to partition noise arising from fluctuations in the modal distribution from pulse to pulse. Much effort has therefore been devoted to the development of single frequency, or dynamic single mode lasers, stable under high-speed modulation. The key point is the ability to provide adequate gain or loss discrimination between the desired modes and all the unwanted modes of the laser resonator. For an

FP cavity, all longitudinal modes have nearly equal losses and are equally spaced (in a 250 μm cavity the spacing is about 1 nm at 1.4 μm). Thus, the broad gain spectrum of a 1.3 μm laser operated well above threshold can support oscillation over a wavelength range of several tens of nanometers, which leads to multimode operation. Conceptually, the simplest way to increase mode discrimination is to shorten the cavity. For instance by reducing L from 250 to 25 μm, the mode spacing is increased to 10 nm.

Recently, a lot of interest has been directed toward surface-emitting microcavity lasers, where a resonator of a few micrometers is formed in the direction normal to the active region [4, 7]. Multiple-element resonators, or resonators with distributed reflectors, provide a loss function with a frequency selectivity sufficiently sharp to produce single-mode oscillation under most operating conditions. Cleaved coupled cavity (C3) lasers belong to this category.

The best way to achieve single-frequency operation involves the use of distributed resonators, fabricated into the laser structure to give integrated wavelength selectivity. Figure 1.5 compares the operation of a standard FB laser and a laser employing a distributed Bragg diffraction grating. The gratings provides a periodic variation of the refractive index in the laser heterostructure waveguide along the direction of wave propagation. Thus feedback of optical energy is obtained through Bragg refraction rather than by the usual cleaved mirrors. Hence the corrugated grating structure shown in Fig. 1.5a determines the wavelength of the longitudinal mode emission instead of the FB gain curve shown in Fig. 1.5b. When the period of the corrugation is equal to $l\lambda_B/2n_e$, where l is the integer order of the grating, λ_B the Bragg wavelength and n_e the effective refractive index of the waveguide, then only the mode near the Bragg wavelength is reflected

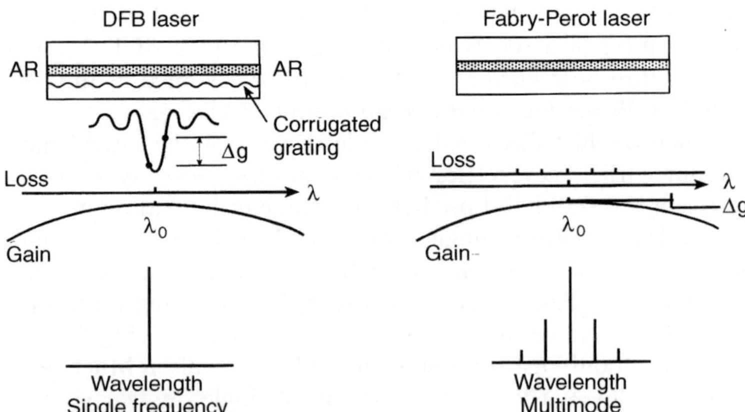

Fig. 1.5 Illustration showing the single frequency operation of: (a) the distributed feedback (DFB) laser in comparison with (b) the Fabry–Perot.

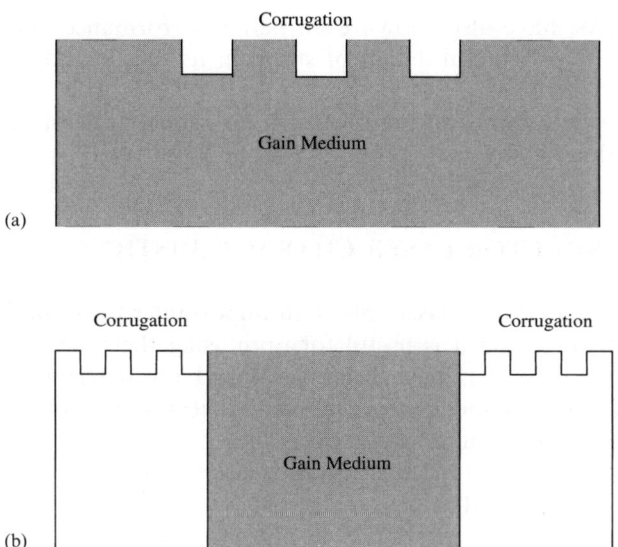

Fig. 1.6 Schematic of typical single-mode lasers: (a) DFB laser; (b) DBR laser.

constructively. This particular mode will lase (Fig. 1.5a) whereas the others exhibiting higher losses will be suppressed from oscillation.

From the viewpoint of device operation, semiconductor lasers employing the distributed feedback mechanism can be classified into two broad categories, the distributed feedback (DFB) and the distributed Bragg reflector (DBR) lasers. A schematic representation of the two devices is presented in Fig. 1.6. In the DFB laser, the optical grating is usually applied over the entire active region which is pumped, whereas in the DBR laser the grating is etched only near the cavity ends. Thus, the DBR laser presents a central active region without distributed feedback and two unpumped corrugated end regions, which effectively act as mirrors whose wavelength-dependent reflectivity results from the distributed feedback mechanism. Presently, most gigabit optical systems employ DFB lasers.

The performance of semiconductor lasers can improve when the size of the active region is reduced to nanometer dimensions. In this situation, quantisation of the electronic states occurs in the direction perpendicular to the heterointerfaces. The optical transitions link states in the conduction and in the valence band that are two-dimensional in nature. As a consequence, quantum well (QW) or multiple quantum well (MQW) lasers display reduced temperature dependence, higher frequency modulation, narrower gain spectrum, lower linewidth enhancement factor and lower threshold than DH lasers. Commercially available laser diodes with threshold current less than 20 mA are available nowadays. In this context, strained QW structures, where an InGaAs layer is pseudomorphically grown on GaAs, have also shown even better performance and potential for improvement by careful design of strain in its tensile and compressive manifestation [8].

System considerations and requirements for semiconductor lasers to be used in optical networks can be found in in the literature [9].

1.3 SEMICONDUCTOR LASER CHARACTERISTICS

Analysis of semiconductor lasers plays an important role in understanding their characteristics, and it is useful for optimising their performance and efficiency. Several models have been developed for describing electrical, optical and thermal characteristicsof semiconductor lasers'. Here we briefly review some of the most important ones.

The output power of the laser inside a laser cavity due to an applied current I can be expressed as

$$P_O = \frac{hv}{q}[\eta_i \cdot (I - I_{th})] \qquad I > I - th \qquad (1.5a)$$

$$P_O = \frac{hv}{q}\eta_i \cdot I \qquad I < I_{th} \qquad (1.5b)$$

where the term $(I\text{-}I)$ represents the number of injected electrons, η_i is the internal quantum efficiency of the laser, which is a measure of the total photons emitted inside the laser cavity per second, and I_{th} is the threshold current density, which is temperature dependent according to the relationship

$$I_{th} = I_{th}(0)\exp(\Delta T/T_C) \qquad (1.6)$$

where $I_{th}(0)$ is the current at a reference temperature, ΔT is the temperature difference between the laser and the reference value, and T_C is the characteristic temperature of the laser, which ranges from 100 to 200 K for AlGaAs lasers and 40 to 80 K for InGaAsP lasers. Therefore, InGaAsP lasers have much stronger dependence on temperature than AlGaAs lasers.

The external differential quantum efficiency of the laser is defined as the number of photons generated for each electron-hole pair injected into the p–n junction, above the threshold current. Experimentally, it can be calculated from the straight line approximation of the $P\text{-}I$ curve as

$$\eta_{ext} = \frac{q\Delta P}{hv\Delta I} \qquad (1.7)$$

where ΔP is the incremental change in the emitted power over the corresponding change in the current ΔI. The external quantum efficiency varies by 30–40% for a typical semiconductor laser.

One of the most important characteristics of a laser diode is its modulation. It depends on the device geometry and structure. A semiconductor laser, being a p–n junction coupled to an optical cavity, would normally modeled by combining mature two-dimensional simulation tools developed for electronic devices (for instance based on drift-diffusion or hydrodynamic models) with a numerical solution of the wave equation for the photons in the laser cavity (Appendix 1A). It turns out that such an approach is computationally very heavy and has been followed only rarely [10].

A more standard approach is to use a set of rate equations describing, in a one dimensional fashion, the temporal evolution of the carrier and photon populations.

The equations, for the single-mode case, take the general form

$$\frac{dN}{dt} = \frac{I}{q} - \gamma_e N - GP \tag{1.8}$$

$$\frac{dP}{dt} = (G - \gamma_p)P + R_{sp} \tag{1.9}$$

In the carrier rate equation (1.8) q is the electron charge, N is the number of carriers in the active layer, I is the current density flowing through the device and γ_e is the carrier recombination rate that can be used to define the spontaneous carrier lifetime τ_e. Both radiative and non-radiative recombination processes contribute to γ_e [1]. In particular, it can be expressed in the following way:

$$\gamma_e = (A_{nr} + Bn + Cn^2) = \tau_e^{-1} \tag{1.10}$$

where A_{nr} is the recombination rate due to mechanisms such as trap or surface recombination, and B is the radiative recombination coefficient; C is related to the Auger recombination processes. The last term in (1.8) is due to stimulated recombination and leads to a non-linear coupling between photons and charge carriers.

In the photon rate equation (1.9) P is the cavity photon number (proportional to the light intensity), γ_p is the photon decay rate (i.e. the reciprocal of the photon lifetime in the cavity) and G is the net rate of stimulated emission. The last term in (1.9) accounts for spontaneously emitted photons. If is assumed that a fraction β_{sp} of spontaneously emitted photons goes into the lasing mode, R_{sp} is given by

$$R_{sp} = \beta_{sp}\eta_{sp}\gamma_e N \tag{1.11}$$

where $\eta_{sp} = Bn/\gamma_e$ and is the (internal) spontaneous quantum efficiency showing the fraction of carriers that emit photons through spontaneous recombination. The parameter β_{sp} is called the spontaneous-emission factor it and it depends on the lateral guiding mechanisms and is enhanced for gain-guided lasers due to wavefront curvature [1]. It can be shown [1] that (1.8) and (1.9) can be derived, with a series of approximations, from more physical equations. The laser rate equations and examples of their use will be presented in Chapter 2.

Equations (1.8) and (1.9) are the single-mode rate equations. For a discussion of multimode phenomena, these equations should be generalised to include the number of possible longitudinal modes for which the gain G is positive. This number depends on the width of the gain spectrum and the frequency spacing between different longitudinal modes [1].

APPENDIX 1A: MODELING OF SEMICONDUCTOR DEVICES

Two approaches can be adopted to model semiconductor devices: the drift-diffusion (DD) model, and the hydrodynamic (HD) model. A detailed analysis of the two methods is certainly beyond this book (a very good discussion can be found in the literature [11, 12]. A third has been proposed in the literature involving Monte Carlo (MC) procedures [13, 14]. This book makes some general remarks about these methods.

All methods are based on the Boltzmann transport equation.

By taking the first moments of the Boltzmann equation, it is possible to obtain, after few assumptions, three coupled equations [11–14]:

$$\frac{\partial n}{\partial t} + \nabla \cdot (n\mathbf{v}) = 0 \tag{A.1}$$

$$\frac{\partial \mathbf{p}}{\partial t} + (\mathbf{v} \cdot \nabla)\mathbf{p} + e\mathbf{E} + \frac{1}{n}\nabla(nk_B T_e) = -\frac{\mathbf{p}}{\tau_p} \tag{A.2}$$

$$\frac{\partial \omega}{\partial t} + \mathbf{v} \cdot \nabla\omega + e\mathbf{E} \cdot \mathbf{v} + \frac{1}{n}\nabla \cdot (k_B T_e n\mathbf{v}) - \frac{1}{n}\nabla \cdot (K\nabla T_e) = -\frac{\omega - \omega_0}{\tau_e} \tag{A.3}$$

where n, $\mathbf{p}$, and ω represent the carrier density, average momentum and average energy, respectively; $\mathbf{v}$ represents the velocity (with $\mathbf{p} = m\mathbf{v}$), T_e denotes the carrier temperature, k_B the Boltzmann constant, $\mathbf{E}$ the electric field and K the thermal conductivity.

Equation (1) is the usual current continuity equation, and expresses charge conservation (here no generation and recombination process is considered). Equations (1A.2) and (1A.3) express, respectively, momentum conservation and energy conservation in the relaxation time approximation.

The DD approach assumes that carriers are always in equilibrium with the lattice ($T_e = T$), which leads to the usual equation, for steady-state conditions:

$$v = -\mu \mathbf{E} - \frac{1}{n} D \nabla n \qquad (A.4)$$

where the carrier mobility and diffusivity (i.e. μ and D) can be field-dependent quantities related through a generalised Einstein relation.

The HD approach uses all three equations and relies on reasonable models for the relaxation times. Clearly, the HD scheme is far superior to the DD scheme, since it can account (when all terms of the third equation are considered) for carrier heating and non-homogeneous distributions of the carrier tememprature.

The MC procedure stands on an even higher level, since it provides (even in non-homogeneous, non-stationary conditions) an exact solution of the Boltzmann equation. It correctly describes non-local effects. Unfortunately, the complexity and costs of each approach are inversely proportional to the refinement of the physical model it is based on. Therefore, the use of one approach or another depends on the specific device under investigation. This book always refers to the DD approach.

REFERENCES

1. G.P. Agrawal and N.K. Dutta, *Long-wavelength semiconductor lasers*, Van Nostrand Reinhold, New York (1986).
2. J.E. Bowers and M.A. Pollack, in *Optical Fiber Communications II*, Eds. S.E. Miller and I.P. Kaminow, p. 509, Academic Press, Boston (1988).
3. Y. Yamamoto, (ed.), *Coherence, Amplification and Quantum Effects in Semiconductor Lasers,* Wiley, New York (1991).
4. K. J. Ebeling, *Integrated Optoelectronics*, Springer-Verlag, Berlin (1995)
5. N.K. Dutta, *J. Appl. Phys.* **51**, 6095 (1980).
6. G.P. Agrawal, IEEE J. *Quantum Electron.* **QE23**, 860 (1987).
7. Y. Suematsu, K. Iga, and S. Arai, *Proc. IEEE* **80**, 383 (1992).
8. H. Morkoc, B. Sverdlov, and G. Gao, *Proc. IEEE* **81**, 493 (1993).
9. D.J.H. Maclean, *Optical line systems*, Wiley, Chichester (1996).
10. Ohtoshi *et al.*, *Solid State Electron.* **30**, 627 (1987).
11. S. Selberherr, *Analysis and simulation of semiconductor devices*, Springer-Verlag (1984).
12. C.M. Snowden, *Introduction to semiconductor device modelling*, World Scientific, Singapore, (1986).
13. P. Lugli, C. Jacoboni, Monte Carlo Simulation of semiconductor devices: a

critical review, in *Solid State Devices*, Eds. P.U. Calzolari, G. Soncini, pp. 861–865 Elsevier, Amsterdam (1987).

14. Jacoboni and P. Lugli, *The Monte Carlo Method for Semiconductor Device Simulation*, Springer-Verlag, Wien (1989).

2

Noise and Dynamic Behaviour of Semiconductor Lasers

One of the most attractive characteristics of a single-mode laser is the spectral purity of the lasing mode. The linewidth and the lineshape were of interest from the beginning of laser physics. Furthermore, the dynamic behavior of the single-mode laser is a relevant issue that has to be treated to better understand how to employ such laser in practical high-speed optical communications systems. In this chapter a concise discussion of phase noise, line broadening and the dynamic behvior of semiconductor lasers is presented. The modeling reported in this chapter can favourably be used to investigate the performance of lasers in fiber-optic communication.

2.1 LASER NOISE FROM VARIOUS SPECTRA

The evaluation of the noise characteristics can be accomplished through the definition of three types of noise spectra: AM-noise spectrum, FM-noise spectrum, and field spectrum (i.e. the spectrum of the electric field of light).

In the semiclassical approximation, the electric field $E(t)$ including noise is expressed as

$$E(t) = [A_0 = a_n(t)]e^{j[\phi_0 + \phi_n(t)]} \tag{2.1}$$

where A_0 and ϕ_0 are stationary values of the amplitude and phase, respectively, and $a_n(t)$ and ϕ_n denote fluctuations of these quantities.

2.1.1 AM-noise spectrum: $S_A(f)$

The AM noise spectrum is defined as the power spectral density function of the amplitude fluctuation. If we denote the Fourier transform of $a_n(t)$, for $-T/2 < t < T/2$, by $A_n(f)$, the AM-noise spectrum is defined as

$$S_A(f) = \lim_{t \to \infty} \frac{1}{T} \langle A_n^*(f) A_n(f) \rangle \tag{2.2}$$

where $< >$ denotes the ensemble average, f the frequency, and T the measurement time.

Instead of $S_A(f)$, the power spectrum of the relative intensity noise (RIN) is commonly used to express the amplitude fluctuations. RIN is defined as the ratio of the intensity fluctuation to the average intensity. Its spectrum $S_{RIN}(f)$ is therefore given as

$$S_{RIN}(f) = \lim_{T \to \infty} \frac{1}{T} \frac{4}{A_0^2} \langle A_n^*(f) A_n(f) \rangle \qquad (2.3)$$

2.1.2 FM-noise spectrum: $S_F(f)$

The FM-noise spectrum is defined as the power spectral density function of the instantaneous frequency fluctuation. It is given by using the Fourier component $F_n(f)$ of the instantaneous frequency fluctuations $f_n(t) = (1/2\pi)(d\phi_n/dt)$, as

$$S_F(f) = \lim_{T \to \infty} \frac{1}{T} \langle F_n^*(f) F_n(f) \rangle \qquad (2.4)$$

2.1.3 Field spectrum: $S(f)$

The field spectrum is the actual power spectral density function of the electric phasor, which is nothing but what is measured by a monochromator. In a semiconductor laser, at bias levels well above threshold, we can assume that the AM noise is negligibly small, since the module of the electric phasor resulting from the sum of stimulated and spontaneous emission is practically coincident with the one corresponding to stimulated emission, while its phase change can be perceptible [1, 2]. In such a case the field spectrum is determined solely by the FM noise spectrum.

In the following, first the autocorrelation function of $E(t)$ is derived. The assumption of neglecting AM noise leads to the following expression of the phasor $E(t)$

$$E(t) = A_0 e^{j[\phi_0 + \phi_n(t)]} \qquad (2.5)$$

The normalised autocorrelation function $R(\tau)$ of $E(t)$ is defined as

$$R(\tau) = <E^*(t)E(t+\tau)> / A_0^2 \qquad (2.6)$$

From these last two equations we have

$$R(\tau) = < \exp(j\Delta\phi_n(\tau) > \qquad (2.7)$$

where

$$\Delta\phi_n(\tau) = \phi_n(t+\tau) - \phi_n(t) = \int_t^{t+\tau} f_n(t\prime)dt\prime. \tag{2.8}$$

Generally speaking, a zero-mean Gaussian noise $G(t)$ satisfies the following relation:

$$< exp[j2\pi G(t)] >= \exp < -2\pi^2 G^2(t) > . \tag{2.9}$$

Therefore, the autocorrelation function of $E(t)$ is given as

$$R(\tau) = exp(-\sigma_\phi^2/2) \tag{2.10}$$

where $\sigma_\phi{}^2$ denotes the mean square fluctuation of the phase noise:

$$\sigma_\phi^2 =< \Delta\phi_n(\tau)^2 >= 4\pi^2 \int_t^{t+\tau} \int_t^{t+\tau} Q(t' - t'')dt'' \tag{2.11}$$

where $Q(t)$ being the autocorrelation function of the instantaneous frequency fluctuation $f_n(t)$. Equation (2.11) can be transformed into a form

$$\sigma_\phi^2 = 8\pi^2 \int_0^\tau (\tau - t)Q(t)dt. \tag{2.12}$$

On the other hand, the FM-noise spectrum $S_F(f)$ is related to the autocorrelation function $Q(t)$ by the relation

$$Q(t) = \int_0^\infty S_F(f)cos(2\pi ft)df \tag{2.13}$$

Note that the FM-noise spectrum is a one-sided spectrum. Substituting equation (2.13) into equation (2.12) and performing some computations, the final expression for $\sigma_\phi{}^2$ can be obtained:

$$\sigma_\phi^2 = 4 \int_0^\infty S_F(f)[sin(\pi f\tau/f)]^2 df. \tag{2.14}$$

The field spectrum $S(f)$ is given as the Fourier transform of $R(\tau)$. Therefore, it can easily computed from equations (2.10) and (2.14).

The 3 dB spread of the field spectrum $S(f)$, i.e. the so-called full width at half maximum (FWHM), is widely used as a measure of the temporal coherence. It should be noted that the field spectrum $S(f)$, and hence its

FWHM, can be determined from the FM noise spectrum $S_F(f)$, whereas the FM noise spectrum cannot be obtained from the field spectrum.

2.2 SEMICONDUCTOR LASER LINESHAPE: THE PHASE NOISE

Before analysing the noise properties and dynamical behavior of lasers under modulation, this section deals with the lineshape of lasers in CW and reports the analytical expression of the lineshape.

The main mechanism which determines the laser spectrum is phase noise. It is caused by randomly occurring spontaneous emission events, which are an inevitable aspect of laser operation. Each event causes a sudden jump (of random magnitude and sign) in the phase of the electromagnetic field generated by the device. As time evolves, the phase executes a random walk away from the value it would have had in the absence of spontaneous emission. The mean squared phase deviation grows with time, and since the average time between steps in the random walk becomes vanishingly small, the random phase $\theta(t)$ becomes in the limit a Wiener process characterised by a zero-mean, white Gaussian frequency noise $\mu(t)$ with two-sided spectral density N_0 [3–5].

Thus, the phase process is represented as

$$\theta(t) = 2\pi \int_0^\tau \mu(t)dt, \tag{2.15}$$

and the mean-squared phase deviation is

$$E\theta^2(t) = E\left[2\pi \int_0^\pi \mu(t)dt\right]^2 = (2\pi)^2 N_0 \tau, \tag{2.16}$$

where E denotes mathematical expectation.

Hence the sine wave random process can be represented as

$$s(t) = A\cos(2\pi f_0 t + \theta(t) + \varphi), \tag{2.17}$$

where the inclusion of the uniform phase φ renders $s(t)$ a stationary process with a correlation function

$$R(\tau) = Es(t)s(t+\tau) = \frac{A^2}{2}\mathrm{Re}\left\{\exp\left[j\,2\pi f_0\tau - \frac{(2\pi)^2}{2}N_0|\tau|\right]\right\}. \tag{2.18}$$

In equation 2.17 f_0 represents the laser emission frequency. A simple calculation reveals that the Fourier transform of $R(\tau)$, i.e. the power spectrum, is given by

$$G(f) = \frac{A^2}{4\pi^2 N_0}\left\{\left[1 + \left(\frac{f+f_0}{\pi N_0}\right)^2\right]^{-1} + \left[1 + \left(\frac{f-f_0}{\pi N_0}\right)^2\right]^{-1}\right\}, \qquad (2.19)$$

with

$$\int_{-\infty}^{\infty} G(f)\,df = \frac{A^2}{2}, \qquad (2.20)$$

as it should. As a result the spectrum has a Lorentzian shape.

A figure of merit for the laser phase noise is its FWHM. Schawlow and Townes proposed the following formula for the emission linewidth [1, 2, 6]:

$$(\delta f)_{ST} = \frac{\pi h f(\Delta f)^2 n_{sp}}{P_0}, \qquad (2.21)$$

where f denotes the oscillation frequency, Δf the passive resonator bandwidth, P_0 the total emitted power, and n_{sp} the spontaneous emission factor representing the population inversion $N_2 - N_1$ between the lasing levels:

$$n_{sp} = \frac{N_2}{N_2 - N_1}. \qquad (2.22)$$

However, several experiments revealed three important results:

(1) The measured linewidth is much wider than the value given by equation (2.21) [7].
(2) The theory predicts that the linewidth is inversely proportional to the output power; therefore, the linewidth-versus $-P^{-1}$ line should pass through the origin. However, experiments showed that this line crosses the ordinate at a certain positive value. This fact suggests that a power-independent contribution exists in the linewidth [8].
(3) A relaxation resonance of the FM noise spectrum was observed in semiconductor lasers [9–11], and the spectrum broadened when the modulation current increased. In other words, the spectrum did not have a Lorentzian shape (Fig. 2.8).

Henry pointed out [12] that the linewidth of semiconductor lasers is enhanced by carrier density fluctuation. In order to account for this effect, a new parameter can be introduced, the linewidth enhancement factor

$$\alpha = \frac{\Delta n'}{\Delta n''} \qquad (2.23)$$

where $\Delta n'$ indicates the change in the real part of the refractive index due to an increment of the carrier density, whereas $\Delta n''$ is the corresponding change in its imaginary part. The linewidth δf is then given by the following equation[12]:

$$\delta f = (1 + \alpha^2)(\delta f)_{ST} \qquad (2.24)$$

where $(\delta f)_{ST}$ denotes the linewidth given by the Shawlow-Townes formula (2.21).

As a result, the effect of the linewidth enhancement factor is to couple intensity and frequency of the field through the carrier density, so that a dynamically evolution of the intensity reflects on a corresponding dynamic change of the frequency. When a laser operates at steady state, only spontaneous emission fluctuations affect the intensity, and the coupling induces a broadening of the emission line. However, when the driving current of the laser is modulated, the carrier modulation induces both an intensity and a frequency modulation which cannot be separated.

The power-independent spectral broadening, which determines the ultimate lower limit of the linewidth, was found to be related to the $1/f$ noise component appearing in the FM noise laser spectra [13, 14].

At present, it is believed that noise phenomena of semiconductor lasers can be described successfully within the framework of the semiclassical theory of laser noise [15–17], once the following three effects are considered: spontaneous emissions, carrier fluctuations and $1/f$ noise.

2.3 SEMICLASSICAL THEORY OF SEMICONDUCTOR LASER NOISE

In this section the semiclassical theory of laser noise is described for a single-mode semiconductor laser, following the approach based on the Langevin equation. This approach deals with the 'equations of motion' for the electric field and the carrier density in a laser. The noise driving force (Langevin force) due to spontaneous emission events is simply regarded as a white Gaussian noise.[1] The equations of motion can be written as

$$\frac{dE}{dt} = -j\Delta\omega(n)E + \frac{1}{2}[G(n) - \Gamma_c]E = F(t), \qquad (2.25)$$

---

[1]In semiconductor lasers the light intensity is strong enough even below threshold; therefore the creation and annihilation operators describing the electric field quantum mechanically can be replaced by the classical phasor having definite values of the amplitude and phase.

$$\frac{dn}{dt} = -\gamma n - G(n)|E|^2 + I/q \qquad (2.26)$$

where

$\Delta\omega(n)$ is the deviation of oscillation angular frequency from the resonant angular frequency ω_c of the cavity resonator
$G(n)$ is the power gain function
Γ_c is the damping constant of the cavity
γ is the damping constant of the carrier density
$F(t)$ is the Langevin force caused by spontaneous emission events;
I/q is the rate of carrier injection (intensity divided by unit charge)

The first term on of the right-hand side of equation (2.25) represents the deviation of the oscillation frequency from the resonance frequency of the resonator (whose stationary value is zero), which fluctuates in accordance with carrier fluctuations, thus inducing a phase fluctuation in the electric

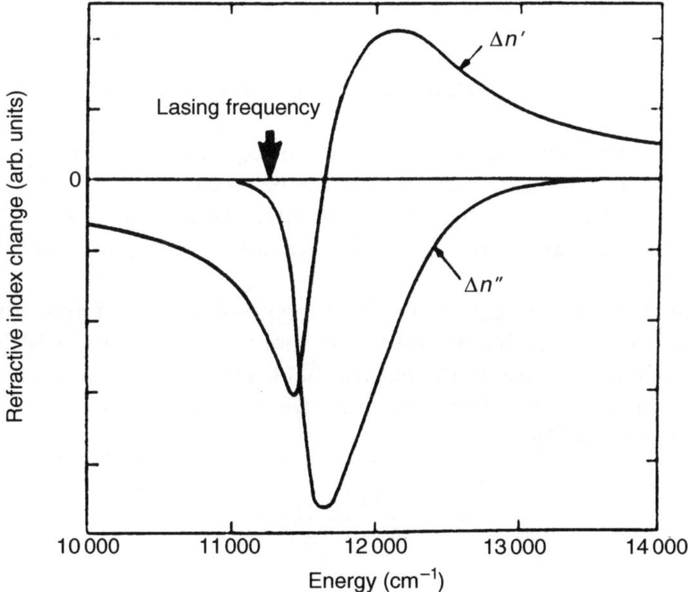

Fig. 2.1 Dispersion of the real and imaginary changes in refractive index in GaAs. This is a measure achieved by Henry [12] for a GaAs active layer of a buried heterostructure laser. The lasing frequency is indicated by the arrow. (After [1] with permission of John Wiley & Sons).

field. In fact, the carrier fluctuation causes a variation of both the real and imaginary parts of the refractive index n' and n'', respectively. As an example, the real and imaginary changes in the refractive index $\Delta n'$ and $\Delta n''$, respectively, due to small carrier density increase, in the case of a GaAs active layer of a buried heterostructure laser [17], have the dispersion illustrated in Fig. 2.1.

Since the propagation constant k is related to the real part of the refractive index by $k = (\omega/c)n'$, when changes the mode frequency ω will also change by

$$\Delta\omega = -\frac{\omega}{c}\Delta n' v_g, \tag{2.27}$$

where v_g is the group velocity. The corresponding change of the imaginary part

$$\Delta n' = \alpha\Delta n'' \tag{2.28}$$

produces in turn a laser gain variation:

$$\Delta g(t) = -2\frac{\omega}{c}\Delta n''(t). \tag{2.29}$$

This effect plays an important role in spectral broadening of semiconductor lasers.

The second term on the right-hand side of equation (2.25) represents the net gain per unit time, i.e. the difference between gain and losses in the cavity. In the absence of other terms, the intensity will increase as $e^{\Delta G t}$ due to this term, since $\Delta G = G - \Gamma_c = \Delta g\, v_g$, and $G = g\, v_g$, g being the gain per unit length.

The third term in equation (2.25) represents the Langevin force accounting for the stochastic nature of spontaneous events, which affect the amplitude and phase of the electric field. Here $F(t)$ is assumed to be a white Gaussian noise, whose spectral density can be expressed by the following equation [2]:

$$C_n = \frac{4\pi N_2}{N_2 - N_1} hf\,\Delta f, \tag{2.30}$$

where Δf denotes the passive resonator bandwidth given by $\Delta f = \Gamma_c / 2\pi$.

As far as equation (2.26) is concerned, the terms on the right hand side represent, respectively, the decay of the carrier due to the spontaneous emission, the decay due to the induced emission, and the injection current I.

2.4 EVALUATION OF SEMICONDUCTOR LASER NOISE

It is relevant to analyse the noise characteristics when the laser operates well above threshold, as in most practical applications. Both $E(t)$ and $n(t)$ have small fluctuations $a_n(t)$ and $\Delta n(t)$ from their stationary values A_0 and n_0, respectively. In such a case, $\Delta\omega(n)$ and $G(n)$ are expressed in terms of $\Delta n(t)$ as

$$\Delta\omega(n) = -\frac{\partial\omega_c}{\partial n}\Delta n, \tag{2.31}$$

$$G(n) = G(n_0) + \frac{\partial G}{\partial n}\Delta n. \tag{2.32}$$

By substituting (2.31) and (2.32) in (2.25) and (2.26), considering only the linear terms in n, and by taking separately intensity and phase of the electric field, we have

$$\frac{da_n}{dt} = \frac{A_0}{2}\frac{\partial G}{2}\Delta n + F_r, \tag{2.33}$$

$$\frac{d(\Delta n)}{dt} = -\left(\gamma + A_0^2\frac{\partial G}{\partial n}\right)\Delta n - [2A_0 G(n)]a_n, \tag{2.34}$$

$$\frac{d\phi_n}{dt} = \frac{\partial\omega_c}{\partial n}\Delta n + \frac{F_i}{A_0}, \tag{2.35}$$

where F_r and F_i are the real and imaginary parts of F, respectively. They can be assumed to be white Gaussain noises, being uncorrelated with each other, and to have the same spectral density, equal to $C_n/2$ [2]. As a result, equation (2.33) to (2.35) are the linearised version of the semiclassical Langevin equations.

2.4.1 Laser noise far below resonance: the adiabatic approximation

The dynamical behavior of a laser strongly depends on the operating frequency. Semiconductor lasers show a resonance frequency, usually appearing at several gigahertz, when the laser is well above threshold. In order to better understand the laser noise mechanisms, it is convenient to consider separately the case in which the laser operates in a frequency range far below the resonance, and then its behavior close to the resonance itself.

In the first case, the semiclassical equations can be solved within the adiabatic approximation, i.e. by neglecting all time derivatives. Hence from (2.34) and (2.33) we have

$$\Delta n = -\frac{F_r}{\frac{1}{2}\frac{\partial G}{\partial n}A_0},$$

(2.36)

$$a_n = \frac{\gamma + A_0^2\frac{\partial G}{\partial n}}{A_0^2 G(n_0)\frac{\partial G}{\partial n}}F_r,$$

(2.37)

It is worth noticing that fluctuations of carrier density and amplitude have a phase difference of π. This is due to the fact that when the amplitude increases, the carrier density has been utilised for the photon emission. In other words, a positive amplitude fluctuation leads to a positive fluctuation in stimulated emission, which in turn causes a negative fluctuation in density.

Equation (2.35), regarding phase fluctuation, can be rewritten in a more significant way as

$$\frac{d\phi_n}{dt} = -\alpha\frac{F_r}{A_0} + \frac{F_i}{A_0}$$

(2.38)

where the linewidth enhancement factor can be expressed as

$$\alpha = \frac{\Delta n'}{\Delta n''} = \frac{\frac{\partial \omega_c}{\partial n}}{\frac{1}{2}\frac{\partial G}{\partial n}}.$$

(2.39)

As a result, it is possible to illustrate the noise mechanisms (Fig. 2.2).

The origin of all the noise processes is in spontaneous emission events. The phase of the electric field is directly affected by spontaneous emission. On the other hand, amplitude fluctuation and carriers density fluctuation are coupled with each other. The carrier density flucuation induces a refractive index fluctuation which ultimately affect the phase.

The FM noise spectrum and the AM noise spectrum can be derived from (2.38) and (2.37)

$$S_f(f) = (1 + \alpha)(\delta f)_{ST}/\pi,$$

(2.40)

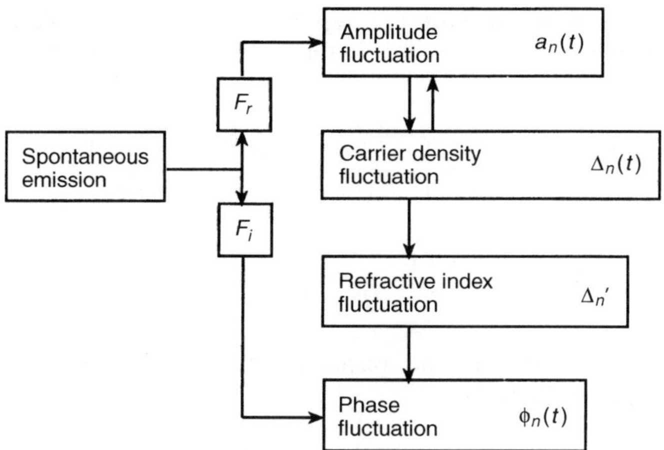

Fig. 2.2 Laser noise mechanisms: spontaneous emission events are the origin of laser noise.

$$S_a(f) = \frac{4\pi\left(\gamma + A_0^2\dfrac{\partial G}{\partial n}\right)^2}{\left(G(n_0)\dfrac{\partial G}{\partial n}\right)^2 A_0^2}(\delta f,)_{ST}. \tag{2.41}$$

The field spectrum can be evaluated, according to the criterium in section 2.1, showing a Lorentzian shape:

$$S(f) = \frac{(1+\alpha^2)(\delta f)_{ST}}{2\pi\{f^2 + [(1+\alpha^2)(\delta f)_{ST}/2]^2\}}. \tag{2.42}$$

It is now possible to find the linewidth of the field spectrum:

$$\delta f = (1+\alpha^2)(\delta f)_{ST}, \tag{2.43}$$

which is the relationship previously anticipated.

Obviously when carrier effects are negligible, i.e. $\alpha = 0$, we reobtain the Shawlow-Townes formula (2.21). In semiconductor lasers the value of α ranges between 4 and 7, so the linewidth broadening is much larger than predicted by equation (2.21).

2.4.2 Laser noise under the relaxation resonance effect: exact solution

The semiclassical Langevin equations can be solved analytically without using the adiabatic approximation. For the sake of brevity, only the expressions for noise spectra are reported, omitting the mathematical calculations, which can be found in the literature [2]. The FM and AM noise spectra are given by

$$S_F(f) = \frac{(\delta f)_{ST}}{\pi} \left[1 + \frac{\alpha^2 f_R^4}{(f_R^2 - f^2)^2 + (\gamma_e/2\pi)^2 f^2} \right], \tag{2.44}$$

$$S_A(f) = \frac{A_0^2 (\delta f)_{ST}}{\pi} \frac{f^2 + (\gamma_e/2\pi)^2}{(f^2 - f_R^2)^2 + (\gamma_e/2\pi)^2 f^2}, \tag{2.45}$$

where F_R and γ_e denote the resonance frequency and the damping constant, respectively, with

$$f_R = \frac{1}{4\pi^2} A_0^2 G(n_0) \frac{\partial G}{\partial n}, \tag{2.46}$$

$$\gamma_e = \gamma + A_0^2 \frac{\partial G}{\partial n}. \tag{2.47}$$

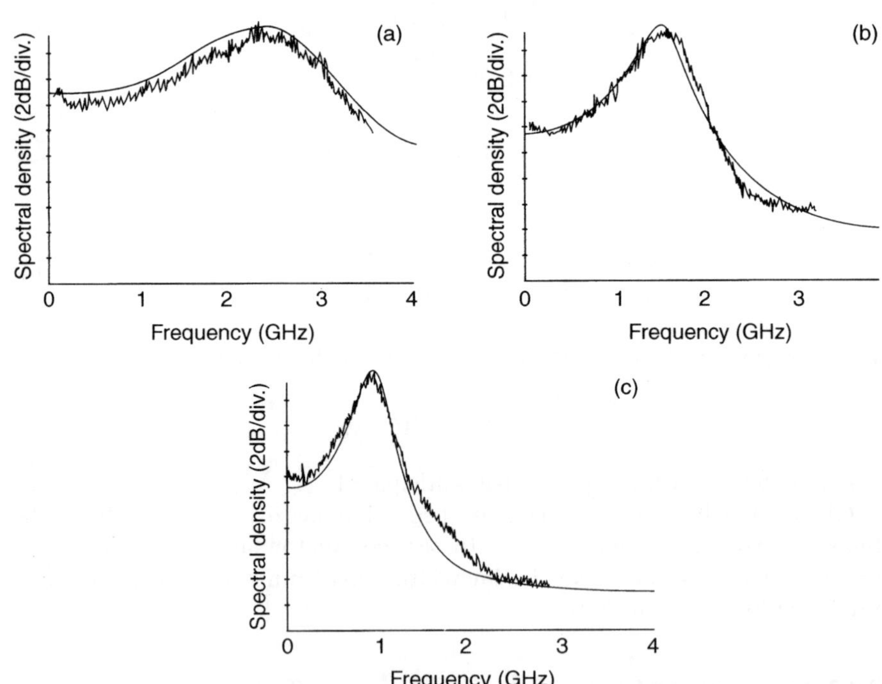

Fig. 2.3 FM noise spectrum. (After [2] with permission of Kluwer Academic Publishers).

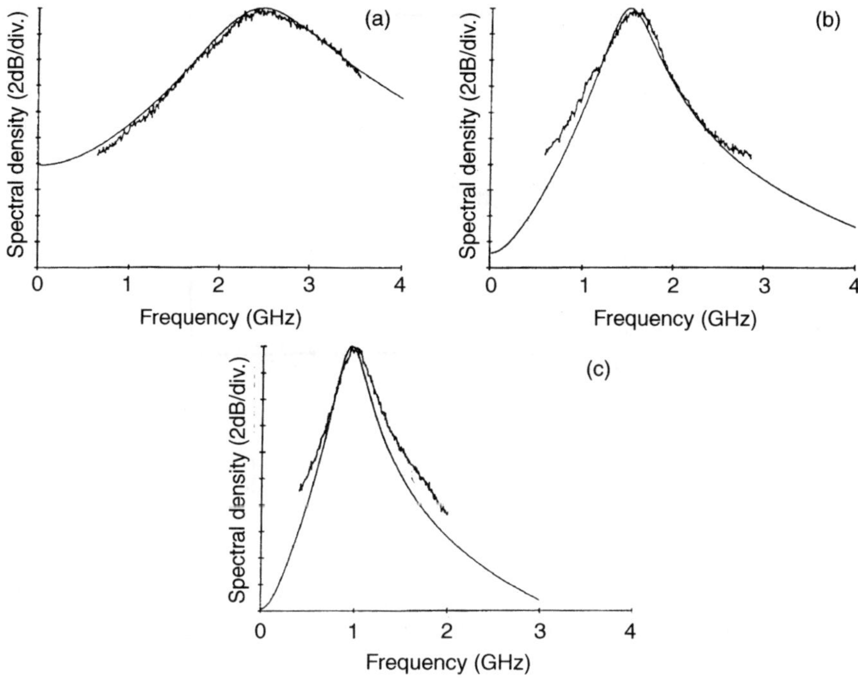

Fig. 2.4 AM noise spectrum. (After [2] with permission of Kluwer Academic Publishers).

It is worth noting that for $f \ll f_R$, the noise spectra assume the form obtained under the adiabatic approximation. Figures 2.3 and 2.4 show typical FM and AM noise spectra [18]. Both spectra display a maximum near to the resonance frequency, which becomes higher as the output power increases. Figures 2.5 and 2.6 plot the resonance frequency and the damping constant versus the normalised bias current [2].

As reported in the literature [2, 18], it is possible to extract some fundamental noise parameters from FM noise spectra, such as α, f_e and γ_e. In particular, α can be easily evaluated in the limits $f = 0$ and $f \gg f_R$.

To evaluate the field spectrum, it is necessary to evaluate $R(\tau)$ (section 2.1) and then find its Fourier transform.

Figure 2.7 shows a typical example of the calculated field spectrum [18]. The lineshape is not a Lorentzian, and it shows additional maxima which correspond to the relaxation resonance in the FM noise spectrum. When the laser operates well above threshold these maxima are far from the central peak. But the field spectrum can still be approximated as a Lorentzian lineshape, having a linewidth increase of $(1 + \alpha^2)$. The maxima shown by the spectrum can also be viewed as the spectral components of

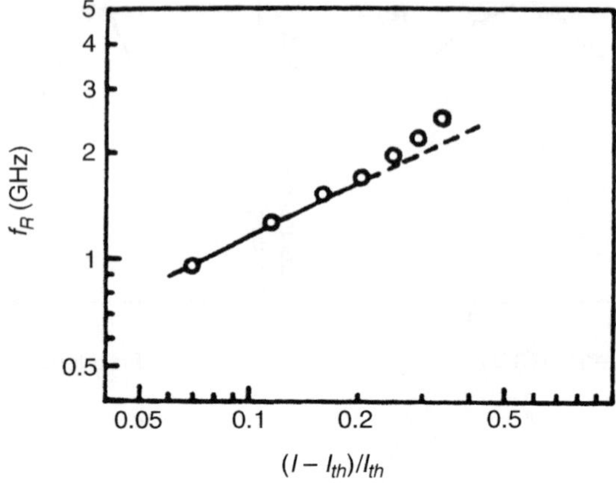

Fig. 2.5 Resonance frequency f_R as a function of the normalised bias current $(I-I_{\text{th}})/I_{\text{th}}$. After [2] with permission of Kluwer Academic Publishers

an FM modulated signal. However, a more detailed analysis on the modulation response is reported in section 2.5.

2.4.3 Effect of $1/f$ noise in the FM noise spectrum

As mentioned in section 2.2, experimental results have shown a power independent contribution to the FM noise spectrum [13, 14]. The noise

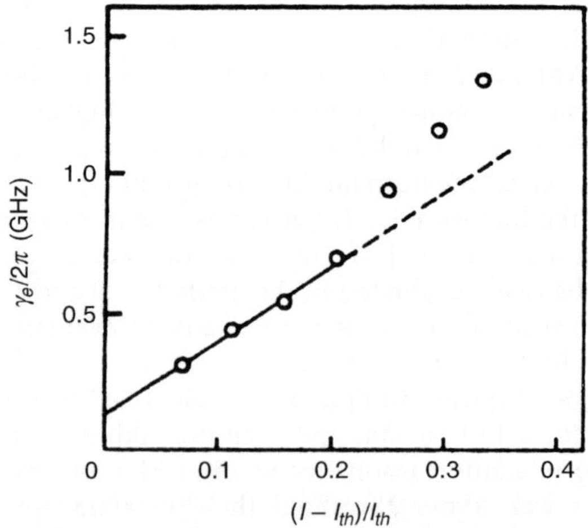

Fig. 2.6 Damping constant γ as a function of the normalised bias current $(I-I_{\text{th}})/I_{\text{th}}$. After [2] with permission of Kluwer Academic Publishers

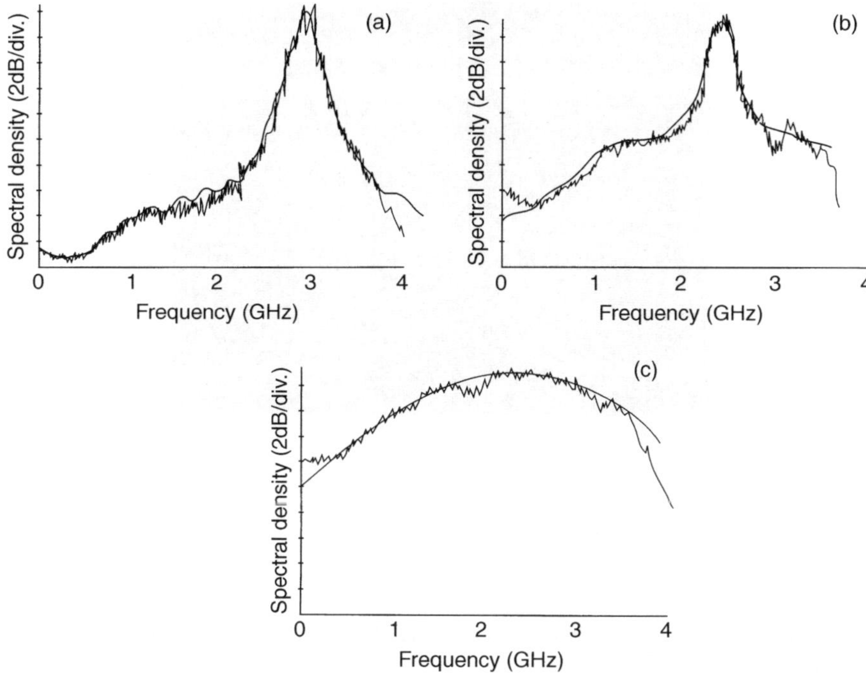

Fig. 2.7 A field spectrum. (After [2] with permission of Kluwer Academic Publishers).

responsible for such a contribution can be schematised as $1/f$ noise. In fact, in the adiabatic approximation, the FM noise can be written as

$$S_F(f) = \frac{K}{f} + \frac{\partial f}{\pi},\tag{2.48}$$

where the first term depicts the $1/f$ contribution, and K is the power-independent term. Thus the $1/f$ noise component shows a marked difference from the white noise component $\delta f/\pi$, whose spectral density is inversely proportional to the output power.

The existence of $1/f$ noise in the FM noise spectrum gives birth to two effects. The first is that the linewidth has a non-zero value even at the high-power limit. Then it gives the predominant contribution to spectral broadening. The second effect is related to the measurement time: when the measurement time becomes longer, the linewidth increases because $1/f$ contributions become more and more significant.

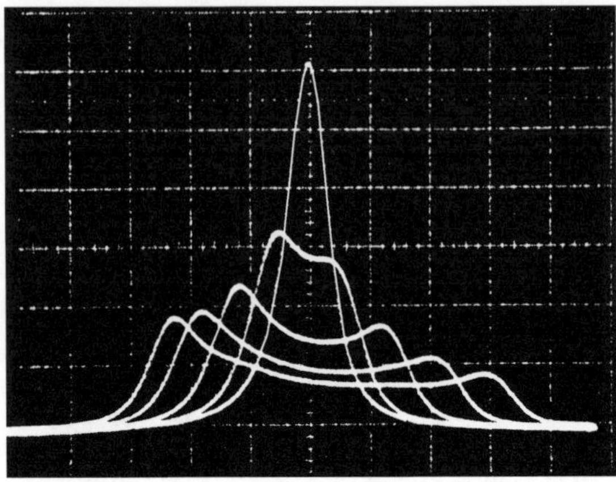

Fig. 2.8 Emission spectrum of a single-mode laser for different modulation currents.

2.5 MODULATION RESPONSE: THE FREQUENCY CHIRPING EFFECT

The effect of the linewidth enhancement factor is to couple the field intensity and frequency through the carrier density, so that a dynamic evolution of the intensity is reflected in a corresponding dynamic change of the frequency. When the laser operates in the steady state, only spontaneous emission fluctuations affect the intensity and the coupling induces a broadening of the emission line. However, when the driving current of the laser is modulated, carrier modulation induces both an intensity and a frequency modulation which cannot be separated. This is shown in Fig. 2.8 where the emission spectrum of a single-mode laser is reported for different values of the modulation current [19].

In order to model the modulation response of a laser diode, taking into account the frequency chirping effect, it is convenient to use a rate equation approach in which the light signal is considered through the photon density function $p(t)$.

2.5.1 Modeling of DFB lasers under modulation

The dynamics of a single-mode laser can be described by the rate equations, which depict the interactions between photon density $p(t)$, the carrier density $n(t)$, and the phase of the electric field $\phi(t)$. The first equation regarding the photon density can be written as follows [20]:

$$\frac{dp}{dt} = (G - \gamma)p + R_{sp}, \tag{2.49}$$

where G is the net gain of stimulated emission, $\gamma = 1/\tau_p$ the photon decay rate that can be used to define the photon lifetime τ_p inside the laser cavity, and R_{sp} is rate of spontaneously emitted photons.

In order to take into account the gain saturation effect, the net gain G can be expressed as follows [20]:

$$G = \frac{G_0}{1 + \varepsilon p} \approx G_0(1 - \varepsilon p) \tag{2.50}$$

where G_0 is the linear gain given by $G_0 = \Gamma v_g g$, with Γ the mode confinement factor and g the local gain; ε is the gain compression factor.

In order to write the complete set of rate equation, the local gain can be approximated as

$$g = a_0(n - n_t), \tag{2.51}$$

being a_0 the gain coefficient, n the carrier density and n_t the carrier density at transparency. Furthermore, the spontaneous emission rate can be expressed as

$$R_{sp} = \frac{\beta \Gamma n}{\tau_n}. \tag{2.52}$$

where β is the fraction of spontaneous emission coupled into the lasing mode, and τ_n is the carrier recombination lifetime. Therefore, the complete set of rate equations, with explicit time dependence of the variables p, n and ϕ, is given as [21]

$$\frac{dp(t)}{dt} = \Gamma \frac{v_g a_0}{1 + \varepsilon p(t)}[n(t) - n_t]p(t) - \frac{p(t)}{\tau_p} + \frac{\beta \Gamma n(t)}{\tau n}, \tag{2.53}$$

$$\frac{dn(t)}{dt} = \frac{I_p(t)}{q V_a} - \frac{v_g a_0}{1 + \varepsilon p(t)}[n(t) - n_t]p(t) - \frac{n(t)}{\tau_n}, \tag{2.54}$$

$$\frac{d\phi(t)}{dt} = \frac{\alpha}{2}\left\{\Gamma v_g a_0[n(t) - n_t] - \frac{1}{\tau_p}\right\}, \tag{2.55}$$

where q is the electron charge and V_a is the active layer volume.

Equation (2.55) holds under the assumption that the physical mechanism behind nonlinear gain is spectral-hole burning. More complex expressions apply if carrier heating also plays a role in the saturation mechanism [22].

Several expression have been obtained in the literature for the frequency chirping [23–25]. A meaningful expression for the instantaneous frequency shift $\Delta f(t)$, i.e. the frequency chirp, can be obtained from (2.55) and (2.53):

$$\Delta f(t) = \frac{1}{2\pi}\frac{d\phi}{dt} = \frac{\alpha}{4\pi}\left(\frac{1}{p}\frac{dp}{dt} - \frac{R_{sp}}{p} + \varepsilon p\right). \tag{2.56}$$

The last two terms in the bracket correspond to the DC shift, the *adiabatic chirp* (related to the frequency offset between the ON and OFF power levels during modulation, when digital modulation is applied), and arises from spontaneous emission and gain suppression. The first term, the *transient chirp*, corresponds to a dynamic frequency shift (it occurs when the laser is turned ON and OFF in digital modulation).

The first two equations can be numerically integrated, taking as an input the current signal $I_p(t)$. Then the optical power emitted per facet $P(t)$ is directly obtained by

$$P(t) = \frac{p(t)}{2}\frac{V_a\eta_0 hv}{\Gamma\tau_p}. \tag{2.57}$$

The module of the electric field can be directly evaluated by making the square root of the optical power $P(t)$, while its instantaneous phase $\phi(t)$ can be obtained integrating equation (2.55). The effect of frequency chirping on the performance of optical communication systems has been analysed in different papers [26–28]. In the next section a typical example of this kind of analysis is reported.

2.5.2 Simulated behavior of DFB dynamics under modulation

The modeling reported in the previous section allows the simulation of the dynamic behavior of a directly modulated DFB laser to be achieved.

The driving current $I_p(t)$ represents an NRZ (non-return-to zero) signal pattern. Such a current can be expressed as follows:

$$I_p(t) = \begin{cases} I_{bias} + I_m\left(1 - e^{-\frac{2.2t}{\tau_r}}\right) \text{ if current bit} = 1, \text{ previous bit} = 0 \\ I_{bias} + I_m e^{-\frac{2.2t}{\tau_r}} \text{ if current bit} = 0, \text{ previous bit} = 1 \\ I_{bias} \text{ if current bit} = 0, \text{ previous bit} = 0 \\ I_{bias} + I_m \text{ if current bit} = 1, \text{ previous bit} = 1 \end{cases} \tag{2.58}$$

where I_{bias} is the bias current (with respect to 0, and close to the lasing threshold), I_m is the modulation current, and τ_r is the rise time and the fall

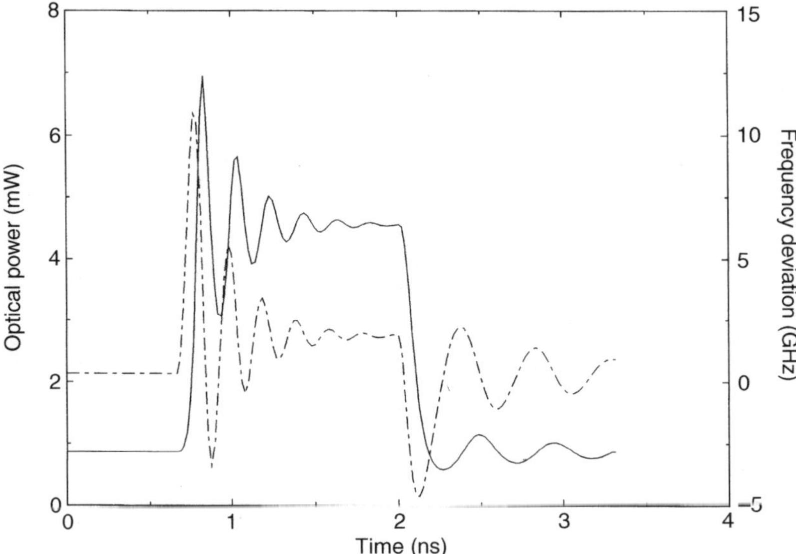

Fig. 2.9 Optical power and frequency chirping for the input current pattern 0011110000. The bit rate of the input waveform is 3 Gb/s, the threshold current of the laser is 33.5 mA, the bias current is 38.86 mA and the modulation current is 23.12 mA. The rise time of the driving time is half the period T and the extinction ratio is 8 dB.

time (both of which are measured between 10% and 90%) of the modulation current.

A relevant example of direct modulation of the laser is shown in Fig. 2.9 and 2.10, where the simulated output of the instantaneous optical power and frequency deviation are plotted for two different input patterns.

Looking at the optical power waveforms, it is possible to argue that the laser is working not far from its relaxation oscillation frequency. As a matter of fact, the relaxation oscillation frequency is 5 GHz for the ONE bit, and 2 GHz for the ZERO bit.

As far as the chirping is concerned, it is worth observing the impact of adiabatic and transient chirp contributions. In fact, Fig. 2.9 clearly reveals the adiabatic chirp: after stabilisation of the waveform (following a long sequence of ONE bits), the deviation frequency set to a stable value which is different from zero (about 2 GHz). This means that different emission frequencies exist for different values of the driving current.

The transient chirp effect can be seen from Fig. 2.10, where a fast variation of the signal pattern induces a fast oscillation of the deviation frequency too. In particular, the transient chirp spans the range from − 5 GHz to 12 GHz during the first cycle oscillation.

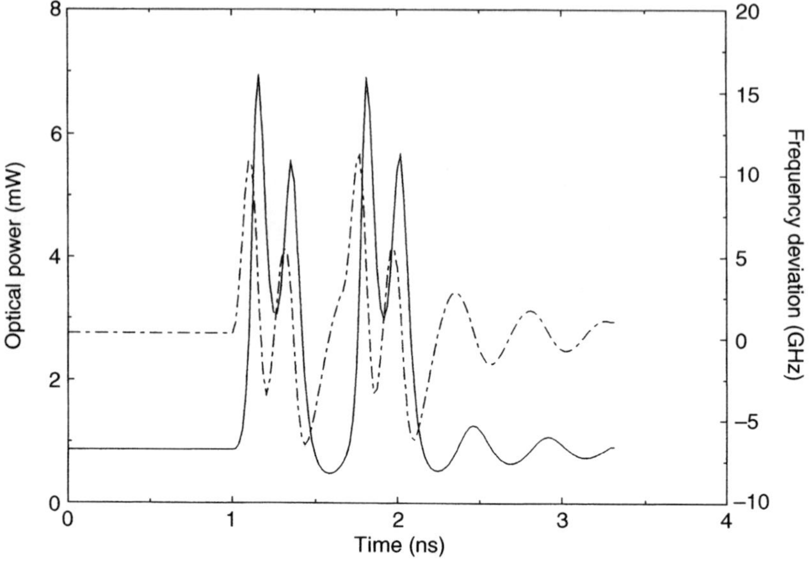

Fig. 2.10 Optical power and frequency chirping for the input current pattern: 0001010000. The conditions are the same as in Fig. 2.9.

If the same patterns were transmitted at much lower bit rates, both optical power oscillations and frequency chirping would be significantly reduced.

APPENDIX 2A: Kramers-Krönig relations in optics

Linear dispersion relations were first derived for X-rays in the mid 1920s by Kramers [29] and Krönig [30] by considering scattering from a continuum of resonators. Krönig subsequently went on to prove the equivalence of causality and dispersion, showing that the dispersion relation is the necessary and sufficient condition for strict causality to be satisfied [31]. It was thus recognised that the same dispersion relations can be applied to optics in general.

Causality refers to the statement that no output can occur before the input. An interesting way of viewing the necessity of dispersion relations was given by Toll [32]. A wave train consisting of a superposition of many frequencies, arrives at a medium that acts as a perfect filter. Thus, one frequency component is completely absorbed and the output should be given by the difference between them. However, this output would violate causality, with an output signal occurring at times before the incident wave train arrives. In order for causality to be satisfied, the absorption of one frequency component must be accompanied by a phase shift in all of the remaining components in such a fashion that, when the components are summed, zero output results for times before the arrival of the wave train.

Determination of linear Kramers-Krönig relations

In a dielectric medium the optical polarisation $P(t)$ can be obtained from the electric field $E(t)$, by means of a response function:

$$P(t) = \int_{-\infty}^{\infty} R(\tau)E(t - \tau)d\tau. \qquad (A.1)$$

The response function $R(\tau)$ is equivalent to a Green's function, as it gives the response (polarisation) resulting from a delta function input (electric field). This equation is more often stated in terms of its Fourier transform, where the convolution is transformed into a product

$$P(\omega) = \chi(\omega)E(\omega) \qquad (A.2)$$

where $\chi(\omega)$ is the susceptibility defined in terms of the response function as

$$\chi(\omega) = \int_{-\infty}^{\infty} R(\tau)e^{i\omega t}d\tau \qquad (A.3)$$

Note that this is not the true Fourier transform as the factor $1/(2\pi)$ is omitted.

Causality states that the effect cannot precede the cause. In the above case this requires that $E(t - \tau)$ cannot contribute to $P(t)$ for $t < (t - \tau)$. Therefore, in order to satisfy causality, $R(\tau) = 0$ for $\tau < 0$. An easy way to see it is to consider the response to a delta function $E(T) = E_0\delta(T)$, where the polarisation would then follow $R(\tau)$. This has important consequences for the relation between susceptibility $\chi(\omega)$, and the response function $R(\tau)$, since the integration needs to be performed only for positive times. Therefore, the lower limit of the integral in equation (2A.3) can in general be replaced by zero.

To determine the Kramers-Krönig relation, we first state the principle of causality as

$$R(T) = R(T)\Theta(T) \qquad (A.4)$$

i.e. the response to an impulse at $t = 0$ must be zero for $t < 0$. Here $\Theta(T)$ is the step function defined as $\Theta(T) = 1$ for $T > 0$ and $\Theta(T) = 0$ for $T < 0$. On Fourier transforming this equation, the product in the time domain becomes a convolution in frequency space, that is

$$\chi(\omega) = \chi(\omega) * \left(\frac{\delta(\omega)}{2} + \frac{i}{2\pi\omega}\right) = \frac{\chi(\omega)}{2} + \frac{i}{2\pi} \Phi \int_{-\infty}^{\infty} \frac{\chi(\Omega)}{\omega - \Omega} d\Omega$$
$$= \frac{1}{i\pi} \Phi \int_{-\infty}^{\infty} \frac{\chi(\Omega)}{\Omega - \omega} d\Omega \qquad (A.5)$$

which is the Kramers-Krönig relation for the linear optical susceptibility. In (2A.5) ϕ denotes the Cauchy principal value. Thus, the familiar Kramers-Krönig relation is simply a restatement of the causality condition, i.e. (2A.4), in the frequency domain.

This relation can also be obtained by starting from (2A.3), multiplying both sides by $1/(\omega - \omega)$ and integrating over all ω. Using the identity

$$\Phi \int_{-\infty}^{\infty} \frac{e^{i\omega} d\omega}{\omega - \Omega} = i\pi e^{i\Omega\tau} \qquad (A.6)$$

which is true only for $\tau > 0$ (which is required by causality) and again (2A.5) is obtained.

Non-linear Kramers-Krönig relation

The Kramers-Krönig relation can be used to evaluate the change in refractive index from the change in absorption due to some external perturbation. The linear Kramers-Krönig relation can be applied both in the presence and in the absence of the perturbation, and the difference is taken between the two cases. Doing this, we can write a modified form of Kramers-Krönig relation (which we also derive below specifically for an optical perturbation):

$$\Delta n(\omega; \zeta) = \frac{c}{\pi} \Phi \int_0^{\infty} \frac{\Delta n(\omega'; \zeta)}{\omega'^2 - \omega^2} d\omega' \qquad (A.7)$$

where ζ denotes the perturbation. An equivalent relation also exists whereby the change in absorption coefficient can be calculated from the change in the refractive index, but this is rarely used for the reasons described below. It is essential that the perturbation is independent of the frequency of observation ω' in the integral (i.e. the excitation must be held constant). The main reason for calculating the refractive index change in this way rather than directly calculating the expectation value of the real part of the polarisation is that calculating the absorption via transition rates is generally far easier. Furthermore, this form of calculation of the refractive index for non-linear optics is often more useful than for linear optics as absorption changes (which can be evaluated or measured) usually occur only over a limited frequency range hence the integral in (2A.7) need to be calculated only over this finite frequency range.

It is possible to use eq. (A.7) in non-linear optics under resonant conditions, where the material is excited into some real state. This excitation can be treated as a perturbation itself. The change in refractive index is calculated from the change in absorption, treating this perturbation as constant, and only afterwards is the optical source of the perturbation considered.

REFERENCES

1. C.H. Henry, Line broadening of semiconductor lasers, in *Coherence, Amplification, and Quantum Effects in Semiconductor Lasers*, Ed.Y. Yamamoto, Wiley, New York (1991).
2. T. Okoshi and K. Kikuchi, *Coherent Optical Fiber Communications*, Kluwer, Tokyo (1988).
3. J. Salz, Coherent lightwave communications, AT&T Tech. J. vol. **64**, no. 10, 2153–2209 (1985).
4. A. Papoulis, *Probability, Random Variables, and Stochastic Processes*, McGraw-Hill, New York (1965).
5. S.O. Rice, Mathematical analysis of random noise, *Selected Papers on Noise and Stochastic Processes*, Ed. W. Nelson Dover, New York (1954).
6. M. Sargent III, M.O. Scully, and W.E. Lamb, Jr, *Laser Physics*, Addison-Wesley, London (1974).
7. M.W. Fleming and A. Mooradian, Fundamental line broadening of single-mode (GaAl)As diode lasers, *Appl. Phys. Lett.* **38**, 511–513 (1981).
8. D. Weldford and A. Mooradian, Output power and temperature dependence of the linewidth of single-frequency CW (GaAl)As diode lasers, **Appl. Phys. Lett.**, **40**, 865–867 (1982).
9. Y. Yamamoto, S. Saito, and T. Mukai, AM and FM noise in semiconductor lasers – Part II: Comparison of theoretical and experimental results for AlGaAs lasers, *IEEE J. Quantum Electron.*, **QE-19**, 47–58 (1983).
10. B. Daino, P. Spano, M. Tamburrini, and S. Piazzolla, Phase noise and spectral line shape in semiconductor lasers, *IEEE J. Quantum Electron.*, **QE-19**, 266–270 (1983).
11. K. Kikuchi and T. Okoshi, FM- and AM-noise spectra of 1.3 μm InGaAsP DFB lasers in 0–3 GHz range and determination of their linewidth enhancement factor α, *Electron. Lett.*, **20**, no, 1044–1045 (1984).
12. C.H. Henry, Theory of the linewidth of semiconductor lasers, *IEEE J. Quantum Electron.*, **QE-18**, 259–264 (1982).
13. M.J. O'Mahony and I.D. Henning, Semiconductor laser linewidth broadening due to $1/f$ carrier noise, *Electron. Lett.*, **19**, 1000–1001 (1983).
14. K. Kikuchi and T. Okoshi, Dependence of semiconductor laser linewidth on measurement time: evidence of predominance of $1/f$ noise, *Electron. Lett.*, **21**, 1011–1012 (1985).
15. P. Spano, S. Piazzolla, and M. Tamburrini, Phase noise in semiconductor

lasers: a theoretical approach, *IEEE J. Quantum Electron.*, **QE-19**, 1195–1199 (1983).

16. K. Vahala and A. Yariv, Semiclassical theory of noise in semiconductor lasers – Part I, *IEEE J. Quantum Electron.*, **QE-19**, 1096–1101 (1983).

17. C.H. Henry, Phase noise in semiconductor lasers, *IEEE/OSA J. Lightwave Technol.*, **LT-4**, 298–311 (1986).

18. K. Kikuchi and T. Okoshi, Measurement of FM noise, AM noise and field spectra of 1.3 mm InGaAsP DFB laser and determination of the linewidth enhancement factor, *IEEE J. Quantum Electron.*, **QE-21**, 1814–1818 (1985).

19. P. Spano, A. Mecozzi, A. Sapia, and A. D'Ottavi, Noise and transient dynamics in semiconductor lasers, in *Nonlinear Dynamics and Quantum Phenomena in Optical Systems*, Eds. R. Vilaseca and R. Corbalan, pp. 259–292, Springer-Verlag, Berlin (1991).

20. G.P. Agrawal and N.K. Dutta, *Long-Wavelength Semiconductor Lasers*, ch.6, Van Nostrand Reinold, New York (1986).

21. C.-S. Li, F.F.-K. Tong, K. Liu, and D.G. Messerschmitt, Channel capacity optimization of chirp-limited dense WDM/WDMA system using OOK/FSK modulation and optical filters, *IEEE/OSA J. Lightwave Technol.*, **10**, 1148–1161 (1992).

22. A. Mecozzi and J. Mørk, Saturation induced by picosecond pulses in semiconductor optical amplifiers, J. Opt. Soc. Amet. B., **14**, 761–770 (1997).

23. D. Weldford, A rate equations analysis for the frequency chirp to modulated power ratio of a semiconductor diode laser, *IEEE J. Quantum Electron.*, **QE-21**, 1749–1751 (1985).

24. T.L. Koch and R.A. Linke, Effect of nonlinear gain reduction on semiconductor laser wavelength chirping, *Appl. Phys. Lett.*, **48**, 613–615 (1986).

25. T.L. Koch and J.E. Bowers, Nature of wavelength chirping in directly modulated semiconductor lasers, *Electron. Lett.*, **20**, 1038–1040 (1984).

26. P.J. Corvini and T.L. Koch, Computer simulation of high bit-rate optical fiber transmission using single-frequency lasers, *IEEE/OSA J. Lightwave Technol.*, **LT-5** 1591–1595 (1987).

27. S. Yamamoto, M. Kuwazuru, H. Wakabayashi, and Y. Iwamoto, Analysis of chirp power penalty in 1.55-μm DFB-LD high-speed optical fiber transmission systems, *IEEE/OSA J. Lightwave Technol.*, **LT-5**, 1518–1524 (1987).

28. J.C. Cartledge and G.S. Burley, The effect of laser chirping on lightwave system performance, *IEEE/OSA J. Lightwave Technol.*, **7**, 568–573 (1989).

29. H.A. Kramers, *Atti Congr. Int. Fis. Como*, **2**, 545 (1927).

30. R. De L. Krönig, *J. Opt. Soc. Amet. Rev. Sci. Instrum.*, **12**, 547 (1926).

31. R. De Krönig, *Ned. Tijdschr. Natuurk*, **9**, 402 (1942).

32. J.S. Toll, *Phys. Rev.*, **104**, 1760 (1956).

3
Optical Modulators

This chapter deals with the devices which act as intensity modulators at high modulation speed. Interest in coherent systems has subsided, so frequency and phase modulators are not treated here. Nevertheless, most issues relating to modulator technology are common to intensity as well as coherent modulators.

The strong interest in intensity modulation at high speed is due to the following reason. The direct current modulation of single-longitudinal mode semiconductor lasers causes a dynamic variation of the peak emission wavelength, especially at high bit rates, as discussed in the previous chapter. The intermingling of such linewidth broadening (chirping) with the chromatic dispersion characteristics of single-mode optical fibres causes signal distortion and eventually generates inter-symbol interference with consequent transmission penalties. As a result, bit-rate and span product are reduced. Therefore, achieving multiple gigabit transmission over large distances requires the use of low-chirp light sources as provided by an external modulator coupled to a laser operating under CW conditions.

3.1 TECHNOLOGIES FOR INTENSITY MODULATORS

Different types of modulator have been proposed in the literature. It is possible to distinguish intensity modulators in two families: loss modulators and electro-optic modulators [1].

The former relates to the modulation of a loss which is introduced in some way into the optical radiation. In simple words, introducing a loss or not induces a space or a mark, respectively, at the output of the modulator. Two different types of loss modulators exist: electroabsorption and carrier injection modulators. The first type is based on the principle that, to change the loss in a modulator, the electroabsorption effect in a waveguide or thin film structure bulk crystal, or in a quantum well structure [2], can be utilised. The second type exploits the change of injected carrier density [2].

The electro-optic modulator can be classified in three types: directional coupler, Mach-Zehnder interferometer and total internal reflection. The first type works by electrically switching a directional coupler from the crossover state to the straight-through state [2]. The Mach-Zehnder interferometer exploits the principle of constructive and destructive interference that can be originated by applying proper voltages on the electrodes put on the two branches of the interferometer. Constructive and destructive interference correspond to having at the device output a mark and a space, respectively. Finally, the principle of a total internal reflection type is that, in a device formed by three zones with different refractive index, a reflected light or a transmitted light is modulated by varying the amplitude reflectivity at the boundary between two materials, due to the refractive change in the inner region.

Two successful candidates have emerged: electroabsorption (using either bulk semiconductor or quantum wells structures) and electro-optic modulators.

3.2 ELECTRO-ABSORPTION MODULATORS

The basic principle of an electroabsorption (EA) modulator is the following. A semiconductor waveguide is realised with a core semiconductor material (usually InGaAsP) sandwiched between two InP materials doped p^+ and n, respectively. This waveguide is actually a p-i-n structure; which, biased, it absorb the optical radiation (OFF modulation) if not, the radiation reaches the output of the device (ON modulation). A figure of merit is the extinction ratio, that is the ratio between the output powers relative to the ON state and the OFF state, respectively.

Electro-absorption modulators show the following advantages with respect to their counterparts based on electro-optic effects:

- monolithic integration with laser diodes lower the production costs;
- low driving voltages;
- low insertion losses (practically negligible for monolithic integration);
- wide bandwidth (with respect to $LiNbO_3$ devices);
- great working stability.

Depending on the physical phenomenon responsible for the electroabsorption of the optical radiation, these types of device can be classified into three kinds:

- Franz-Keldysh modulators (FK): the absorbing material is a bulk semiconductor [3, 4].
- Stark shift modulators: the absorbing layer is constituted by a multi-quantum well structure [5, 6].
- Wannier-Stark modulators: the absorbing layer is an undoped superlattice [7].

In the following we mainly treat the FK modulator, since it is widely used and considered in most applications.

3.2.1 Absorbing spectrum

This is a basic characteristic for a proper design of an FK modulator. It represents the variation of the absorbing coefficient with the wavelength of the incident optical radiation, and with the electric field in the active region, imposed by the reverse bias voltage.

In the transparent range, i.e. the wavelength range in which the light is not absorbed by the material ($\lambda < \lambda_{gap}$), the absorbing coefficient decreases almost exponentially with the energy gap (ΔE) of the semiconductor which has been used in the active region of the modulator, according to the following expression [3]:

$$\alpha(hv) \approx \alpha_0 \cdot \exp\left(-\frac{E_0 - hv}{\Delta E(E)} \right) \tag{3.1}$$

where E is the electric field and $v = c/\lambda$. When $\lambda > \lambda_{gap}$ the absorption coefficient increases.

The absorption coefficient versus the optical wavelength, using different values of the applied electric field, is reported in Fig. 3.1 [3], ralting to the quaternary compound InGaAsP; whereas fig. 3.2 shows the relative variation of the index change Δn, evaluated by the Kramers-Kroenig relation. Moreover, Fig. 3.3 reports the absorption coefficient versus the electric field for two different wavelength.

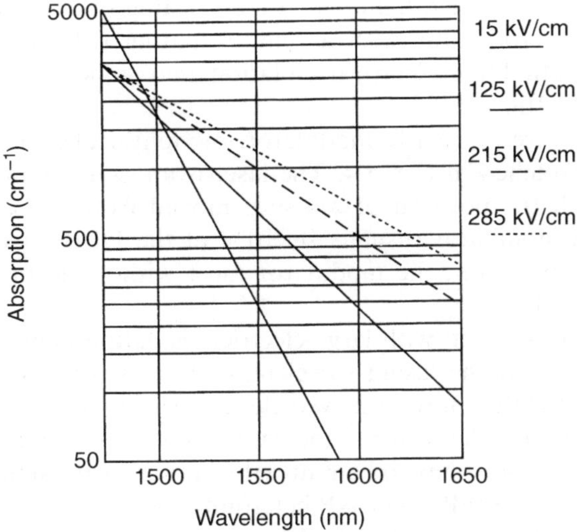

Fig. 3.1 Variation of the absorption coefficient for the quaternary compound InGaAsP versus the optical wavelength at different applied electric fields.

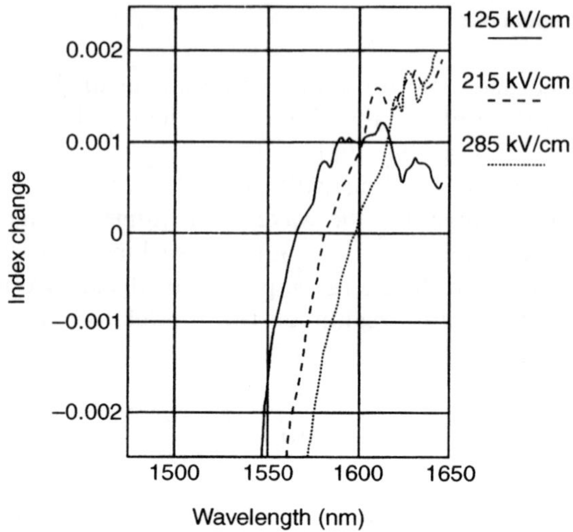

Fig. 3.2 The relative variation of the index change \Delta$n$. The curve has been evaluated using the Kramers-Kroenig relation.

3.2.2 Design characteristics

Several attempts were initially made to integrate a laser with an EA modulator, but it was difficult to keep both large electrical isolation and good optical coupling between the laser and the modulator. Then, at the beginning of the 1990s, integrated light sources, consisting of a $\lambda/4$-shifted DFB laser and an EA modulator with a semi-insulating (SI) InP buried BH structure with both large electrical isolation and good optical coupling were proposed [8–10].

A typical structure of an FK modulator, monolithically integrated with a DFB laser is sketched in Fig. 3.4. The laser is an asymmetric $\lambda/4$-shifted DFB device, with the position of $\lambda/4$-shift moved from the centre of the laser toward the modulator (usually by 10% of the DFB region) in order to get higher output from the modulator facet without losing the single-mode property [8–10].

Such devices are realised with large electrical isolation between laser and modulator, to prevent the lasing wavelength shift associated with biasing the modulator. Furthermore, the wavelength shift which is caused by reflection back to the laser from the laser/modulator boundary and/or from the modulator facet has to be reduced as much as possible.

A buffer layer at the InP/InGaAsP heterointerface can be used to reduce the hole pile-up at that interface, which causes degradation of the modulation characteristic of the device, when the input power into the absorption layer is large. In fact, a degradation of the extinction ratio occurs when the

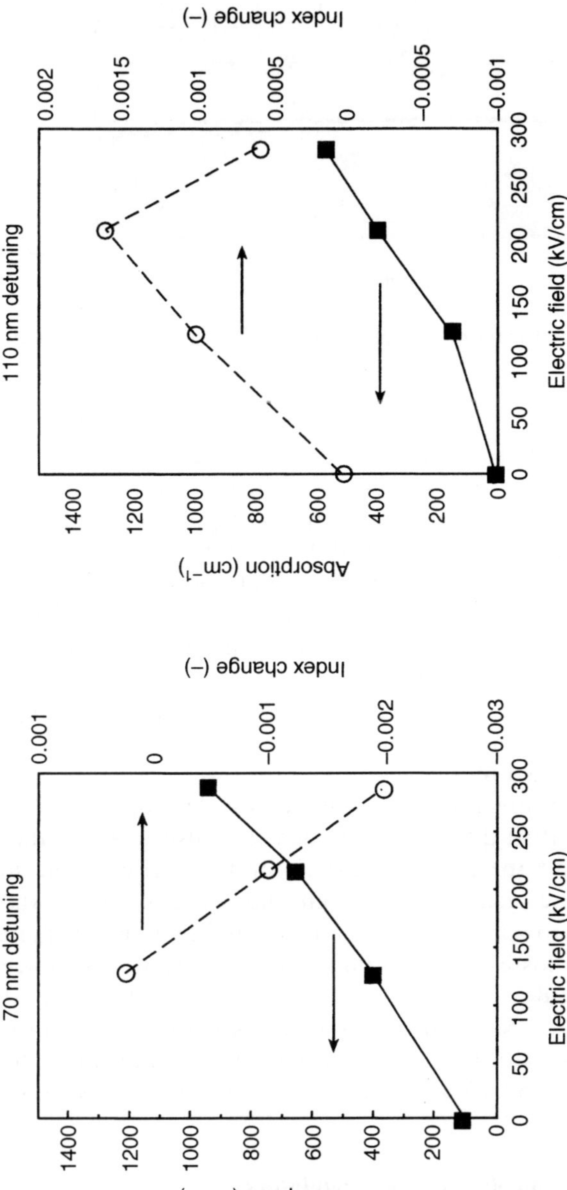

Fig. 3.3 Change of the absorption coefficient versus the electric field for two different wavelengths. The variation is almost linear over the ranges considered.

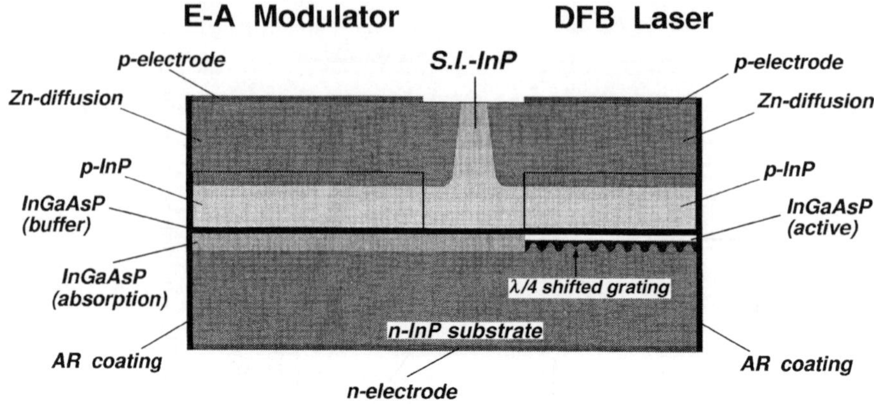

Fig. 3.4 An FK modulator monolithically integrated with a \lambda/4-shifted DFB laser.

optical input power increases. Besides the hole pile-up, another effect can degrade the performance: the space-charge effect. The hole pile-up is due to the large number of photogenerated holes trapped at the valence band discontinuity of the p-InP/i-InGaAsP hetero-interface in the depletion region. The accumulation of trapped holes induces a large potential drop at the hetero-interface, thus reducing the electric field (and therefore the total electroabsorption) of the other region. Hence the waveguide becomes more transparent. The latter effect relates to the large amount of space charge generated in the waveguide by electroabsorption of high input optical power that instantly produces an internal electric field which tends to screen out the applied bias.

A quasi-two-dimensional numerical model was developed by the authors [11, 12] based on a drift-diffusion approach that also includes the hetero-structure models and the Fermi statistics for the carriers. The model is described in the next section. The design and optimisation of the device characteristic have to take into account the different device parameters, such as insertion losses, modulation index, chirping parameter and capacity. In particular, the last two parameters affect the dynamic behavior of the device, since the chirping limits the bit rate-span product, whereas the capacity directly influences the device bandwidth. Of course, the design should try to keep both these parameters as low as possible.

3.2.3 Mathematical modeling of FK modulators

The modulator waveguide scheme is shown in Fig. 3.5.

The quasi-two-dimensional model adopted here is based on a self-consis-

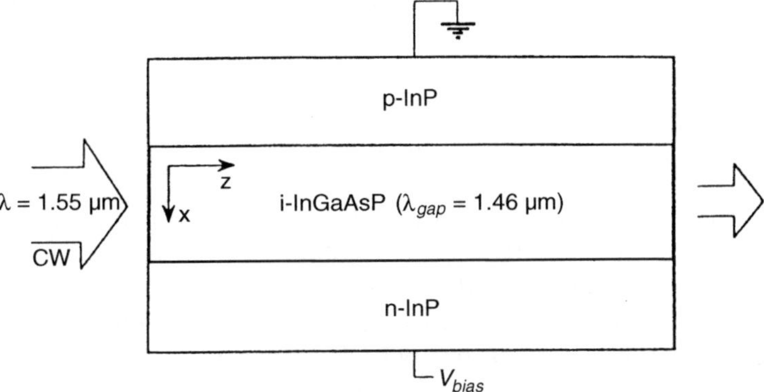

Fig. 3.5 A modulator waveguide.

tent solution of the coupled Poisson, continuity and transport equations along the x-direction:

Poisson equation:

$$\frac{d}{dx}\left(\varepsilon\frac{d\Psi}{dx}\right) = -q(p - n + N_D - N_A) \tag{3.2}$$

Continuity equation for electrons

$$\frac{1}{q}\frac{dJ_n}{dx} + G - R = 0 \tag{3.3}$$

Continuity equation for holes

$$-\frac{1}{q}\frac{dJ_p}{dx} + G - R = 0 \tag{3.4}$$

where ψ denotes the electrostatic potential, p and n, respectively the hole and electron concentrations, q the electron charge, ε is the dielectric constant, N_A and N_D the acceptor and donor impurity concentrations respectively, G the carrier radiative generation term and R the Shockley-Read-Hall net recombination rate (Auger recombination can be omitted because the carrier concentrations are not sufficiently high). Carrier transport in inhomogeneous media including heterostructures can be originally described by the Boltzmann transport equation and the current is given by the gradient of the quasi-Fermi level. The electron/hole current densities in equations (3.3) and (3.4) can be written as

$$J_n = -q\mu_n(E)n\frac{d\varphi_n}{dx} \tag{3.5}$$

$$J_p = -q\mu_p(E)p\frac{d\varphi_p}{dx} \tag{3.6}$$

Here φ_n and φ_p are the electron and hole quasi-Fermi potentials, while $\mu_n(E)$ and $\mu_p(E)$ denote the carrier mobilities for electrons and holes, respectively, determined from the carrier velocities which can be found in the literature [13, 14] as functions of the electric field. Only thermoionic contributions to the current through the heterolayers were considered. Tunneling contributions were neglected since a calculation based on the WKB approximation for triangular barriers showed that they are indeed negligible.

Carrier degeneracy is considered by defining the electron and hole densities as follows:

$$n = N_C \cdot F_{1/2}\left(\frac{V_C - \varphi_n}{V_T}\right) \tag{3.7}$$

$$p = N_V \cdot F_{1/2}\left(\frac{\varphi_p - V_V}{V_T}\right) \tag{3.8}$$

where N_C and N_V are the effective densities of states in the conduction and valence bands, V_C and V_V are the potentials of conduction and valence band edges, V_T is the thermal voltage, and $F_{1/2}$ the Fermi-Dirac integral of order 1/2 defined as

$$F_{1/2}(x) = \frac{2}{\sqrt{\pi}} \cdot \int_0^\infty \frac{\sqrt{y}}{1 + e^{y-x}}\,dy \tag{3.9}$$

The relation between φ_p, φ_n, ψ, the electron affinity χ, and the potential gap V_{gap} is schematically depicted in Fig. 3.6. All material parameters are supposed to vary with position, according to the As percentage in the quaternary layer [13–16].

Since the light intensity decreases as it propagates along the z-direction, according to the absorption profile, the device can be divided in longitudinal sections, each of length z. The above equations are solved for each section, and the output from one section is used as input to the subsequent section. In each longitudinal section, the generation term is evaluated as

$$G(x,z) = \frac{\alpha(E)}{h\nu}I(x,z) \tag{3.10}$$

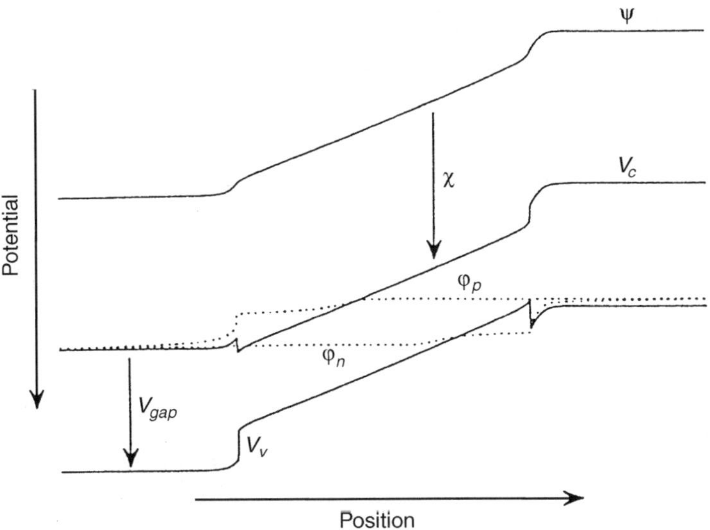

Fig. 3.6 Schematic band diagram of a p-i-n heterostructure.

where the intensity profile $I(x,z)$ at z is calculated assuming fundamental mode propagation [17], and hv is the photon energy. The absorption coefficient α as a function of the electric field has been treated in the literature [18–20].

In a truly two-dimensional model the semiconductor equations and the Poisson equation would directly contain the propagation direction so as to account for the corresponding gradient in carrier concentration; here the model presented above is a quasi-two-dimensional analytical representation of the device, so it neglects carrier motion in the z-direction.

The boundary conditions, needed for the system solutions, are determined by imposing thermal equilibrium in the semiconductor at the border of the quasi-neutral regions of the device. In this hypothesis both quasi-Fermi potentials for electrons and holes must coincide with the equilibrium Fermi potential φ_e, that is

$$\varphi_n = \varphi_p = \varphi_e \tag{3.11}$$

At the same time, for the same hypothesis of thermal equilibrium, the difference between the quasi-Fermi potentials at the border of the quasi-neutral regions, must coincide with the external applied voltage V, that is

$$\varphi_e(0) = 0; \varphi e(x_M) = V \tag{3.12}$$

0 and x_M being the device extremities in the x-direction.

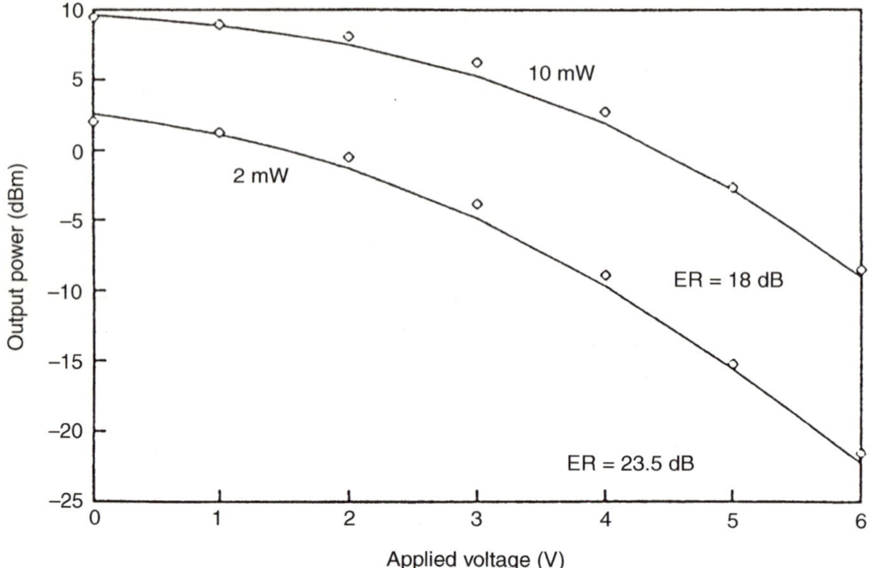

Fig. 3.7 Performance of an FK modulator. The optical output power is plotted versus the applied voltage for two optical intensities. The continuous line indicates the simulation results. The rhombuses depict the measured values. The extinction ratio in both cases has been evaluated considering 0 and 6 V as the levels for space and mark, respectively. (After [12] with permission of IEEE).

Relationships (3.11) and (3.12) represent, as a whole, the boundary conditions of the problem. The solution of the system of equations can be accomplished through the Gummel method [21].

To test this model, we reported a comparison between simulations and measurements, taking into consideration an experimental device [12]. In particular, we reported the output optical power versus the applied voltage, for different optical intensities; Fig. 3.7 shows these results. The agreement between simulation and experiment is quite good. A significant figure of merit of this modulator is the extinction ratio. It is defined as the ratio between the output powers in the ON state (the light exits the modulator) and in the OFF state (the light is essentially absorbed by the modulator). As one might expect, the extinction ratio degradation is stronger at higher input optical intensities [12].

3.2.4 Chirping characteristics in electroabsorption modulators

In direct modulation of a semiconductor laser, frequency chirping is caused by the refractive index change of the active layer due to carrier density modulation. This is represented by the linewidth enhancement factor α_m and the modulation waveform (see also Chapter 2). To eliminate chirping, an external modulation technique represents a valid solution. Nevertheless, chirp-like spectral broadening is still present when employing external modulation. One possible cause is the wavelength shift of the semiconductor laser due to an external reflection outside of the external modulator. The coupled power of the externally reflected light to the laser cavity is in fact varied with intensity modulation, which changes the laser wavelength. A theoretical analysis of this effect was reported in the literature [22]. This sort of chirping can be eliminated by adopting an optical isolator and/or reducing the reflection by means of antireflection (AR) coating to the endface of the modulator. Another possible cause is phase modulation due to the refractive index change of the medium inside an external modulator. This is intrinsic to the device and has to be analysed in detail.

An FK modulator (or more generally an EA modulator) can be regarded as a waveguide in which a loss is changed. In general, if one varies the loss of of a medium, which is related to the imaginary part n'' of the refractive index, the real part n', related to the phase change, will also suffer some modulation, according to Kramers-Krönig relation. Therefore, this causes phase variation of transmitted light through an external modulator together with intensity modulation. Assuming that the input light into the modulator has a constant amplitude $|E_1|$ and constant angular frequency ω_0, the amplitude $|E_2|$ and the phase change ϕ at the output are given by:

$$|E_2| = |E_1| \cdot exp(-k_0 n'' L) \tag{3.13}$$

$$\phi_{out} - \phi_{in} = -k_0 n' L \tag{3.14}$$

where k_0 is the propagation constant in free space and L is the length of an external modulator. Combining these two equations, we obtain the following relationship which relates the instantaneous intensity $S(=|E_2|^2)$ and the phase ϕ of the output light:

$$\frac{d\phi}{dt} = \frac{\alpha_m}{2} \frac{1}{S} \frac{dS}{dt} \tag{3.15}$$

where $\alpha_m = \Delta n'/\Delta n''$, being $\Delta n'$ and the temporal change of the real and imaginary parts of the refractive index, respectively. The derivative on the left side of (3.15) represents the instantaneous angular frequency shift. It is worth observing that this relation is identical to the transient chirp expression of a directly modulated laser (Chapter 2), where the linewidth enhancement factor is around 5–7 in a bulk laser and around 1.7 in a quantum well laser; the chirping parameter α in FK modulators is around 0.5 [23]. From a physical point of view, Δn in lasers depends on the variation of carrier concentration, whereas in modulators it is related to variation in the electric field.

3.3 MACH-ZEHNDER MODULATORS

The basic principle of the Mach-Zehnder (MZ) modulator concern the MZ interferometer function and the electro-optic properties of $LiNbO_3$ material [1]. In fact, a bias voltage applied to an $LiNbO_3$ waveguide varies the waveguide refractive index. This effect can be regarded as a phase modulation operating on the lightwave propagating through the guide. Such a phase modulation can be converted in amplitude modulation exploiting the MZ interferometer feature. A sketch of an MZ amplitude modulator is shown in Fig. 3.8. The electrodes are placed in a travelling wave configuration. The voltages are chosen in order to obtain either a 0° shift or a 180° shift between the two signals propagating in the two branches of the interferometer. In these two possible states the interferometer operates in constructive and destructive interference respectively, which correspond to states ON and OFF modulation (for digital intensity modulation).

This type of device show higher insertion losses and driving voltages, than EA modulators. Furthermore, it requires control systems for stabilising the operating conditions with time. On the other hand, it allows the minimasation or even the disappearence of chirping effects. A very important feature is that MZ modulators can be employed as external modulators in fiber-optic distribution systems for analog signals (e.g. cable television signals), whereas EA modulators cannot, due to their strong non-linear characteristics [24].

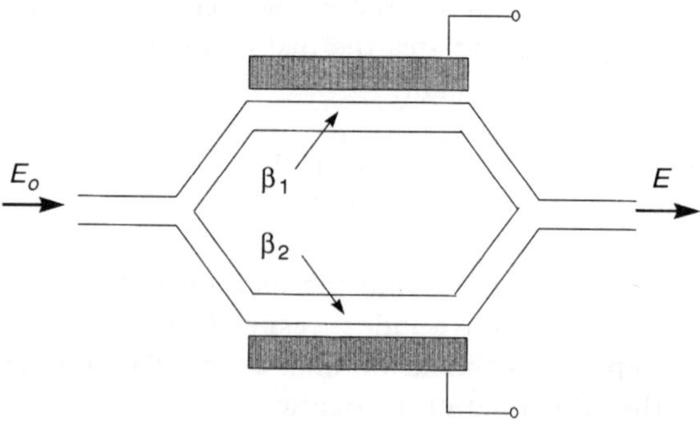

Fig. 3.8 An MZ amplitude modulator in a traveling wave configuration.

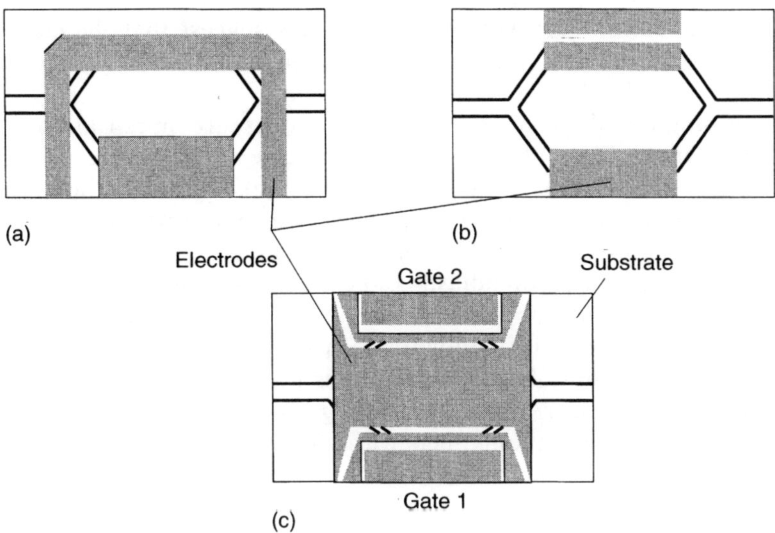

Fig. 3.9 Three possible schemes for an MZ modulator.

3.3.1 MZ modulator structures

Three possible schemes can be realised to achieve MZ modulators, as illustrated in Fig. 3.9. In any case the modulator output power can be expressed as

$$P_0 = P_{in} cos^2 \left(\frac{\pi}{2} \frac{V(t)}{V_\pi} \right) \tag{3.16}$$

where P_o and P_{in} are the output and input optical powers, respectively, and $V\pi$ is the switch voltage of the modulator, that is the required voltage needed to switch the optical output power from zero to its maximum value.

The optical phase φ is given by

$$\varphi = \alpha_{MZ} \frac{\pi}{2} \frac{V(t)}{V_Z} \tag{3.17}$$

being α_{MZ} the modulator chirping parameter of the MZ modulator, which is defined as

$$\alpha_{MZ} = \frac{\eta_1 + \eta_2}{\eta_1 - \eta_2} \tag{3.18}$$

where η_1 and η_2 are the index change per volt (including sign) for the different branches of the MZ modulator. This suggest that the choice of operating point is crucial for correct device operation; henceforth it has to be carefully stabilised to avoid signal distortions.

The position and geometry of the electrodes are important for the chirping characteristics of the device. In scheme (c) there are two input ports instead of one, as in (a) and (b). In scheme (c), because of the electrode symmetry, the driving circuit has to provide two complementary signals for the correct operation; this leads to greater complexity. Anyway, this structure has the advantage of adjusting the chirp as desired, and to allow 'redshift' and 'blueshift' operations, to improve the performance of a given fiber-optic link. In fact, the modulator chirping can been partially compensate for fiber dispersion. For instance, it has was shown [25] that, by operating an MZ modulator in the blueshifting mode, it is possible to realise 10 Gb/s links with 75 km long fibers without any other means for compensating fiber dispersion. This is obtained despite greater modulation voltages.

3.3.2 Chirping characteristics of MZ modulators

The chirping characteristics in the previous section are now described analytically.

The output electric field E is given by

$$E = \frac{E_0}{2}\exp(-j\beta_1 L) + \frac{E_0}{2}\exp(-j\beta_2 L) = E_0\cos(\Delta\beta L)\exp(j\overline{\beta}L) \qquad (3.19)$$

where E_0 is the amplitude of the electric field in the input waveguide $\Delta\beta$ and $\overline{\beta}$ are given by

$$\Delta\beta = \frac{\beta_1 - \beta_2}{2}, \qquad (3.20)$$

$$\overline{\beta} = \frac{\beta_1 + \beta_2}{2}. \qquad (3.21)$$

Since equation (3.15) holds even for MZ modulators, we can simply rewrite it as

$$\alpha_{MZ} = \frac{d\varphi}{(1/E)(dE/dt)} \qquad (3.22)$$

If $d\varphi/dt = 0$, the chirping parameter is zero. This is possible if the propagation constants of the two waveguides are changing by the same amount $\Delta\beta$ with an opposite sign, hence completely compensating for phase modulations. Thus, by properly designing or controlling the device, it is possible to obtain $\alpha = 0$.

APPENDIX 3A

The flowchart for the numerical procedure is shown in Fig. A3.1. In the input step, the data for the device structure, including dimensions, compositions and doping concentrations as well as the material parameters, are read by the program. Then the modulator is analysed, in a double loop, for different bias voltages and for several optical input power levels. The

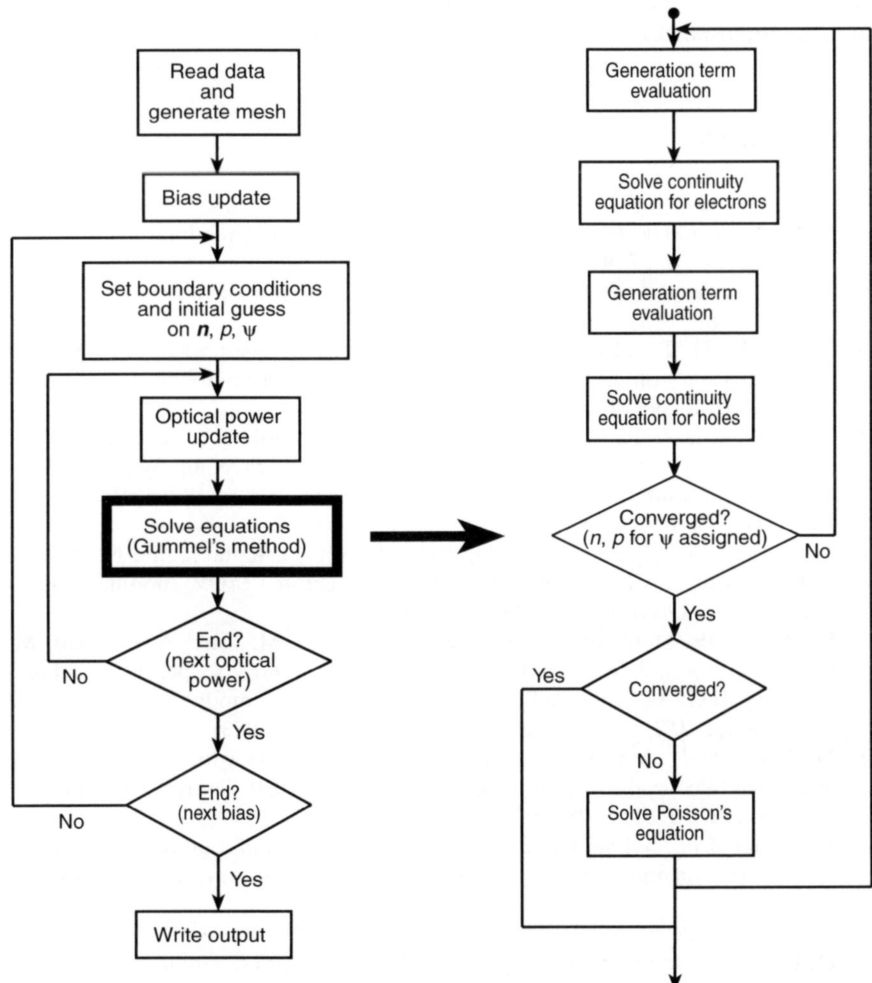

Fig. 3A.1 Flowchart for the numerical procedure.

kernel of the solution procedure is represented by the Gummel loop (on the right-hand side of the figure), where the two continuity equations are solved iteratively with Poisson's equation. In fact, the procedure starts from an initial guess for the carrier concentrations and the potential, and it ends when the stationary condition is reached (within a fixed accuracy). A number of special algorithms have been included in the procedure in order to improve the convergence property of the solution algorithm in the case of high radiative generation, such as the introduction of an inner loop that solves only the two continuity equations for an assigned potential profile, and the gradual increase of the optical input power, which avoids the repeated calculation of the initial guess for the carrier concentrations and the potential distributions for a specified bias.

REFERENCES

1. A. Yariv and P. Yeh, *Optical Waves in Crystals*, John Wiley, New York (1984).
2. F. Koyama and K.Iga, Frequency chirping in external modulators, *J. Lightwave Technol.*, **6**, 87–92 (1988).
3. Internal documents from Fiber Optic Research Center, Ericsson Component AB, Kista, Sweden.
4. M. Suzuki, H. Tanaka, and S. Akiba, Effect of hole pile-up at heterointerface on modulation voltage in GaInAsP electro-absorption modulators, *Electronics Letters*, **25**, 88–89 (1989).
5. T.H. Wood, J.Z. Pastalan, C.A. Burrus, Jr., B.C. Johnson, B.I. Miller, J.L. deMiguel, U. Koren, and M.G. Young, Electric field screening by photogenerated holes in multiple quantum wells: a new mechanism for absorption saturation, *Appl. Phys. Lett.*, **57**, 1081–1083 (1990).
6. T. Kataoka, Y. Miyamoto, K. Hagimoto, K. Sato, I. Kotaka, and K. Wakita, 20 Gbit/s transmission experiments using an integrated MQW modulator/DFB laser module, *Electron. Lett.*, **30**, 872–873 (1994).
7. F. Devaux, F. Dorgeuille, A. Ougazzden, F. Huet, M. Carr, A. Carenco, M. Henry, Y. Sorel, J.-F. Kerdiles, and E. Jeanny, 20 Gbit/s operation of high-efficiency InGaAsP/InGaAsP MQW electroabsorpion modulator with 1.2 -V drive voltage, *IEEE Phot. Techn. Lett.*, **5**, 1288–1290 (1993).
8. M. Goto et al., A 10 Gb/s optical transmitter module with a monolithically integrated electroabsorption modulator with a DFB laser, *IEEE Phoont. Technol. Lett.*, **2**. 896–898 (1990).
9. M Suzuki *et al.*, $\lambda/4$-shifted DFB laser/electroabsorption modulator integrated light source for multigigabit transmission, *IEEE J. Lightwave Technol.*, **10**, 90–95 (1992).
10. P. Ojala, C. Petterson, B. Stoltz, A.C. Moerner, M. Janson, and O. Sahlén, DFB laser monolithically integrated with an absorption modulator with low residual reflectance and small chirp, *Electron.Lett.*, **29**, 859–860 (1993).
11. R. Sabella, P. Lugli, D. Meglio, O. Sahlén, Modeling of nonlinear effects in GaInAsP electroabsorption modulators, in *Design Simulation and Fabrication*

of Optoelectronic Devices and Circuits, OE-LASE'94, Los Angeles CA, pp. 125–137, SPIE (1994).

12. D. Meglio, P. Lugli, R. Sabella, and O. Sahlén, Analysis and optimization of InGaAsP electro-absorption modulators, *IEEE Journal of Quantum Electron.*, **31**, 261–268 (1995).

13. M.A. Littlejohn, T.H. Glisson, and J.R. Hauser, Hot electron transport in n-type $Ga_{1-x}In_xAs_yP_{1-y}$ alloys lattice matched to InP, in *GaInAsP alloy semiconductors* ed. T.P. Pearsall, pp. 243–274, John Wiley, New York (1982).

14. T. Gonzalez Sanchez, J.E. Velazquez Perez, P.M. Gutierrez Conde, and D. Pardo Collantes, Electron transport in InP under high electric field conditions, *Semicond. Sci. Technol.*, **7**, 31–36 (1992).

15. S. Adachi, Materials parameters of $In_{1-x}Ga_xAs_yP_{1-y}$ and related binaries, *J. Appl. Phys.*, **53**, 12 (1982).

16. A.G. Milnes and D. L. Feucht, *Heterojunctions and metal-semiconductor junctions*, pp. 1–13 Academic Press, New York (1972).

17. B.E.A. Saleh and M.C. Teich, Fundamentals of photonics, pp. 248–258, John Wiley, New York (1991).

18. K. Tharmalingam, Optical absorption in the presence of a uniform field, Phys. Rev. **130**, 2204–2206 (1963).

19. R.H. Kingston, Electro-absorption in InGaAsP, Appl. Phys. Lett., **34**, 744–746 (1979).

20. B.R. Bennet and R.A. Soref, Electrorefraction and electro-absorption in InP, GaAs, GaSb, InAs, and InSb, *IEEE Quantum Electron.*, **QE-23**, 2159–2166 (1987).

21. H.K. Gummel, A self-consistent iterative scheme for one-dimensional steady state transistor calculations, *IEEE Trans. Electron. Dev.*, **ED-11**, 455 (1964).

22. K. Matsumoto, Study on integrated external modulators, Tokyo Institute of Technology, Tokyo, Japan, Thesis (1982).

23. O. Sahlén, Optimization of DFB lasers integrated with Franz-Keldysh absorption modulators, *IEEE J. Lightwave Technol.*, ***.

24. R. Sabella, M. Paciotti, and A. Di Fonzo, Impact of non-linear distortion for different modulation schemes in CATV distribution systems, *IEEE LEOS '96*, Boston MA (1996).

25. A. Djupsjöbacka, Residual chirp in integrated-optic modulators, *IEEE Photon. Technol. Lett.*, **4**, 41–43 (1992).

4

Optical Fibers

The objective of this chapter is to provide a basic theory describing the propagation of optical pulses in single-mode optical fibers. Maxwell's equations and basic concepts such as the linear and non-linear parts of the induced polarisation and the frequency-dependent dielectric constant, are introduced in section 4.1. Propagation under linear conditions is treated in section 4.2; section 4.3 reports the theory of pulse propagation in non-linear dispersive media, in the slowly-varying-envelope approximation with the assumption that the spectral width of the pulse is much smaller than the frequency of the incident radiation.The numerical techniques to solve the resulting propagation equation are desribed in section 4.4. Finally, in section 4.5 the basic non-linear effects during fiber propagation, are reviewed.

4.1 BASIC THEORY OF PROPAGATION IN FIBERS

Maxwell's equations, governing all the electromagnetic phenomena, can be considered as the starting point for describing propagation in fibers. They can be written as

$$\nabla \times \mathbf{E} = -\frac{\partial \mathbf{B}}{\partial t}, \tag{4.1}$$

$$\nabla \times \mathbf{H} = \mathbf{J} + \frac{\partial \mathbf{D}}{\partial t}, \tag{4.2}$$

$$\nabla \cdot \mathbf{D} = \rho_f, \tag{4.3}$$

$$\nabla \cdot \mathbf{B} = 0, \tag{4.4}$$

where $\mathbf{E}$ and $\mathbf{H}$ are the electric and magnetic field vectors, respectively, and $\mathbf{D}$ and $\mathbf{B}$ are the corresponding electric and magnetic flux densities. The current density vector $\mathbf{J}$ and the charge density ρ_f depict the sources for the electromagnetic field. In the absence of free charges in a medium such as optical fibers, they are both equal to zero.

The flux densities $\mathbf{D}$ and $\mathbf{B}$ arise in response to the electric and magnetic fields $\mathbf{E}$ and $\mathbf{H}$ propagating inside the medium and are related to them through the constitutive relations given by

$$\mathbf{D} = \varepsilon_0 \mathbf{E} + \mathbf{P} \tag{4.5}$$

$$\mathbf{B} = \mu_0 \mathbf{H} + \mathbf{M} \tag{4.6}$$

where ε_0 and μ_0 are the vacuum permittivity and permeability respectively, and $\mathbf{P}$ and $\mathbf{M}$ are the induced electric and magnetic polarisations. For a non-magnetic medium $\mathbf{M} = 0$.

To obtain the wave equation that describes light propagation in optical fibers, the curl of (4.1) can be taken and (4.2), (4.5) and (4.6) can be used to eliminate $\mathbf{D}$ and $\mathbf{B}$:

$$\nabla \times \nabla \times \mathbf{E} = -\frac{1}{c^2}\frac{\partial^2 \mathbf{E}}{\partial t^2} - \mu_0 \frac{\partial^2 \mathbf{P}}{\partial t^2}, \tag{4.7}$$

where $\varepsilon_0 \mu_0 = 1/c^2$, and c is the velocity of light in a vacuum.

To accomplish the description a relation between the induced polarisation P and the electric field $\mathbf{E}$ is needed. On the whole, the evaluation of P requires a quantum mechanical approach, especially when the optical frequency is near a medium resonance of the medium. However, a phenomenological relation can be used to relate them far from the medium resonance, as in the considered case. The induced polarisation can in general be divided into two parts:

$$\mathbf{P}(\mathbf{r}, t) = \mathbf{P}_L(\mathbf{r}, t) + \mathbf{P}_{NL}(\mathbf{r}, t). \tag{4.8}$$

The linear component, $\mathbf{P}_L$, and the non linear component $\mathbf{P}_{NL}$ are related to the electric field by the following general relationships:

$$\mathbf{P}_L(\mathbf{r}, t) = \varepsilon_0 \int_{-\infty}^{\infty} \chi^{(1)}(t - t') \cdot \mathbf{E}(\mathbf{r}, t) dt' \tag{4.9}$$

$$\mathbf{P}_{NL}(\mathbf{r}, t) = \varepsilon_0 \int\int_{-\infty}^{\infty} \chi^{(3)}(t - t_1, t - t_2, t - t_3) \cdot \mathbf{E}(\mathbf{r}, t_1)\mathbf{E}(\mathbf{r}, t_2\mathbf{E}(\mathbf{r}, t_3) dt_1 dt_2 dt_3.$$

$$\tag{4.10}$$

These relations are valid in the electric dipole approximation such that the medium response is local. The last four equations provide a general method for treating the lowest-order non-linear effects in optical fibers. Because of their complexity, it is necessary to make several simplifying approximations.

As a first approximation, the non-linear polarisation $\mathbf{P}_{NL}$ can be treated as a small perturbation to the total induced polarisation. This is justified because $|\mathbf{P}_{NL}(\mathbf{r},t)| \ll |\mathbf{P}_L(\mathbf{r},t)|$ for optical fibers. So the first step is to solve (4.7) with $\mathbf{P}_{NL} = 0$. Because (4.7) is linear in $\mathbf{E}$, it takes simple form in the Fourier domain. Introducing $\tilde{E}(\mathbf{r},\omega)$ as the Fourier transform of

$E(\mathbf{r},t)$, it become:

$$\nabla \times \nabla \times \widetilde{\mathbf{E}}(\mathbf{r},\omega) + (\omega)\frac{\omega^2}{c^2}\widetilde{\mathbf{E}}(\mathbf{r},\omega) = 0 \qquad (4.11)$$

where the frequency dependent-dielectric constant is defined by

$$\varepsilon(\omega) = 1 + \widetilde{\chi}^{(1)}(\omega) \qquad (4.12)$$

$\widetilde{\chi}^{(1)}(\omega)$ is the Fourier transform of $\chi^{(1)}(t)$. Note that both the terms in (4.12) are complex. In particular, the dielectric constant can be related to the refractive index $\varepsilon(\omega)$ and absorption coefficient $\alpha(\omega)$ by using the following definition:

$$\varepsilon = (n + i\alpha c/2\omega)^2 \qquad (4.13)$$

Two further simplifications can be made:

1) Because of low optical losses in fibers, the imaginary part of $\varepsilon\omega$ is negligible in comparison to the real part; thus ε can be replaced by $n(\omega)^2$;
2) Since $n(\omega)$ is independent of the spatial coordinates in both core and cladding of a step-index fiber, it is possible to write

$$\nabla \times \nabla \times \mathbf{E} = \nabla(\nabla \cdot \mathbf{E}) - \nabla^2\mathbf{E} = -\nabla^2\mathbf{E} \qquad (4.14)$$

since $\nabla \cdot \mathbf{D} = \varepsilon\nabla \cdot \mathbf{E} = 0$.

With these simplifications the propagation equation can be written:

$$\nabla_2\widetilde{\mathbf{E}}(\mathbf{r},\omega) + n^2(\omega)\frac{\omega_2}{c^2}\widetilde{\mathbf{E}}(\mathbf{r},\omega) = 0 \qquad (4.15)$$

4.2 ANALYSIS OF LINEAR FIBER PROPAGATION

For any frequency there is a finite number of modes whose spatial distribution $\widetilde{E}(\mathbf{r},\omega)$ is a solution of the propagation equation and which satisfy the pertinent boundary conditions.

To find the propagation modes, it is useful to assume that the fiber has a perfect cylindrical symmetry. So the propagation equation can be re-expressed in cylindrical coordinates ρ, ϕ, and z:

$$\frac{\partial^2 \widetilde{\mathbf{E}}}{\partial \rho^2} + \frac{1}{\rho}\frac{\partial \widetilde{\mathbf{E}}}{\partial \rho} + \frac{1}{\rho^2}\frac{\partial^2 \widetilde{\mathbf{E}}}{\partial \varphi^2} + \frac{\partial^2 \widetilde{\mathbf{E}}}{\partial z^2} + n^2 k_0 \widetilde{\mathbf{E}} = 0 \qquad (4.16)$$

where $k_0 = \omega/c = 2\pi/\lambda$.

Similar expressions can be written for the magnetic field $\mathbf{H}$. Since $\mathbf{E}$ and $\mathbf{H}$ satisfy Maxwell's equations, only two components out of six are independent. It is common to choose $\tilde{E}_z$ and $\tilde{E}_z$ as the independent components and obtain $\tilde{E}_\rho$, $\tilde{E}_\phi$, $\tilde{H}_\rho$, $\tilde{H}_\phi$ in terms of the first two. Naturally, both $\tilde{E}_z$ and $\tilde{H}_z$ satisfy (4.16), which can be solved by making the substitution

$$\widetilde{E}_z(r,\omega) - A(\omega)F(\rho)\exp(im\phi)\exp(i\beta z) \qquad (4.17)$$

where A is a normalisation constant, β is the propagation constant, and m is an integer. By substituting (4.17) in (4.16), the function F is found to satisfy

$$\frac{\partial^2 F}{\partial \rho^2} = \frac{1}{\rho}\frac{\partial F}{\partial \rho} + \left[\kappa^2 - \frac{m^2}{\rho^2}\right]F = 0, \qquad (4.18)$$

where

$$\kappa^2 = n^2 k_0^2 - \beta^2 \qquad (4.19)$$

In a step-index fiber of core radius a, the refractive index is given by

$$n = \begin{cases} n_1 & \text{for } \rho \leq a \\ n_2 & \text{for } \rho > a \end{cases} \qquad (4.20)$$

Equation (4.18) is the known equation for Bessel functions. Its general solution in the core can be expressed as a linear combination of Bessel functions $J_m(\kappa\rho)$ and Neuman functions $N_m(\kappa\rho)$.

However, since $N_m(\kappa\rho)$ has a singularity at $\rho = 0$, the physical solution is

$$F(\rho) = J_m(\kappa\rho), \rho \leq a. \qquad (4.21)$$

In the cladding region ($\rho \geq a$), the solution should be such that it decays exponentially for large ρ. The modified Bessel function K_m represents such a solution. Therefore

$$F(\rho) = K_m(\gamma\rho), \rho \geq a, \qquad (4.22)$$

where

$$\gamma = (\beta^2 - n_2^2)^{1/2} \tag{4.23}$$

The same procedure can be adopted to find $\tilde{H}_z$. The boundary condition that the tangential components of the electric and the magnetic fields should be continuous across the core-cladding interface requires that $\tilde{E}_z$, $\tilde{H}_z$, $\tilde{E}_\phi$, and $\tilde{H}_\phi$ be the same at $\rho = a$. This condition leads to an eigenvalue equation whose solution determines the propagation constant β for the fiber modes. Since the whole procedure is well-known [1], we write the eigenvalue equation directly:

$$\left(\frac{J'_m(\kappa\alpha)}{\kappa J'_m(\kappa\alpha)} + \frac{K'_m(\gamma\alpha)}{\gamma K'_m(\gamma\alpha >)} \right) \cdot \left(\frac{J'_m(\kappa\alpha)}{\kappa J'_m(\kappa\alpha)} + \frac{n_2^2}{n_1^2} \frac{K'_m(\gamma\alpha)}{\gamma K'_m(\gamma\alpha)} \right)$$
$$= \left(\frac{m\beta k_0(n_1^2 - n_2^2)}{\alpha \kappa^2 \gamma^2 n_1} \right)^2 \tag{4.24}$$

where the prime denotes differentiation with respect to the argument.

Equation (4.24) may have, in general, several solutions for each integer value of m. It is common to express these solutions by β_{mn}, where both m and n take integer values. Each eigenvalue β_{mn} corresponds to one mode which the fiber can support. The fiber modes are hybrid, i.e. all six components of the electromagnetic field are non-zero. So there are two types of fiber mode, designated as HE_{mn} and EH_{mn}, according to the fact the transverse magnetic or electric field gives the biggest contribution.

Equation (4.24) can be used to determine the values of the frequencies at which different modes reach cut-off. Since in actual fiber-optic systems we are essentially interested in single-mode fibers, we limit the discussion to the first mode HE_{11}. The field distribution corresponding to the HE_{11} mode has three non-zero components. In this respect, even a single-mode fiber is not truly single mode since it can support two modes of orthogonal polarisation. The notation LP_{mn} is often used to denote the linearly polarised modes which are approximate solution of (4.16). The fundamental mode HE_{11} corresponds to LP_{01} in this notation.

Under ideal conditions, the two orthogonally polarised modes of a single mode fiber are degenerate that is, they have the same propagation constant. In practice, imperfections such as random variations in the core diameter along the fiber length break the degeneracy slightly, mix the two polarisation components randomly, and scramble the polarisation of the incident light as it propagates down the fiber.

Assuming that the incident light is polarised along the principle axis (e.g. the x-axis), the electric field for the fundamental fiber mode is approximately given by the following equation [1]:

$$\tilde{\mathbf{E}}(\mathbf{r}, \omega) = \hat{\mathbf{x}}\{A(\omega)F(x, y)\exp[j\beta(\omega)z]\} \tag{4.25}$$

where $\hat{\mathbf{x}}$ is the polarisation unit vector of the light. The propagation constant $\beta(\omega)$ is obtained by solving (4.25). Its frequency dependence results not only from the frequency dependence of the refractive indeces n_1 and n_2, but also from the frequency dependence of κ. The former is called the material dispersion and the latter is called the waveguide dispersion. Material dispersion generally dominates unless the light wavelength is close to the zero-dispersion wavelength.

In practical cases, when linear propagation can be assumed, it is useful to express the fiber transfer function as

$$H_{fiber}(f) = \exp\left(\frac{\pi\lambda^2 D(f)f^2 L}{c}\right) \cdot (-\alpha_L L) \tag{4.26}$$

L being the length of the considered fiber, λ the optical wavelength in vacuum, c the velocity of light, and $D(f)$ the fiber dispersion parameter, which is generally frequency dependent, α_L is the attenuation per unit length.

Usually for standard single-mode fibers, D is a constant and its value is around -17 ps/km/nm. For dispersion-shifted fibers it is about 4 ps/km/nm.

4.3 ANALYSIS OF NON-LINEAR FIBER PROPAGATION

This section derives a basic equation governing the propagation of optical pulses in non-linear dispersive fibers. By using (4.8) and (4.14), the wave equation (4.7) becomes

$$\nabla^2 \mathbf{E} - \frac{1}{c^2}\frac{\partial^2 \mathbf{E}}{\partial t^2} = -\mu_0 \frac{\partial^2 \mathbf{P}_L}{\partial t^2} - \mu_0 \frac{\partial^2 \mathbf{P}_{NL}}{\partial t^2}. \tag{4.27}$$

In order to solve this equation, some simplifying assumptions are made:

(1) $\mathbf{P}_{NL}$ is treated as a small perturbation with respect to $\mathbf{P}_N$
(2) The optical field maintains its polarisation along the fiber length so that a scalar approach is valid
(3) The optical field is quasi-monochromatic, i.e. its spectrum, centered at ω_0, has a spectral width $\Delta\omega$ such that $\Delta\omega/\omega_0 \ll 1$.

For the considered frequencies, the last assumption is valid for pulses whose width is greater than 0.1 ps. In the slowly-varying-envelope approximation, it is useful to separate the rapidly varying part of the electric field by writing it in the form

$$\mathbf{E}(\mathbf{r},\omega) = \frac{1}{2}\hat{\mathbf{x}}\{\overline{E}(\mathbf{r},t)\exp[j\omega_0 t] + c.c.\} \tag{4.28}$$

where c.c. stands for complex conjugate, and $E(\mathbf{r},t)$ is a slowly varying function of time (relative to the optical period). The polarisation components can also be expressed in a similar form:

$$\mathbf{P}_L(\mathbf{r},\omega) = \frac{1}{2}\hat{\mathbf{x}}\{\overline{P}_L(\mathbf{r},t)\exp[j\omega_0 t] + c.c.\} \tag{4.29}$$

$$\mathbf{P}_{NL}(\mathbf{r},\omega) = \frac{1}{2}\hat{\mathbf{x}}\{\overline{P}_{NL}(\mathbf{r},t)\exp[j\omega_0 t] + c.c.\} \tag{4.30}$$

In order to obtain the wave equation for the slowly varying amplitude $\overline{E}(\mathbf{r},t)$, it is better to work in the Fourier domain. This is generally not possible as (4.27) is non-linear, ε_{NL} is intensity dependent. It is possible to treat ε_{NL} as a constant during the derivation of the propagation equation for $\overline{E}(\mathbf{r},t)$. The approach is justified in view of the slowly-varying-envelope approximation and the perturbative nature of $\overline{P}_{NL}(\mathbf{r},\omega)$. By substituting eqs. (4.28)–(4.30) in Eq. (4.27), the Fourier transform of $\overline{E}(\mathbf{r},t)$ is found to satisfy

$$\nabla^2\overline{E} + \varepsilon(\omega)k_0^2\overline{E} = 0 \tag{4.31}$$

where $k_0 = \omega/c$, and

$$\varepsilon(\omega) = 1 + \widetilde{\chi}_{xx}^{(1)}(\omega) + \varepsilon_{NL} \tag{4.32}$$

is the dielectric constant whose non-linear part is ε_{NL}.

Again, the dielectric constant can be used to define the refractive index $\overline{n}$ and the absorption coefficient α. However, $\overline{n}$ is intensity dependent because of ε_{NL}. It is usual to adopt the definition of Agrawal [1]

$$\overline{n}(\omega) = n(\omega) + n_2 \mid \overline{E} \mid^2 \tag{4.32a}$$

where the non-linear index coefficient n_2 can be shown to be $n_2 = 3 \,/\, 8n$ $\chi_{xxxx}^{(3)}$ [1].

Equation (4.31) can be solved by separation of variables, assuming the following relation:

$$\widetilde{E}(\mathbf{r},\omega - \omega_0) = \widetilde{A}(\omega - \omega_0)F(x,y)\exp[j\beta_0 z] \tag{4.33}$$

where $\widetilde{A}(\omega - \omega_0)$ is a slowly varying function of z and β_0 is the wave number to be determined later. Equation (4.31) leads then to the following expression for $\widetilde{A}(\omega - \omega_0)$ and $F(x,y)$:

$$\frac{\partial^2 F}{\partial x^2} + \frac{\partial^2 F}{\partial y^2} + [\varepsilon(\omega)k_0^2 - \overline{\beta}^2]F = 0 \tag{4.34}$$

$$2j\beta_0 \frac{\partial \widetilde{A}}{\partial z} + (\overline{\beta}^2 - \beta_0^2)\widetilde{A} = 0 \tag{4.35}$$

In obtaining equation (4.35), the second derivative $\partial^2 \widetilde{A}/\partial z^2$ has been neglected. The wavenumber β is determined by solving the eigenvalue equation (4.34) for the fiber modes using a procedure similar to that used in the previous section. The dielectric constant can be approximated by

$$\varepsilon = (n + \Delta n)^2 \cong n^2 + 2n\Delta n \tag{4.36}$$

where Δn is a small pertubation given by

$$\Delta n = n_2 \, | \, \overline{E} \, |^2 + \frac{j\alpha}{2k_0} \tag{4.37}$$

In particular, the solution of (4.34) can be carried out using first-order pertubation theory [1] by replacing ε with n^2 and obtain the modal distribution $F(x,y)$ and the corresponding wavenumber $\beta\omega$. For a single mode fiber, $F(x,y)$ corresponds to the modal distribution of the fundamental fiber mode HE_{11}. The fundamental fiber mode is often approximated by a Gaussian distribution [1]. Then the effect of Δn can be included in equation (4.34). In first order pertubation theory, Δn does not affect the modal distribution $F(x,y)$. However, the eigenvalue β is given by

$$\overline{\beta}(\omega) - \beta(\omega) + \Delta\beta \tag{4.38}$$

where

$$\Delta\beta = \frac{k_0 \iint_{-\infty}^{\infty} \Delta n \, | \, F(x,y) \, |^2 \, dxdy}{\iint_{-\infty}^{\infty} | \, F(x,y) \, |^2 \, dxdy} \tag{4.39}$$

As far as (4.35) is concerned, it can be rewritten as follows:

$$\frac{\partial \widetilde{A}}{\partial z} = j[\beta(\omega) + \Delta\beta - \beta_0]\widetilde{A} = 0 \tag{4.40}$$

where we approximated $(\overline{\beta}^2 - \beta_0^2)$ by $2\beta_0(\overline{\beta} - \beta_0)$. The inverse Fourier transform (4.40) provides the propagation equation for $A(z,t)$. For this purpose, it is favourable to expand $\beta(\omega)$ in a Taylor series around the carrier frequency ω_0., i.e.:

$$\beta(\omega) = \beta_0 + (\omega - \omega_0)\beta_1 + \frac{1}{2}(\omega - \omega_0)^2\beta_2 + \frac{1}{6}(\omega - \omega_0)^3\beta_3 + ... \qquad (4.41)$$

where

$$\beta_n = \left(\frac{d^n\beta}{d\omega^n}\right)_{\omega=\omega_0} \qquad (4.42)$$

The cubic and higher-order terms in thes series are generally negligible when $\Delta\omega \ll \omega_0$. Their omission is consistent with the quasi-monochromatic assumption used in the derivation of (4.40), and limits the validity to pulse widths longer than 0.1 ps. Substituting (4.41) into (4.40) and taking the inverse Fourier transform, we obtain

$$\frac{\partial A}{\partial z} = -\beta_1\frac{\partial A}{\partial t} - \frac{j}{2}\beta_2\frac{\partial^2 A}{\partial t^2} + j\Delta\beta A. \qquad (4.43)$$

where the term proportional to $\Delta\beta$ contains the effect of fiber losses and non-linearity. By using (4.37) and (4.39), $\Delta\beta$ can be evaluated and substituted in (4.43). The result is

$$\frac{\partial A}{\partial z} + \beta_1\frac{\partial A}{\partial t} + \frac{j}{2}\beta_2\frac{\partial^2 A}{\partial t^2} + \frac{\alpha}{2}A = j\gamma \mid A \mid^2 A \qquad (4.44)$$

where the non-linearity coefficient γ is defined by

$$\gamma = \frac{n_2\omega_0}{cA_{eff}}. \qquad (4.45)$$

The parameter A_{eff} is known as the effective core area and is given by

$$A_{eff} = \frac{\left(\iint_{-\infty}^{\infty} \Delta n \mid F(x,y) \mid^2 dxdy\right)^2}{\iint_{-\infty}^{\infty} \mid F(x,y) \mid^4 dxdy} \qquad (4.46)$$

Being dependent on the modal distribution F, A_{eff} depends on the fiber parameters (radius and index of core and cladding). Typically $A_{eff}= 50 - 80$ μm^2 in the 1.5 μm region. As a result, the non-linearity coefficient γ can vary from 2 to 30 W/km, depending on wavelength.

The meaning of the different terms in the propagation equation (4.44) is easy to understand. It includes the effects of fiber losses through α, of chromatic dispersion through β_1 and β_2 , and of fiber non-linearity through γ. In particular, the pulse envelope moves at the group velocity

$v_g = 1 / \beta_1$, while group whereas velocity dispersion (GVD) is taken into account by β_2. Specifically, the GVD parameter β_2 can be positive or negative depending on whether the pulse wavelength is below or above the zero-dispersion wavelength λ_D. In the anomalous-dispersion regime $(\lambda > \lambda_D)$ β_2 is negative.

Equation (4.44) does not include the effects of stimulated inelastic scattering such as Raman scattering (SRS) and Brillouin scattering (SBS). In most practical situations it is easy to design the communication system in order to avoid both SBS and SRS, which in any case can be taken into account for ultrashort pulses or Raman gain, as reported in the literature [1].

4.4 NUMERICAL METHODS FOR SOLVING NON-LINEAR PROPAGATION

In the previous section the non-linear propagation equation has been obtained. However, this is a non-linear partial differential equation that does not generally lead to analytical solutions, except for some specific cases in which the inverse scattering method can be adopted [1]. Hence a numerical approach is often needed. Different methods have been reported in the literature, mainly based on two approaches: (1) the finite difference methods and (2) pseudospectral methods. The latter approach is almost an order of magnitude faster than to the former. Here we describe the so-called split-step Fourier method.

First, it is useful to employ a frame of reference moving with the pulse at the group velocity v_g: i.e. transforming the time reference as

$$T = t - z/v_g = t - \beta_1 z. \tag{4.47}$$

Consequently, the term proportional to β_1 disappears from (4.44). Hence, it is opportune to rewrite (4.44) formally as

$$\frac{\partial A}{\partial z} = (\hat{D} + \hat{N})A \tag{4.48}$$

where $\hat{D}$ and $\hat{N}$ are differential operators that account for dispersion and absorption in a linear medium, and for fiber non-linearities on pulse propagation, respectively. That is

$$\hat{D} = -\frac{j}{2}\beta_2\frac{\partial^2}{\partial t^2} - \frac{\alpha}{2} \tag{4.49}$$

$$\hat{N} = j\gamma \mid A \mid^2 \tag{4.50}$$

Equations (4.49) and (4.50) contain more terms if other effects are considered, such as higher-order dispersion effects, and the self-steepening of the pulse edge.

In general, dispersion and non-linearity operate together during signal propagation in the fiber. A mathematical simplification can be obtained if it is assumed that dispersion and non-linear effects act separately. This assumption can be made realistic if only small section of fibers are considered any time. In other words, if we consider a small span of fiber, say h we can assume that propagation from z to $z + h$, can be regarded as a two-step process. In the first one, the non-linearity acts alone, and dispersion is assumed to be zero. In the second one, dispersion is assumed to operate alone in the absence of non-linearities. From a mathematical point of view, the field can be represented as

$$A(z + h, T) = exp(h\hat{D})\exp(h\hat{N})A(z, T) \tag{4.51}$$

It is clear that the smaller h, the smaller the error due to this approximation. In particular, the split-step method is accurate to second order in the step size h [1].

The calculation of the exponential operator $\exp(h D^3)$ is performed in the Fourier domain using the formula

$$\exp(h\hat{D})B(z, T) = \{F^{-1}\exp[h\hat{D}(i\omega)]F\}B(z, T) \tag{4.52}$$

where F denotes the Fourier operator. Since $\hat{D}(i\omega)$ is just a number in the Fourier domain, the evaluation (4.52) is straightforward. The use of an FFT algorithm makes numerical evaluation of (4.52) very fast compared with other methods perhaps based on finite difference schemes.

The accuracy of this method can be improved by adopting a slightly different approach to propagate the optical pulse over one segment from z to $z + h$. In this way the solution (4.51) can be replaced by

$$A(z + h, T) = \exp\left(\frac{h}{2}\hat{D}\right)\exp\left(\int_z^{z+h} \hat{N}(z')dz'\right)\exp\left(\frac{h}{2}\hat{D}\right)A(z, T) \tag{4.53}$$

Thus, the non-linearity effect is included in the middle of the segment rather than at the segment boundary. The integral in the middle exponential is useful to include the z-dependence of the non-linear operator. The most important advantage of using the latter approach is that the leading error is third order in the step size h.

4.5 NON-LINEAR EFFECTS IN FIBER PROPAGATION

In sections 4.3 and 4.4 the pulse propagation in non-linear dispersive media and its numerical solution have been treated, assuming single-channel transmission, in the slowly mvarying envelope approximation, with the assumption that the spectral width of the pulse is much smaller than the frequency of the incident radiation.

Here the main non-linear effects affecting the transmission are reviewed from a physical point of view.

Most non-linear effects in fibers are due to refractive index changes with optical power or they are to scattering phenomena [1–5]. The power dependence of the refractive index is responsible for the Kerr effect whereas scattering phenomena are responsible for Brillouin and Raman effects.

The Kerr effect can give rise to different phenomena: self-phase modulation (SPM), cross-phase modulation (XPM), four-wave mixing (FWM), and modulation instability.

4.5.1 The Kerr effect

The presence of the non-linear susceptibility $\chi^{(3)}$ implies that the refractive index depends on field intensity, according to the expression

$$n = n_0 + n_2 \frac{|\mathbf{E}|^2}{\sqrt{\mu/\varepsilon}} = n_0 + n_2 I \qquad (4.54)$$

where n_0 and n_2 are the linear and non-linear refractive index respectively, and I the field intensity. The presence of Kerr non-linearity gives rise to different effects, depending on the optical field at the fiber input.

Self-phase modulation

Due to n_2, the phase of a signal propagating through the fiber varies with the distance z according to the relation

$$\phi = (n_0 z + \phi_0) + \frac{2\pi}{\lambda} n_2 I(t) z, \qquad (4.55)$$

ϕ_0 being the initial phase. The first term represents the linear phase shift due to signal propagation, the second one depicts the non-linear phase shift. If the optical signal is intensity modulated, a spurious phase modulation occurs due to non-linear phase shift depending on the field intensity (SPM).

When a signal enters the fiber, such time-dependent phase-shift causes a chirp of the transmitted field. The presence of a chirp causes a non-linear broadening of the spectrum, which depends on the bandwidth and on the shape of the injected signal. Note that the SPM spectral broadening enhances pulse broadening due to chromatic dispersion. Thus SPM and GVD strictly interact and their effects generally cannot be separated.

Cross-phase modulation

If N signals, carried by different frequencies, propagate into a fiber, the non-linear phase evolution of any signal depends also on the power of the signals propagating at different frequencies:

$$\Delta\phi_i = \frac{2\pi n_2 z}{\lambda}\left[I_i(t) + 2\sum_{i\neq j} I_j(t)\right].\qquad(4.56)$$

The first term in square brackets depicts the contribution from the SPM, and the second term represents the contribution from all other channels (XPM). The XPM contribution, whose weight is double the SPMweightin, causes a further non-linear spectral broadening so interacting with fiber GVD [4]. Notice that XPM is effective only when the interactive signals are superimposed in time: this means that increasing the GVDwill caused the XPM efficiency to decrease.

Four-wave mixing

Four-wave mixing (FWM) is a parametric interaction among four waves satisfying a particular phase relationship, named *phase matching* [4]. The simplest embodiment of this effect is shown in Fig. 4.1.

Two copropagating waves at frequencies f_1 and f_2 mix and generate sidebands at $2f_1 - f_2$ and $2f_1 - f_2$. These sidebands copropagate with the initial waves and grow at their expense. Similarly, three copropagating waves will generate nine new optical waves (Fig. 4.2) at frequencies $f_{ijk} = f_i + f_j - f_k$ where i, j and k can be 1, 2, or 3. If the channels are equally spaced, some of the generated waves will have the same frequencies as the injected waves. Clearly the appearance of the additional waves as well as the depletion of the initial waves will degrade multichannel systems by crosstalk or excess attenuation.

The efficiency of FWM depends on channel spacing and fiber dispersion

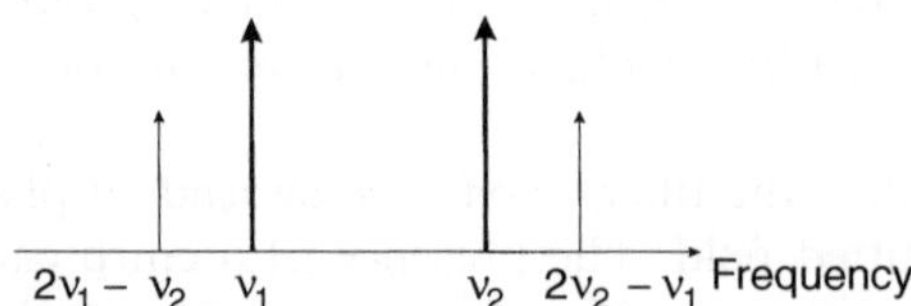

Fig. 4.1 Idealisation of FWM effect.

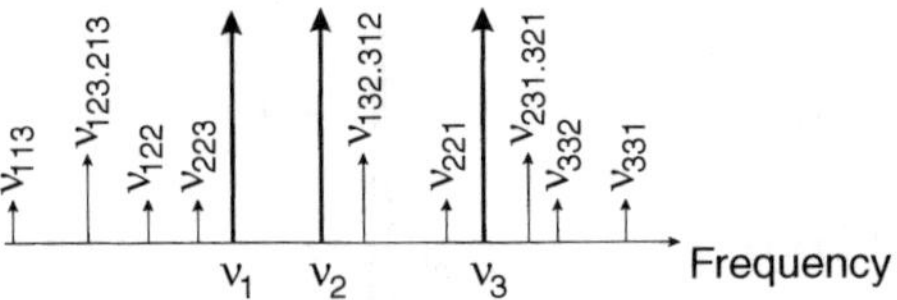

Fig. 4.2 Effect of three copropagating waves: nine new optical waves are generated at frequencies $f_{ijk} = f_i + f_j - f_k$ where i, j and k can be 1, 2, or 3.

[6, 7]. Because of chromatic dispersion, the interacting and generated waves have different group velocities. This destroys the phase matching of the interactive waves and lowers the efficiency of power generation at new frequencies. As a result, FWM efficiency decreases with increasing GVD. Consequently, larger channel spacing and greater GVD lead to lower efficiency.

Modulation instability

Modulation instability (MI) describes different phenomena originated by intermingling between the Kerr effect and chromatic dispersion. All these phenomena generate the exponential growth or attenuation of the sidebands with respect to a central frequency propagating through the fiber [8–10].

In the anomalous dispersion region of the fiber, if a strong CW signal with pulsation ω_0 and power A^2 is injected into the fiber, a probe with an angular frequency satisfying the condition

$$\omega < \omega_c = \sqrt{\frac{3\chi^{(3)}_{xxxx}\omega_0 A^2}{2n_0 c\pi\rho_m \mid \beta_2 \mid}} \tag{4.57}$$

grows exponentially with the distance as a result of MI and at the expense of the signal at ω_0. Due to MI, a signal constituted by a CW carrier with a very small modulation can be transformed in a train of pulses as far as the small sidebands are amplified at the expense of the central carrier. The same mechanism can cause a large spectral broadening of a modulated signal if the MI gain is sufficiently high. In Fig. 4.3 the gain curve of MI is reported versus the detuning $\omega - \omega_0$ for two different values of the chromatic dispersion β_2. The CW power is 1 mW.

A different behavior can be observed, in both the anomalous and the normal regions of the fiber, in the case where a modulated signal with two strong sidebands and a weak central carrier are injected into the fiber. Amplification of the central carrier can then accur at the expense of the sidebands if the following conditions are satisfied on the power A_S^2 of the sidebands:

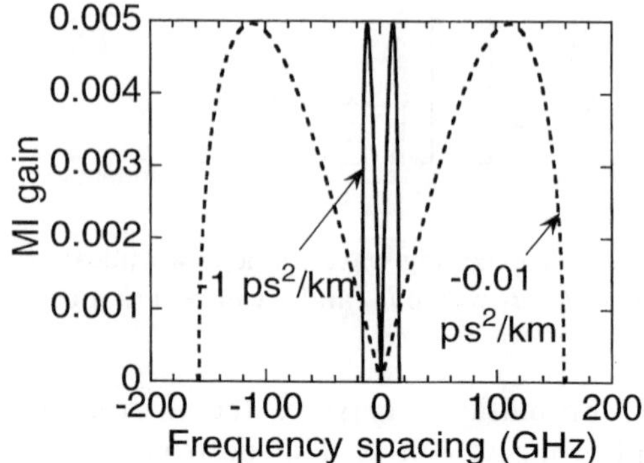

Fig. 4.3 Gain curve of MI versus the detuning $\omega - \omega_0$ for two different values of the chromatic dispersion β_2 and a CW power of 1 mW.

$$A_S^2 > \frac{4n_0^2 \mid \beta_2 \mid \Delta\omega^2}{3\beta_0\chi_{xxxx}^{(3)}} \quad \beta_2 < 0 \tag{4.58}$$

$$A_3^2 > \frac{4n_0^2\beta_2\Delta\omega^2}{9\beta_0\chi_{xxxx}^{(3)}} \quad \beta_2 > 0 \tag{4.59}$$

where β_0 is the propagation constant at the central carrier and $\Delta\omega$ is the modulation frequency, that is the frequency spacing between a sideband and the carrier.

4.5.2 Stimulated Brillouin scattering

Stimulated brillovin scattering (SBS) relates to the interaction of light with sound waves. In particular, it can be described as a three-wave interaction among the original optical wave, the acoustic wave and the so-called Stokes wave [5]. The three waves must fulfil the energy conservation law that, in the analysed case, write $\omega_0 = \omega_S + 2\pi f_a$, where ω_0 is the angular frequency of the injected light (pump wave), ω_S is the the angular frequency of the Stokes wave and f_a is the frequency of the acoustic wave, which is about 11 GHz in conventional fibers. In a single-mode optical fiber, optical waves can propagate only along the direction of the fiber axis, so the conservation law is simply a scalar relationship, $\beta_0 = \beta_S + \beta_a$. As an important consequence, no effect takes place in practice if the Stokes and pump waves propagate in the same direction; the maximum efficiency is obtained when the Stokes and pump waves propagate in opposite directions. SBS depletes the incident wave and generates a potentially strong scattered beam propagating back toward the transmitter.

In a single channel the critical power level at which SBS degrades system performance is given by the following equation [5, 11–13]:

$$P_c = \frac{21bA_e}{g_B L_e} \qquad (4.60)$$

where b is factor accounting for the relative polarisation properties of the fiber. In a polarisation-maintaining fiber with identical pump and probe polarisation states, $b = 1$; in a conventional fiber $b = 2$. A_e is the effective area of the propagating waves, which is calculated by evaluating the average modal overlap between the pump and the probe waves [5]. In general, if the pump and the probe wavelengths are comparable and both are slightly longer than the fiber cutoff wavelength, A_e is equal to the core area of the fiber. L_e is the effective fiber length, given by $L_e = (1 - e^{-\alpha L})/\alpha$, with α and L the fiber loss and the fiber length respectively. Finally, g_B is the maximum steady-state Brillouin gain.

In multichannel systems it can be shown [5] that each channel interacts with the fiber independently of other channels. Consequently, the critical power is constant with increasing number of channels.

Moreover, the optical bandwidth for SBS in silica is about 20 MHz at 1550 nm and varies as λ^{-2}. High modulation rates produce broad optical spectra hence they reduce the SBS effect.

4.5.3 Stimulated Raman scattering

Stimulating Raman scattering (SRS) is due to the interaction of photons with molecular vibrations. Incident light scattered by molecules experiences a downshift in optical frequency. The frequency change is just the molecular vibrational frequency (the Stokes frequency) [5]. If two optical waves separated by the Stokes frequency are coinjected into a Raman-active medium, the lower-frequency (probe) wave will experience optical gain generated by, and at the expense of, the higher frequency (pump) wave.

The Raman curve for a typical single-mode fiber is shown in Fig. 4.4 [4]. Unlike SBS, the SRS effect is extremely broadband and the Raman Stokes wave is frequency down shifted by about 12 THz with respect to the pump (about 100 nm in the third window).

A Raman threshold power can be defined [14]. For a fiber longer than a few kilometers, the pump power at which a pump depletion of 3 dB occurs is given by

$$gL_e P/(bA_e) = 16 \qquad (4.61)$$

where g is the gain coefficient. The Raman threshold in conventional fibers

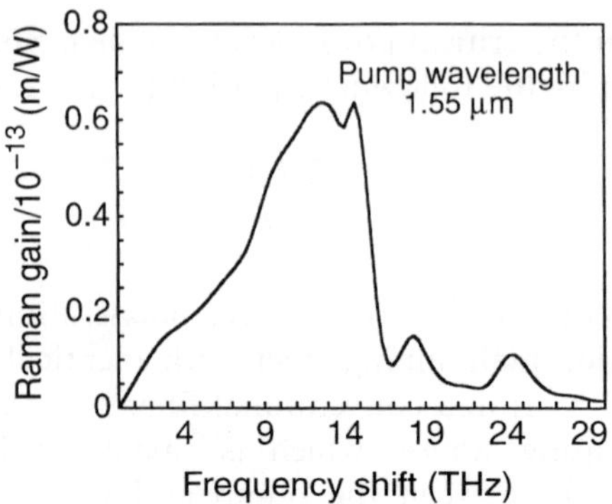

Fig. 4.4 The Raman curve for a typical single-mode fiber.

is of the order of 500 mW for copolarised pump and Stokes waves (about 1 mW for random polarisations), making SRS negligible in practice for single-channel transmission.

However, because of the extremely large ban dwidth of the Raman gain curve, this effect can be important for a signal composed of many waves at different frequencies. In this case the signals at longer wavelengths, are amplified by the Raman effect at the expense of those at shorter wavelengths and a relevant degradation of the worst channel can occur even when the overall optical power is quite a lot smaller than the Raman threshold.

A general result can be obtained by approximating the Raman gain curve versus the pump-Stokes separation as a triangle with its maximum for $\Delta\omega = \Omega_M$ and zero for $\Delta\omega < 0$ and $\Delta\omega > \Omega_M$ [15]. In general, considering that the worst channel accurs at the lowest frequency ($\Delta\omega < 0$), the transmitted channels are all in the bandwidth of the triangular gain curve, giving a size of about 125 nm (15 THz). In this case the limit power per channel P_{Ram} causing a depletion of 0.5 dB on the worst channel is given by

$$P_{Ram} = \frac{8.7 \cdot 10^{15}}{N_C B_{opt} L_{eff}} \tag{4.62}$$

where B_{opt} is the total optical bandwidth and N_c is the channel number. The Raman effect can therefore be avoided by keeping the peak optical power along the link below P_{Ram}.

REFERENCES

1. G.P. Agrawal, Nonlinear Fiber Optics. Academic Press: New York (1989).
2. A. Yariv, *Quantum Electronics*, 3rd edn, John Wiley, New York (1989).

3. M. Schubert and B. Wilhelmi, *Nonlinear Optics and Quantum Electronics*, John Wiley, New York (1986).

4. E. Iannone, F. Matera, A. Mecozzi, and M. Settembre, *Nonlinear Optical Communication Systems*, John Wiley, New York (1998).

5. A.R. Chraplyvy, Limitations on lightwave communications imposed by optical-fiber nonlinearities, *IEEE/OSA J. Lightwave Technol.*, **8**, 1548–1557 (1990).

6. N. Shibata, R.P. Braun, and R.G. Waarts, Phase-mismatch dependence of efficiency of wave generation through four-wave mixing in single-mode optical fiber, *IEEE J. Quantum Electron*, **QE-23**, 1205–1210 (1987).

7. R.W. Tkach, A.R. Chraplyvy, F. Forghieri, A.H. Gnauck, and R.M. Derosier, Four-photon mixing and high-speed WDM systems, *IEEE/OSA J. Lightwave Technol.*, **13**, 841–849 (1995).

8. D. Anderson and M. Lisak, Modulation instability induced by cross phase modulation, *Phys. Rev. Lett.*, **59**, 880–883 (1987).

9. A. Barthelemy and R. La Fuente, Unusual modulation instability in fibers with normal dispersion, *Opt. Commun.*, **73**, 409–412 (1989).

10. G. Cappellini and S. Trillo, Third order three-wave mixing in single-mode fibers: exact solution and spatial instability effects, *J. Of Opt. Soc. Amer. B*, **8**, 824–838 (1991).

11. D. Cotter, Stimulated Brillouin scattering in monomode optical fibre, *J. of Opt. Commun.*, **4**, 10–19 (1983).

12. A. Cosentino and E. Iannone, SBS threshold dependence on line coding in phase modulated coherent optical systems, Electron. Lett., **25**, 1459–1460 (1989).

13. G.C. Valley, A review of stimulated Brillouin scattering excitedwith a broadband pump laser, *IEEE J. of Quantum Electron.*, **QE-22** (1986).

14. J. Auyeung and A. Yariv, Spontaneous and stimulated Raman scattering in long low loss fibers, *IEEE J. of Quantum Electron.*, **QE-14** (1987).

15. A.R. Chraplyvy, Optical power limits in multi-channel wavelength-division multiplexed systems due to stimulated Raman scattering, *Electron. Lett.*, **20**, 58–59 (1984).

5

Optical Amplifiers

In the latter half of the 1980s there was a significant change in the telecommunications industry. Optical fiber networks were installed throughout the developed world and submarine cables containing optical fibers were deployed to link continents. This great mutation in the fundamental technology used in long-haul transmission systems was induced by several advantages of optical fibers. To make better use of their potential, it was necessary to improve receiver sensitivities and to develop practical wavelength division multiplexing (WDM) techniques to combine channels (Chapter 11).

Practical optical amplifiers are critical components in the development of both these capabilities. The sensitivity of direct-detection (non-heterodyne) receivers is limited by thermal noise in the receiver front-end to one or two orders of magnitude worse than the quantum limit (chapters 8 and 9). The use of optical amplifiers allows this limitation to be overcome without the use of more complex coherent (heterodyne or homodyne) systems. Furthermore, given the speed of the electronics used in optical transmitters and receivers, WDM is essential for ultrahigh bit rate systems. In fact, the complexity of the regenerators increases significantly if WDM channels must be separated, i.e. demultiplexed and multiplexed again. One way for greatly simplifying the regenerators used in WDM systems is to replace them with optical amplifier repeaters. The optical multplexers and demultiplexers are then all replaced by a single device.

Finally, optical amplifiers can allow transparent optical networks to be realised, thus avoiding optoelectronic conversion when a signal travels throughout the network itself. Optical amplifiers can also be used to achieved other functions besides signal amplification. In this chapter two different technologies are reported here for the realisation of optical amplifiers, based on semiconductors or based on optical fibers, respectively.

5.1 SEMICONDUCTOR OPTICAL AMPLIFIERS

Even if at the present time EDFAs are widely used for signal amplification along fiber-optic links, semiconductor optical amplifiers (SOA) are intensively studied as an alternative for system applications [1] as preamplifiers, inline amplifiers and booster amplifiers. Furthermore, they are also investigated for applications as optical switching elements [2], as optical modulators [3], as frequency converters [4] and as frequency chirpers for pulse compression [5].

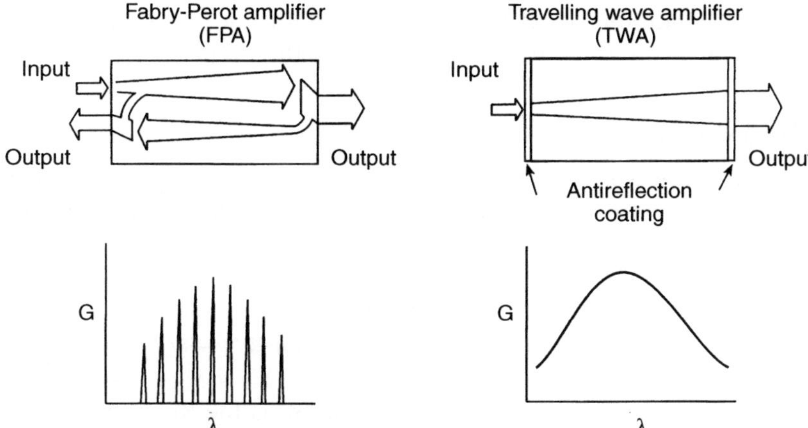

Fig. 5.1 Basic schemes for semiconductor laser amplifiers.

This chapter reports the fundamental device features, the noise characteristics and the dynamic behavior of SOAs.

5.1.1 Fundamental device characteristics

SOAs are based on the same principle of stimulated emission as semiconductor lasers. But they need to avoid any optical feedback and also require a higher pumping rate or longer interaction length than laser oscillators in order to obtain a reasonable signal gain. The comparison between an SOA and a resonant semiconductor laser amplifier (e.g. a Fabry-Perot laser) is sketched in Fig. 5.1. When the reflectivity of both facets are suppressed, for instance by precise antireflection (AR) coating, the device works as a single-pass amplifier, instead of a Fabry-Perot (FP) laser oscillator. This is commonly known as a traveling wave amplifier (TWA). The traveling wave device presents superior gain bandwidth, signal gain saturation and noise figure with respect to the FP device.

In fabricating a TWA, FP cavity resonance in the laser diode must be suppressed through its end-facet reflectivity, whose value R $(= \sqrt{R_1 R_2}$, with R_1 and R_2 the input and output reflectivities, respectively) depends on the single-pass gain G_s and the allowed signal gain ripple ξ due to residual FP resonance [6]:

$$R = \frac{1}{G_s} \cdot > \frac{\sqrt{\xi} - 1}{\sqrt{\xi} + 1} \tag{5.1}$$

When $G_s \sqrt{R_1 R_2} < 0.17$ the signal gain ripple becomes less than 3 dB; that is the TWA condition. Apart from AR coating, the facet reflectivity can be reduced by angled facet and window facet structures [1].

Small-signal gain

Small-signal gain for an SOA is given by the power transmission coefficient of an active FP etalon, which includes a gain medium in the cavity, as a function of input signal frequency ν in the following form [7]:

$$G(\nu) = \frac{(1 - R_1)(1 - R_2)G_s}{(1 - \sqrt{R_1 R_2}G_s)^2 + 4\sqrt{R_1 R_2}G_s \sin^2[\pi(\nu - \nu_0)/\Delta\nu]} \tag{5.2}$$

where ν is the cavity resonant frequency and $\Delta\nu$ is the free spectral range (FSR) of the SOA. The single-pass gain G_s is expressed as

$$G_s = exp[(\Gamma g - \alpha)L] \tag{5.3}$$

where Γ is the optical mode confinement factor for the active layer, g is the material gain coefficient, a is the absorption coefficient, and L is the amplifier length.

From eq. (5.2) the 3 dB bandwidth (FWHM: full width at half maximum) of an FP laser is expressed as

$$B = \frac{2\Delta\nu}{\pi} sin^{-1}\left[\frac{1 - \sqrt{R_1 R_2}G_s}{(4\sqrt{R_1 R_2}G_s)^{1/2}}\right] \tag{5.4}$$

On the other hand, the 3 dB bandwidth of a TWA is three orders for magnitude larger than that for an FP device, since the TWA is determined by the full gain width of the amplifier medium itself, without being restricted by the FP gain profile. Gain ripple ξ, which is defined as the difference between resonant and non-resonant signal gain, is derived from (5.1) as

$$\xi = \left[\frac{1 + \sqrt{R_1 R_2}G_s}{1 - \sqrt{R_1 R_2}G_s}\right]^2 \tag{5.5}$$

It should be less than 3 dB for TWAs over the entire signal-gain spectrum.

Unfortunately signal gain spectra of SOAs are different for the two polarisation states TE and TM. Figure 5.2 shows signal gain spectra of a near-TWA for both TE and TM signal polarisation states [8]. The spectra have different amplitudes (dichroism) and different FSRs (birefringence). The random change over time in the polarisation state at the fiber output is responsible for a reduction in TWA effective bandwidth. Birefringence is

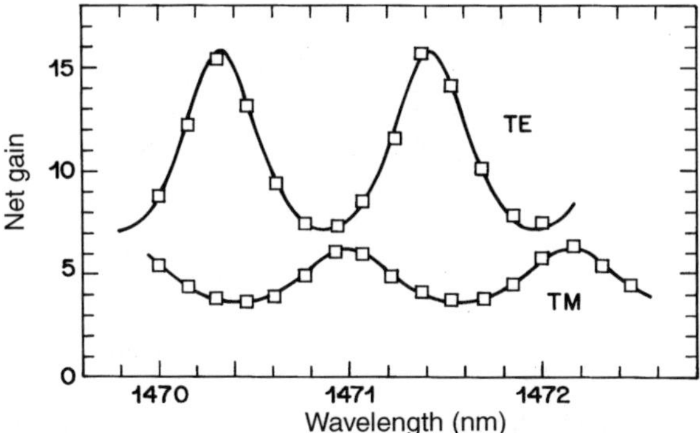

Fig. 5.2 Signal gain spectra for TE and TM polarised input signals. Solid lines are calculated FP resonant curves [8]. (After [1] with permission of John Wiley & Sons).

negligible when the signal gain ripple disappears due to sufficiently low facet reflectivity. Several methods of reducing or compensating for the signal gain difference between polarisations have been demonstrated [1]. Among them, a thick narrow active layer structure, separate confinement heterostructure (SCH), large optical cavity (LOC) strucutres, twin-amplifier configurations or a double-pass configuration.

Ideally, if the active layer width is the same as the active layer thickness, then signal gain will become completely insensitive to polarisation. Problems exist in maintaining the single transverse mode conditions and reproducibility in fabricating narrow active layer structures. However, a TWA with a thick active layer has two drawbacks: low saturation output power if the operating carrier density is low, and thermally related gain saturation if the operating current density is high [1].

Besides the thick narrow active layer structure, the SCH or LOC struc-

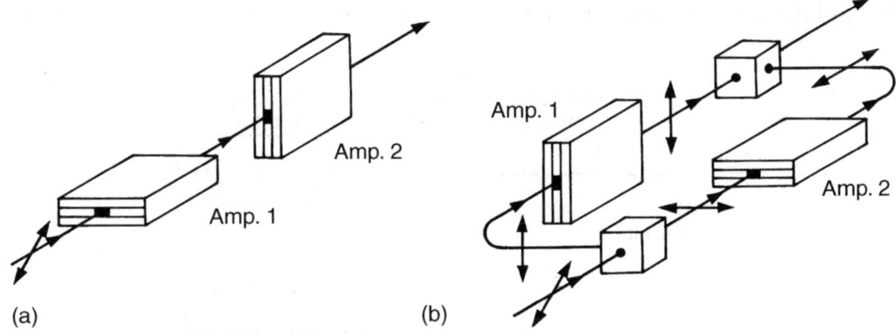

Fig. 5.3 Twin-amplifier configurations: (a) series and (b) parallel settlements [10]. (After [1] with permission of John Wiley & Sons).

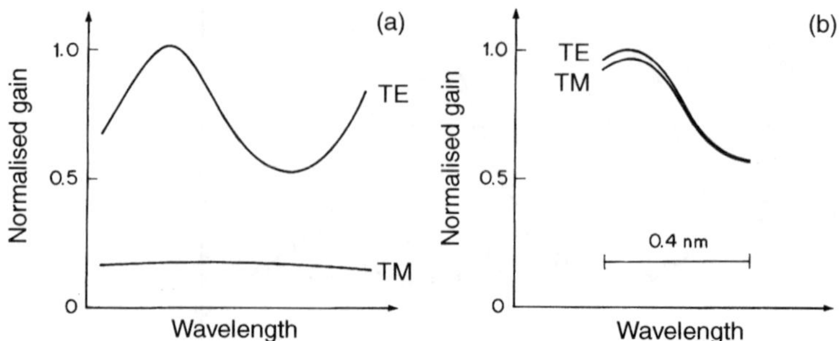

Fig. 5.4 Normalised signal gain for two orthogonal polarisations of input signal: (a) for a single amplifier (TE and TM); (b) for amplifiers in parallel configuration [10]. (After [1] with permission of John Wiley & Sons).

tures are also useful for reducing polarisation sensitivity [9]. They can obtain high saturation output power because they have thin active layers resulting in high operating carrier density.

Twin-amplifier configurations [10] are shown in Fig. 5.3. In series, the TE polarisation in the first amplifier becomes TM polarisation in the second amplifier and vice versa. If both amplifiers exhibit equal gain characteristics, the combined system is expected to exhibit polarisation insensitivity. Even though the series configuration is simple, it suffers from the problem of mutual coupling between amplifiers. In the parallel configuration, each of the two amplifiers sees a TE-polarised signal. Gain spectra for both polarisations are shown in Fig. 5.4, for a single arrangement and the parallel arrangement. The maximum $\xi_{TE/TM}$ of the single amplifier is about 7 dB in the example. The parallel configuration gives $\xi_{TE/TM}$ of less than 0.6 dB at any wavelength.

A double-pass configuration [11] is schematically shown in Fig. 5.5. In this configuration the signals pass through the same amplifier twice, and the polarisation is rotated by 90° between passes. Therefore, this config-

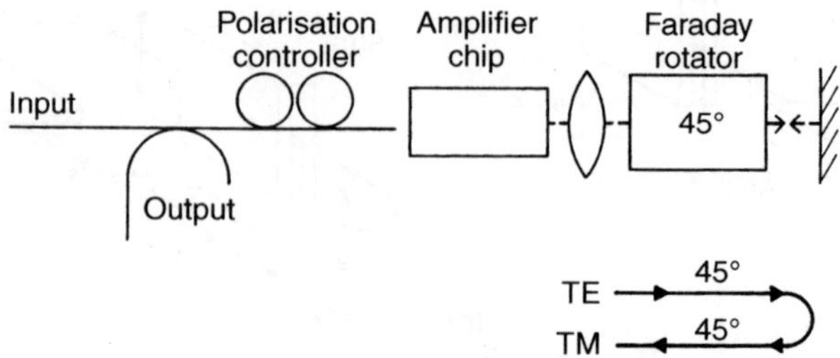

Fig. 5.5 Configuration of a double-pass amplifier [11]. (After [1] with permission of John Wiley & Sons).

uration provides an intermediate value of TE and TM gains for input signals with any polarisation state, similar to the series arrangement. In spite of a 6 dB loss due to the 3 dB fiber coupler, this configuration gives relatively high signal gain as a single amplifier because it operates twice on the same signal. The disadvantages of this configuration is the low saturation output power, as the signal needs to go through the amplifier twice.

Signal gain saturation

Signal-gain saturation of an SOA is caused by a reduction of the population inversion in the active layer due to an increase in stimulated emission. Gain saturation characteristics are especially important in optical repeaters and multichannel amplifiers, which require high-power operation.

The injected carrier density can be determined by using the following single-mode rate equations [12] to which a term for the optical injection is added:

$$\frac{dN_e}{dt} = \frac{J}{qd_a} - \frac{N}{\tau_c} - A_g \nu_g (N_e - N_o)S \tag{5.6}$$

$$\frac{dS}{dt} = -\frac{S}{\tau_p} + \frac{A_g \nu_g \Gamma N_e}{V_0} + A_g \nu_g \Gamma (N_e - N_o)S + S_{in} \tag{5.7}$$

where N_e is the injected carrier density, J the current density, q the electron charge, d_a the active layer thickness, τ_c the carrier lifetime, A_g the differential gain, ν_g the group velocity in the amplifying medium, N_0 the carrier density required for transparency, S the photon density, τ_p the photon lifetime, Γ the optical mode confinement factor, V_0 the optical mode volume, and S_{in} the injected signal photon density. For an accurate analysis, we should use multimode rate equations [13] that can properly include the effect of the gain saturation due total spontaneous emission power. However, the single-mode rate equations are simple and physically sufficient to treat gain saturation due to amplified signal power. By representing the number of photons in the active layer by the optical intensity I ($= h\nu \nu_g S$) and using the linear approximation of the gain coefficient given as $g = A_g(N_e - N_0)$, the following equation is obtained [1]:

$$g = \frac{g_0}{1 + I/I_s} \tag{5.8}$$

where g_0 is the unsaturated gain coefficient and I_s is the saturation intensity defined as the optical intensity that reduces g to $g_0/2$. In (5.8) $I_s = h\nu/(A_g \tau_c)$.

The growth rate for the signal intensity $I(z)$ along the amplifier thus beomes

$$\frac{dI(z)}{dz} = \frac{\Gamma g_0}{1 + I(z)/I_s} I(z) \tag{5.9}$$

Here the absorption coefficient α is omitted. Signal gain G is derived by integrating (5.9) from the input to the output port of the laser amplifier [14]:

$$G \equiv \frac{I_{out}}{I_{in}} = G_0 \exp\left[-\frac{I_{out} - I_{in}}{I_s} \right] = G_0 \exp\left[-\frac{G-1}{G} \frac{I_{out}}{I_s} \right] \tag{5.10}$$

where I_{in} and I_{out} are the input and the output optical intensities and G_0 is the unsaturated signal gain in the steady state. Signal gain is reduced to $1/e$, by 4.34 dB, when the output intensity I_{out} is equal to the saturation intensity I_s. In practical cases the saturation output power P_{3dB}, defined as the power at which the signal gain has fallen by 3 dB from the unsaturated value, ranges from 5 to 9 dBm, which is about 20 dB larger than for an FP laser amplifier. The increase in saturation power of the TWA is attributed to the suppression of the resonant effect; that is, a slight saturation in gain coefficient produces large signal gain saturation in FP amplifiers, and increases the saturation intensity of the amplifying medium due to high operating carrier density [15]. If the device length L, the optical mode confinement factor Γ, and the facet reflectivity R are decreased, then the saturation intensity I_s of the TWA increases because operation takes place in the large-g region. Large I_s can also be obtained if the carrier lifetime is shortened, such as by carrier diffusion from nearby carrier storage regions [16]. If I_s is not degraded, an increase in the mode cross-sectional area ($d_a w/\Gamma$) is another effective way to increase the TWA output power. It has also been shown [17] that the saturation power can be improved by setting the signal wavelength higher than the gain peak wavelength. In fact, P_{3dB} monotonically increases with an increase in the signal wavelength. This is because the gain saturation reduces the injected carrier density and causes the gain peak to shift towards higher wavelength in accordance with the reverse process of band filling, which compensates for the decrease in gain at the longer wavelength. This feature of a semiconductor gain medium has been demonstrated in a TWA, where the saturated signal gain spectrum coincides exactly with the unsaturated spectrum under less biased conditions [18]. This experimental result also clarifies that the semiconductor laser gain saturates homogeneously over the entire gain spectrum, an important feature for understanding saturation-induced crosstalk in multichannel amplification. The choise of operating wavelength is important for both high-power and low-noise operation.

5.1.2 Noise characteristics of semiconductor laser amplifiers

Noise is one of the most important characteristics of any amplifier. Here we present an analysis of noise processes based on the master equation approach [19].

Master equation

The master equation is here reported considering a two-level atomic system. Nevertheless, the results can be applicable to semiconductor laser amplifiers.

The emission and absorption of a photon in a unit time interval are denoted by a and b, respectively. Note that a and b include the population of atomic levels; i.e. $a = AN_2$ and $b = AN_1$, where A is Einstein's A coefficient, and N_1 and N_2 are the population of the lower and the upper states, respectively. The probability that the photon number belonging to a mode increases from $n-1$ to n by spontaneous and induced emission is given by an. On the other hand, the probability that the photon number decreases from $n+1$ to n via an induced absorption is given as $b(n+1)$. Similarly, the probability that photon number increases from n to $n+1$ is $a(n+1)$, and the probability that it decreases from n to $n-1$ is bn. Hence the probability $P_n(t)$ that n photons exist in an amplified wave packet obeys the following differential equation:

$$\frac{dP_n}{dt} = b(n+1)P_{n+1} + anP_{n-1} - a(n+1)P_n - bnP_n \qquad (5.11)$$

which is called the master equation.

Estimation of photon fluctuation

From the master equation, the rth moment of the photon number can be estimated as

$$<n^r> = \sum_n n^r P_n \qquad (5.12)$$

The first and second moment obey, respectively, the following equations:

$$\frac{d<n>}{dt} = (a-b)<n> + a \qquad (5.13)$$

$$\frac{d<n^2>}{dt} = 2(a-b)<n^2> + (3a+b)<n> + a \qquad (5.14)$$

In the following, the fluctuation of photon number is assumed to be a Gaussian noise and we are not concerned with moments higher than the third order.

By solving (5.13) and (5.14) for initial conditions $< n(0) > \overline{n_0}$ and $< n^2(0) > \overline{n_0^2}$, the average and the variance σ^2 of the photon number can be derived as

$$< n >= G_0\overline{n_0} + (G - 1)n_{sp} \tag{5.15}$$

$$\sigma^2 = G\overline{n_0} + (G - 1)n_{sp} + 2\overline{n_0}G(G - 1)n_{sp} + \\ (G - 1)^2 n_{sp} + G^2(\overline{n_0^2} - \overline{n_0}^2 - \overline{n_0}) \tag{5.16}$$

where G and n_{sp} are the gain and the population inversion parameter (spontaneous emission factor), respectively, which are given as $G = \exp[(a-b)t]$ and $n_{sp} = a/(a-b)$.

Here the following assumptions are adopted:

(1) An optical bandpass filter having a bandwidth Δf is inserted at the output end of the amplifier,
(2) The bandwidth of the amplifier gain is much wider than Δf.
(3) The spectral width of the incident light is much narrower than Δf.
(4) The electric field has a single transverse mode.

Both the average and the variance of the photon number are summed up over longitudinal modes. The longitudinal mode density per unit frequency interval of a traveling wave in a suppositious waveguide section having length l is given as follows [20]:

$$m(f) = l/c \cdot \tag{5.17}$$

The outgoing rate of photons is then given simply by

$$< N >= \int_0^\infty < n > (c/l)m(f)df = \int_0^\infty < n > df \cdot \tag{5.18}$$

Therefore, from (5.15), (5.17) and (5.18) we have

$$< N > +G\overline{N_0} + (G - 1)n_{sp}\Delta f, \tag{5.19}$$

where $\overline{N_0}$ denotes the incoming rate of photons:

$$\overline{N_0} = \int_0^\infty \overline{n_0}df. \tag{5.20}$$

The first term of (5.19) expresses the amplified signal, whereas the second term is the spontaneous emission noise.

Similarly, the sum of σ^2, denoted by Σ^2, is given as

$$\Sigma^2 = G\overline{N}_0 + (G-1)n_{sp}\Delta f + 2\overline{N}_0 G(G-1)n_{sp} + $$
$$(G-1)^2 n_{sp}^2 \Delta f + G^2(\overline{N_0^2} - \overline{N_0}^2 - \overline{N}_0) \tag{5.21}$$

where

$$\overline{N_0^2} = \int_0^\infty \overline{n_0^2}\, df. \tag{5.22}$$

The first term of (5.21) expresses the shot noise induced by the amplified signal light, the second term the shot noise induced by spontaneously emitted light, the third term the beat noise between the signal and spontaneous emission, and the fourth term the beat noise among spontaneous emission. The fifth term depends on the state of the incident electric field; however, if we deal with a coherent incident light, this term becomes zero, because the average and for variance of the photon number of the coherent light are equal.

Noise figure of laser amplifier

To evaluate the noise figure of a laser amplifier, we assume an ideal photodetector (quantum efficiency equal to 1), no circuit noise and a coherent incident light. When the signal light is detected with an incoming rate of photons of $\overline{N}_0$, the average signal photocurrent is $q\overline{N}_0$, whereas the variance of the photocurrent (shot noise) is $(2q^2\overline{N}_0 B)$, B being the bandwidth of the noise-measuring system. Then the S/N ratio of the received signal is given as

$$(S/N)_i = \overline{N}_0/(2B). \tag{5.23}$$

This equation can also be derived from the 'photon concept'. The average photon number measured in a time interval T is $\overline{N}_0 T$, whereas the variance is $\Sigma^2 T = \overline{N}_0 T$. The S/N ratio is the ratio of the square of the average photon number to the variance; therefore, $S/N = \overline{N}_0 T$. If we use the effective bandwidth $B = 1/2T$, equation (5.23) is obtained.

On the other hand, when the amplified signal is detected, equation (5.21) the S/N ratio gives

$$\left(\frac{S}{N}\right)_0 = \frac{(G\overline{N}_0)^2}{2[G\overline{N}_0 + (G-1)n_{sp}\Delta f + 2\overline{N}_0 G(G-1)n_{sp} + (G-1)^2 n_{sp}^2 \Delta f]B}. \tag{5.24}$$

When the gain G is sufficiently large, equation (5.24) simplifies to

$$\left(\frac{S}{N}\right)_0 = \frac{\overline{N}_0)^2}{2[2\overline{N}_0 n_{sp} + n_{sp}^2 \Delta f]B}.$$

(5.25)

Furthermore, if the filter bandwidth Δf is so small that the second term of the denominator can be neglected as compared with the first term, the S/N ratio becomes

$$\left(\frac{S}{N}\right)_0 = \frac{\overline{N}_0}{4n_{sp}B}.$$

(5.26)

The S/N ratio given by the last equation is called the beat-noise limited S/N ratio of a laser amplifier output. In an actual TWA laser amplifier, the beat noise among spontaneous emission deteriorates the S/N ratio, because it is difficult to restrict the bandwidth of the spontaneous emission by narrowband filtering.

The 'noise figure' F of a laser amplifier is defined as the ratio of the S/N ratios at the input and output stages:

$$F = \frac{(S/N)_0}{(S/N)_i}$$

(5.27)

By using (5.23), (5.26), and (5.27) in the beat noise limited state, the following expression is obtained:

$$F = 2n_{sp}.$$

(5.28)

This means that even when n_{sp} approaches unity by enhancing the population inversion, the minimum value of the noise figure F is 3 dB. Thus the noise figure of an ideal TWA laser amplifier is 3 dB.

5.1.3 High-speed operation of semiconductor amplifiers: the dynamical model

Amplification of ultrashort optical pulses in semiconductor optical amplifiers produces considerable spectral broadening and distortion of the non-linear phenomenon of self-phase modulation (SPM). The physical mechanism behind SPM is gain saturation, which leads to intensity-dependent changes of refractive index in response to variations in carrier density. The effect of shape and initial frequency chirp of input pulses on the shape and the spectrum of amplified pulses is discussed below. First the basic equations which govern the dynamics of the amplification process are described, indicating the approximations made, and the solution in the relevant case in which the input pulses are much shorter than the carrier lifetime.

Basic equation

Carrier dynamics in the amplifier, assuming that the pulsewidth τ_p is much larger than the intraband relaxation time τ_{in} that governs the dynamics of the induced polarisation, can be described by the carrier density rate equation [21, 22]:

$$\frac{\partial N}{\partial t} = D\nabla^2 N + \frac{1}{qV} - \frac{N}{\tau_c} - \frac{\alpha(N - N)_0)}{h\omega_0} \mid E \mid^2 \qquad (5.29)$$

where N is the carrier density (for electrons as well as holes), D the diffusion coefficient, I the injection current, V the active volume, τ_c the spontaneous carrier lifetime, $h\omega_0$ the photon energy, a the gain coefficient, and N_0 the carrier density required for transparency.

The propagation of the electromagnetic field inside the amplifier is governed by the wave equation

$$\nabla^2 E - \frac{\varepsilon}{c_2}\frac{\partial^2 E}{\partial t_2} = 0 \qquad (5.30)$$

where c is the light velocity. The dielectric constant ε is given by

$$\varepsilon = n_b^2 + \chi \qquad (5.31)$$

where the background refractive index n_b is generally a function of the transverse coordinates x and y to account for the dielectric waveguiding in semiconductor laser amplifiers. The suceptibility χ represents the contribution of the charge carriers inside the active region of the amplifier and is a function of the carrier density N. The exact dependence of χ on N is quite complicated as it depends, among other things, on details of the band structure. A simple phenomenological model has been found very useful in the theory of semiconductor lasers [22]. In this model, χ is assumed to depend on carrier density N linearly and is given by

$$\chi(N) = -\frac{\bar{n}c}{\omega_0}(\alpha + i)\alpha(N - N_0) \qquad (5.32)$$

where $\bar{n}$ is the effective mode index. The carrier-induced index charge, responsible for SPM, is accounted for through the linewitdh enhancement factor α.

Equations (5.29) to (5.32) provide a general theoretical framework for the propagation of optical pulses in semiconductor amplifiers.

In order to obtain a simplified mathematical modeling, the following assumptions are adopted:

(1) The device is an ideal TWA amplifier, whose active region dimensions are such that the amplifier supports a single waveguide mode.
(2) The input light is linearly polarised during propagation.
(3) The width and thickness of the active region are smaller, but the amplifier length is much larger, than the diffusion length.
(4) Carrier diffusion is neglected.

Accordingly, the following three equations are obtained:

$$\frac{\partial P}{\partial z} = (g - \alpha_{int})P \tag{5.33}$$

$$\frac{\partial \phi}{\partial z} = -\frac{1}{2}\alpha g \tag{5.34}$$

$$\frac{\partial g}{\partial \tau} = \frac{g_0 - g}{\tau_c} - \frac{gP}{E_{sat}} \tag{5.35}$$

where $P(z,\tau)$ and $\phi(z,\tau)$ are the power and phase of the electric field, respectively; α_{int} represents the internal losses, and g is the gain defined as

$$g(N) = \Gamma\alpha(N - N_0) \tag{5.36}$$

In equation (5.35) g_0 is the small-signal gain defined as

$$g_0 = \Gamma\alpha N_0(I/I_0 - 1), \tag{5.37}$$

I_0 being the current required for transparency. The term E_{sat} represents the saturation energy of the amplifier, and is given by $E_{sat} = \eta\omega_0\,\sigma\,/\,a$, where σ the cross-section ($= wd/\Gamma$).

Equation (5.34) shows the origin of SPM. The time dependence of the saturated gain $g(z,\tau)$ leads to a temporal modulation of the phase, i.e. the pulse modulates its own phase as a result of gain modulation.

Pulse shape and spectrum

The evolution of the pulse inside the amplifier requires, in general, a numerical solution of (5.33) to (5.35), which represent the dynamical model of the amplifier. However, if the internal losses are much smaller than the gain, as is often the case, the equations can be solved in closed form [21].

Two relevant circumstances are considered here, according to the fact that the input pulsewidth is much smaller than the carrier lifetime (isolated pulse) or that it is comparable to it (repetitive pulse).

Isolated pulse amplification

As an example, Fig. 5.6 and 5.7 show the shape and spectrum of the amplified pulse for several values of the unsaturated gain G_0, when the input pulse is Gaussian with energy such that $E_{in}/E_{sat} = 0.1$. The linewidth enhancement factor may vary from one amplifier to another as it depends on the relative position of the gain peak with respect to the operating wavelength. The value of $\alpha = 5$ has been chosen as a representative valuation for all calculations. In particular, Fig. 5.6 reveals that the amplified pulse becomes asymmetric so that its leading edge is sharper compared with the trailing edge. Sharpening of the leading edge is a common feature of all amplifiers [23, 24], and occurs because the leading edge experiences larger gain than the trailing edge.

In the meantime, the pulse spectra shown in Fig. 5.7 reveal features that are particular to semiconductor laser amplifiers. In general, the spectrum develops a multipeak structure. The dominant spectral peak shifts to the low-frequency side.

Such a redshift increases with amplifier gain G_0 and can be as large as 3–5 times the spectral width of the input pulse. For 10 ps input pulses, the

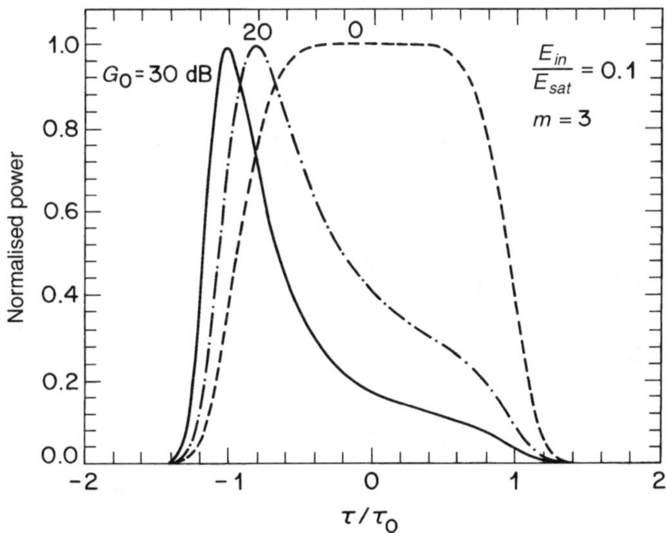

Fig. 5.6 Output pulse shapes for several values of the unsaturated gain G_0 when the input pulse is Gaussian with an energy such that $E_{in}/E_{sat} = 0.1$. The 0 dB curve shows the input pulse shape [21]. (After [21] with permission of IEEE).

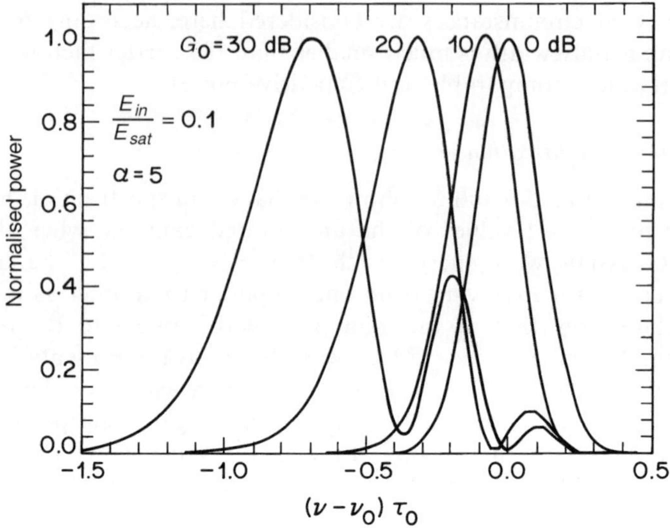

Fig. 5.7 Output pulse spectra corresponding to pulse shapes of Fig. 5.6. Note the spectral shift toward the low-frequency side with increasing G_0. (After [21] with permission of IEEE).

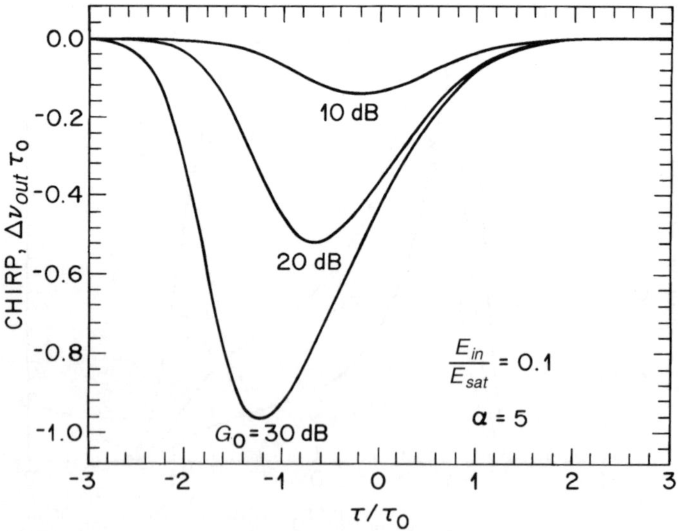

Fig. 5.8 Frequency chirp imposed on the output pulse during amplification of a Gaussian input pulse. The corresponding pulse shape and the spectrum are shown in Figs. 5.6 and 5.7. (After [21] with permission of IEEE).

frequency shift can easily exceed 100 GHz. These spectral changes are due to SPM-induced frequency chirp imposed on the pulse as it propagates through the amplifier.

Figure 5.8 illustrates the frequency chirp across the amplified pulse for the case of an unchirped Gaussian input pulse by using parameter values identical to those used in the previous figures. The instantaneous frequency chirp, $\Delta\nu\,(\tau)$, is negative across the entire pulse, i.e. the instantaneous frequency is downshifted (the redshift) from the incident frequency $\nu = \omega_0/2\pi$ The temporal variation of the chirp is almost identical to that of the output pulse shape. In fact, the structure in the pulse spectra shown in Fig. 5.7 result from an interference phenomenon that is common to SPM in all non-linear media [25]. Physically, the instantaneous frequency is the same at two distinct points within the pulse profile. Depending on the relative phases of the optical fields at those two points, the fields can interfere destructively or constructively. This interference leads to an oscillatory structure in the pulse spectrum. The asymmetry is a direct consequence of the asymmetric shape of the output pulse. It is worth observing that chirp increases almost linearly over the central part ($|\tau| \leq \tau_0$) of the pulse. Such a linear chirp implies that the pulse can be compressed in a dispersive medium, such as an optical fiber, if it experiences anomalous group velocity dispersion (GVD) during propagation in that medium.

The unchirped Gaussian input pulse, discussed up to now, can be far from many real situations, such as directly modulated semiconductor lasers. A super-Gaussian model can be used for studying the pulse shape effects [25, 26]. In this model, the amplitude of the input pulse given by

$$A_{in}(\tau) = \sqrt{P_{in}}\exp\left[-\frac{1 + iC}{2}\left(\frac{\tau}{\tau_0}\right)^{2m}\right] \tag{5.38}$$

where P_{in} is the peak power, C the chirp parameter, τ_0 the width coefficient, and m a parameter controlling tha shape of the super-Gaussian pulse ($m = 1$ corresponds to the Gaussian pulse). The peak power is related to the input pulse energy by the relation

$$E_{in} = \int_{-\infty}^{\infty} |A_{in}(\tau)|^2\, d\tau = \frac{\tau_0}{m}\Gamma\left(\frac{1}{2m}\right)P_{in} \tag{5.39}$$

where $\Gamma\,(x)$ stands for the gamma function of its argument x.

Figure 5.9 illustrates the output pulse shapes when the input pulse in an unchirped ($C = 0$) super-Gaussian pulse with $m = 3$. The input pulse energy E_{in}/E_{sat} is taken to be 0.1. The output pulse has a long tail on the trailing edge and appears to be narrower than the input pulse on the basis of its FWHM. This is in contrast with the Gaussian pulse reported in Fig.

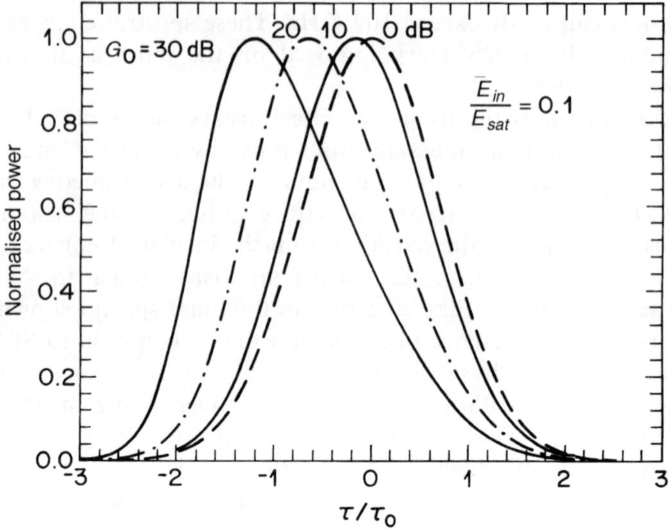

Fig. 5.9 Output pulse shapes for unchirped super-Gaussian input pulses, for $G_0 = 20$ and 30 dB. (After [21] with permission of IEEE).

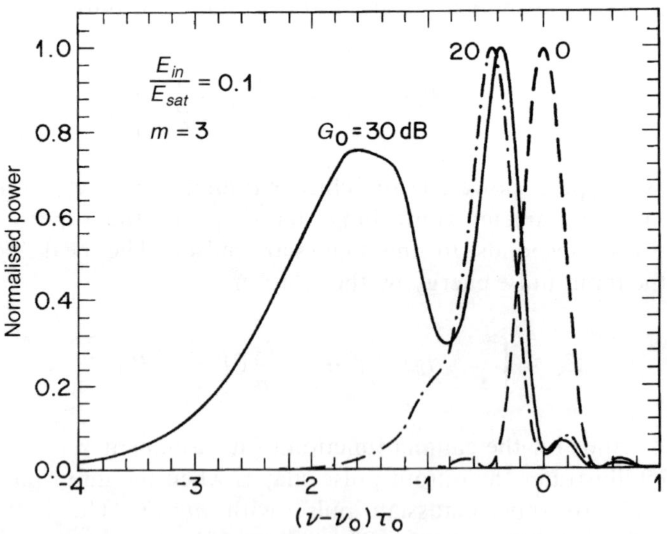

Fig. 5.10 Output spectra corresponding to the pulse shapes of Fig. 5.9. (After [21] with permission of IEEE).

5.6 (the input pulse energy is the same in both the cases for a right comparison), in which the FWHM of the output pulse is larger than the input pulse.

The output spectra corresponding to the pulse shapes of Fig. 5.9 are shown in Fig. 5.10, which can be directly compared with Fig. 5.7. This comparison reveals how much spectral distortion depends on the input pulse shape. In both cases, the spectrum has a multipeak configuration and is redshifted. Nevertheless, the amount of shift and the amplitudes of the peaks are quite different.

Finally, Fig. 5.11 shows the output spectra for a chirped Gaussian pulse. Depending on the sign of the chirp parameter C, the spectral shift can increase or decrease from $C = 0$ (the unchirped input pulse case). The magnitude of C can be determined from the input spectral width which increases by a factor $(1 + C^2)^{1/2}$ for chirped Gaussian pulses. The sign of C depends on whether the frequency increase $(C > 0)$ or decreases $(C < 0)$ with time across the pulse. For $C > 0$ the SPM-induced chirp adds to the input pulse chirp, and the spectrum shifts even more to the red side than $C = 0$. The opposite occurs when $C < 0$. These features are clearly illustrated in Fig. 5.11. A qualitatively similar behavior occurs for other pulse shapes.

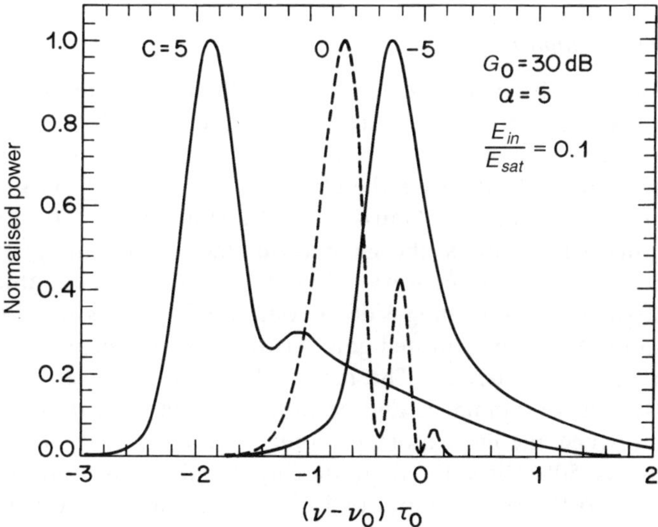

Fig. 5.11 Ourput spectra for chirped Gaussian input pulses for $C = 5$ and -5. The case of unchirped ($C = 0$) Gaussian pulses is also shown for comparison. The other parameters are identical to those of Figs. 5.7 and 5.8. (After [21] with permission of IEEE).

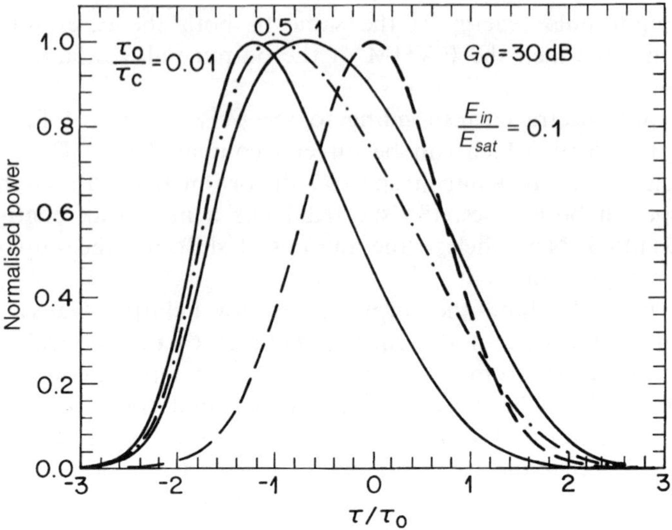

Fig. 5.12 Output pulse shapes for several values of \tau /\tau$_c$, when a Gaussian input pulse (dashed curve) is amplified in an amplifier with G_0 = 30 dB. Changes in the pulse shape are due to partial gain recovery. (After [21] with permission of IEEE).

Repetitive pulse amplification

When the pulsewidth becomes comparable to the carrier lifetime, the saturated gain has time to recover during the pulse. Such a partial gain recovery would affect both the shape and the spectrum of output pulses.

As a result, Fig. 5.12 displays the effect of gain recovery on the output shape when the input pulse is Gaussian and unchirped (C = 0). For τ / τ_c = 0.1 the pulse shape almost the same as obtained by assuming an infinite carrier lifetime (no gain recovery). For τ /τ_c = 0.5 the output pulse becomes broader as its trailing side become intense. This can be understood by noting that the saturated gain has time to recover partially by the time the trailing edge arrives. The net result of gain recovery is that the output pulse is less asymmetric and becomes broader than the input pulse. These features become even more pronounced for τ / τ_c = 1, where the output pulse is 50% broader than the input pulse. The output spectra corresponding to the pulse shapes of Fig. 5.12 are illustrated in Fig. 5.13. For partial gain recovery the spectral shift becomes smaller and the spectrum becomes less asymmetric. In the limit $\tau \gg \tau_c$, the spectrum becomes symmetric as the gain recovery is complete.

Altogether, for input pulses longer than the carrier lifetime the spectrum is broadened on both red and blue sides. The spectral and temporal

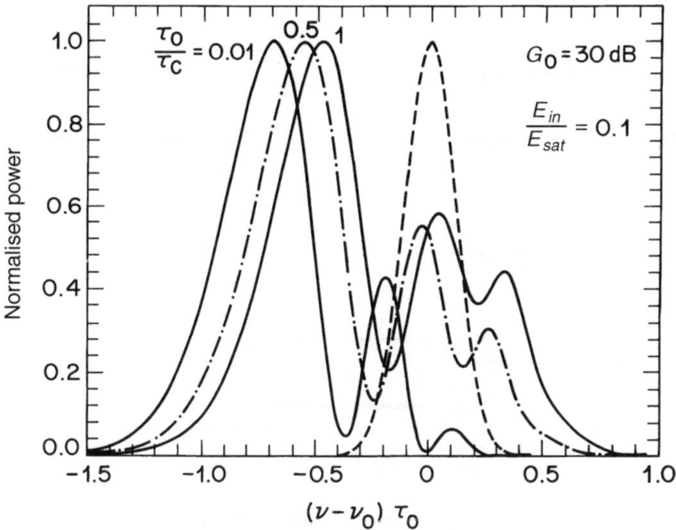

Fig. 5.13 Ouput spectra corresponding to the pulse shape of Fig. 5.12. The spectrum of the Gaussian pulse is shown by a dashed line. (After [21] with permission of IEEE).

changes also depend on the shape and the frequency chirp of the input pulses.

5.2 ERBIUM-DOPED FIBER AMPLIFIERS

With the demonstration of high-gain erbium doped fiber amplifiers (EDFAs) and their application in 1500 nm lightwave transmission systems, the potential for optical communication improved dramatically.

An EDFA is based on a single-mode optical fiber suitably doped with erbium ions, which constitute the optically active elements. Because of the introduction of Er^{3+} ions, the amplifier operation can be dealt with assuming a three-level system, as depicted in Fig. 5.14 for the case of a 980 nm pump. A pump photon is absorbed by a ground state erbium ion that jumps to a higher energy level, then the excited ion is excited to the metastable level through non-radiative decay.

Once in the metastable state, a radiative decay towards the ground state leads to the emission of a photon at the 1550 nm wavelength. In this last transition, a stimulated emission process can occur, leading, in the presence of suitable conditions, to either amplification or lasing action.

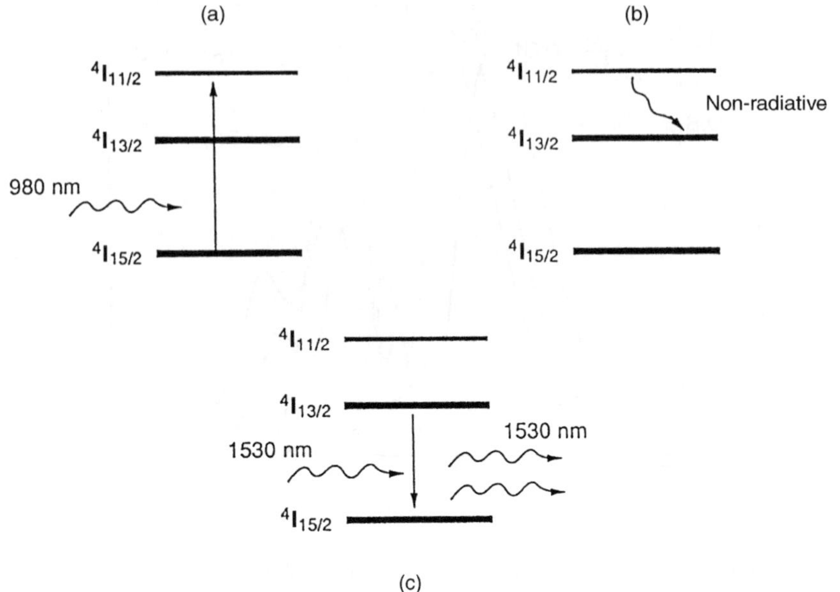

Fig. 5.14 Amplification mechanisms in erbium-doped fibers for a 980 nm optical pump

5.2.1 Structure of an erbium-doped fiber amplifier

The structure of an EDFA, able to provide a small-signal gain in excess of 30 dB, is shown in Fig. 5.15. Besides a doped fiber and a pump laser, the amplifier consists of a power supply line, a pump driving circuit, a monitoring circuit, and an electrically monitoring line. Furthermore, tunable optical filter is usually placed at the amplifier output. The pump can be launched into the fiber in either a copropagating or a counterpropagating direction with respect to the signal, with the two configurations being characterised by different properties. The performance of EDFA was evaluated for both pumping configurations by an accurate physical model [27], taking into account the presence of excited state absorption (ESA). The results are summarised in Table 5.1 [28]. If ESA is not present, the two pumping configurations provide the same gain, but assuming a perfectly coherent state at the amplifier input, the output signal-to-noise ratio (SNR) is improved an increase of about 2–3 dB. If ESA is present, the copropagating configuration provides a better SNR whereas higher gain is obtained in the counterpropagating configuration. The gain improvement strongly depends on the input signal level, usually ranging form a maximum of about 2 dB for an input signal power of 0 dBm, to negligible values for input signal powers below 10 dBm or above 15 dBm.

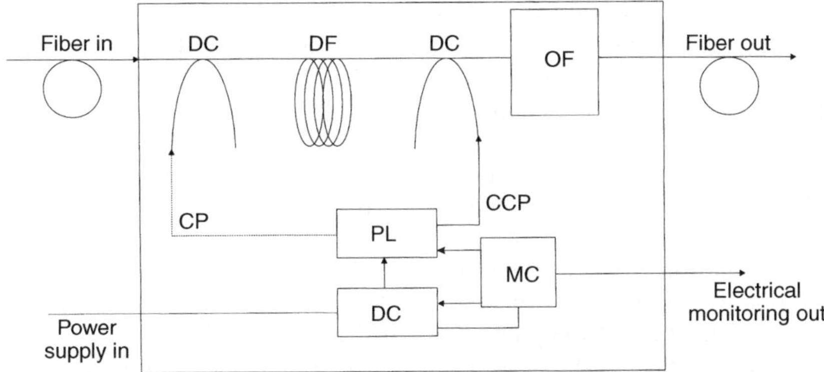

Fig. 5.15 Block scheme of an EDFA. DF = doped fiber; DC = directional coupler; PL = pump laser; DC = driving circuits; MC = monitoring circuit; OF = optical tunable filter; CP = copropagating pumping configuration; CCP = counterpropagating pumping configuration.

In addition to counterpropagating and copropagating pumping configurations, a double-pumping configuration could be used, with the pump injected into the doped fiber from both ends [28].

5.2.2 Gain and noise characteristics of erbium-doped fiber amplifiers

There are different accurate theoretical models that allow the performance of an EDFA to be evaluated [29–32]. In this section a simple and practical model is reported, from which the most important features can be derived [33]. Within this model, an EDFA is assumed to be a three-level one-dimensional system, in which the lifetime of the pumping state is much shorter than the other times of interest, in such a way that the population of this state can be assumed negligible. The assumptions in this model are (i) the transverse doping shape [34] and the dependence of the optical fields on the transverse coordinates have not been taken into

Table 5.1 Properties of different pumping schemes of doped fiber amplifiers

		Copropagating	Counterpropagating
Gain:	ESA	Worse	Better
	No ESA	Does not matter	Does not matter
Noise:	ESA	Better	Worse
	No ESA	Better	Worse

account, (ii) the guiding function of the fiber and consequent effects due to chromatic dispersion and polarisation evolution [35] have been neglected, (iii) both ESA and ion-ion interaction have not been considered, (iv) the possible emission of photons at the pump wavelength has been neglected.

If z is the direction corresponding to the fiber axis, the physical variables which describe the amplifier behavior are as follows:

- The population $N_{gr}(z)$ of the ground state,
- The population $N_{me}(z)$ of the metastable state,
- The intensity $I_p(z)$ of the pump wave at the wavelength λ_p,
- The intensity $I_s(z)$ of the optical wave at the signal wavelength λ_s, propagating in the positive z-direction, which consists of the actual signal and amplified spontaneous emission (ASE).

In these definitions the optical wave propagating at the signal wavelength along the negative z-direction, consisting only of ASE, has been neglected. In steady state, the evolution of these quantities is governed by the following set of relationships:

$$N_{me}(z) = N_{tot} \frac{I_s(z)/I_{ss} + I_p(z)/I_{sp}}{I_p(z)/I_{sp} + 2I_s(z)/I_{ss} + 1} \tag{5.40}$$

$$N_{me}(z) + N_{gr}(z) = N_{tot} \tag{5.41}$$

$$\frac{dI_s(z)}{dz} = 2\pi\Gamma_s \left[\sigma_{se} N_{me}(z) I_s(z) + \frac{\sigma_{se}}{\Gamma_s} N_{me}(z) I_n - \sigma_{sa} N_{gr}(z) I_s(z) \right] \tag{5.42}$$

$$\frac{dI_p(z)}{dz} = \pm 2\pi\Gamma_p \sigma_p N_g(z) I_p(z) \tag{5.43}$$

Equation (5.40) represents the steady-state value of the population of a three-level system, in the presence of pumping and signal waves, if the population of the upper level is set to zero [36]. N_{tot} represents the overall density of erbium ions, while I_{sp} and I_{ss} depict the pump and the signal saturation intensities, respectively, taking into account saturation effects due to pump and signal. These values can be evaluated from the doped fiber parameters, $I_{sp} = h\nu_p/(\sigma_p \tau_{sp})$ and $I_{ss} = h\nu_s/(\sigma_{se} \tau_{ss})$, and τ_{ss} represent the spontaneous emission lifetimes of the metastable state for the pump and the signal, respectively.

Equation (5.41) asserts that all the ions that are not in the metastable state are in the ground state, whereas equation (5.42) describes the propagation equation for the signal wave, in which σ_{se} and σ_{sa} are the emission and absorption cross sections at the signal wavelength, respectively. The three terms within brackets refer to phenomena involving photons at the signal wavelength: the first one represents the stimulated emission, the second the spontaneous emission, and the third one the absorption effect. Actually, the first term is proportional to the occupation number of the metastable state $N_{me}(z)$, whereas the third one is proportional to the occupation number of the ground state $N_{gr}(z)$. Therefore, the signal wave undergoes a net amplification of $N_{gr}(z)$, otherwise it is attenuated. The coupling between the signal mode and the core of the erbium-doped fiber is expressed by Γ_s, which represents the overlapping integral between the normalised transverse signal mode and the erbium concentration in the doped fiber transverse section.

The spontaneous emission term is I_n: it is a random process because the photons are spontaneously emitted at random instants. However, since it is important to determine the value of the signal to noise ratio at the amplifier output, I_n can be substituted with its average value [28]: $I_n = 2h\nu_s B$, where B is the signal bandwidth.

Equation (5.43) depicts the pump wave evolution. The minus sign refers to a copropagating pump, the plus sign to a counterpropagating pump. The parameter σ_p represents the pump absorption cross section, and Γ_p depicts the overlapping integral of the normalised transverse pump mode and the erbium concentration in the doped fiber transverse section.

Typical parameters for a low-noise inline EDFA are reported in Table 5.2. When an EDFA has to operate at high output power, the saturation power for the signal has to be as high as 8–10 mW and the satura-

Table 5.2 Typical parameters for a low-noise in-line EDFA.

Parameter	Symbol	Value	Unit
Signal mode area	$\pi w2$	1.3×10^{-11}	Q
Erbium concentration	N_{tot}	5.4×10^{24}	m^{-3}
Signal overlapping integral	Γ_s	0.4	Non-dimensional
Pump overlapping integral	Γ_p	0.4	Non-dimensional
Signal emission cross-section	σ_{se}	5.03×10^{-25}	m^2
Signal absorption cross-section	σ_{sa}	3.5×10^{-25}	m^2
Pump absorption cross-section	σ_p	3.2×10^{-25}	m^2
Signal saturation intensity	I_{ss}	3.4×10^7	$W\ m^{-2}$
Pump saturation intensity	I_{sp}	4.1×10^7	$W\ m^{-2}$
Signal saturation power	$\pi w^2 I_{ss}$	1.3	mW
Pump saturation power	$\pi w^2 I_{sp}$	1.6	mW

In the case of a power booster, the saturation power for the signal can be as high as 8–10 mW and the saturation power for the pump can be about 20 mW.

tion power for the pump can be around 20 mW. This can be obtained at the expense of noise performance. Starting from equation (5.42), two important parameters of the doped fiber amplifier can be introduced: the local gain per unit length $g_m(z)$ and the attenuation per unit length $\alpha_m(z)$:

$$g_m(z) = 2\pi\sigma_{se}N_{me}(z) \tag{5.44}$$

$$\alpha_m(z) = 2\pi\sigma_{sa}N_{gr}(z) \tag{5.45}$$

Hence the overall amplifier gain G can be calculated by setting $I_n = 0$ in equation (5.42), that is by assuming a noiseless amplifier. After solving the amplifier equations and evaluating the ratio between the output and the input signals, we obtain the following relationship:

$$G = \frac{I_s(L)}{I_s(0)} = exp\left\{ \int_0^L [g_m(z) - \alpha_m(z)]dz \right\} \tag{5.46}$$

where L is the doped fiber length.

To evaluate the performance of an EDFA, have equations (5.40) to (5.43) to be solved numerically. The most significant results are shown in Fig. 5.16 to 5.18. The pumping configuration has no effect on the gain, since ESA has not been considered, unlike in more accurate models [27].

Figure 5.16 shows the gain versus the pump power for different values of fiber length. The pump saturation effect occurs for input powers in the range 1–6 mW. In Fig. 5.17 the gain, shown versus the doped fiber length for different pump powers, presents a maximum corresponding to a given

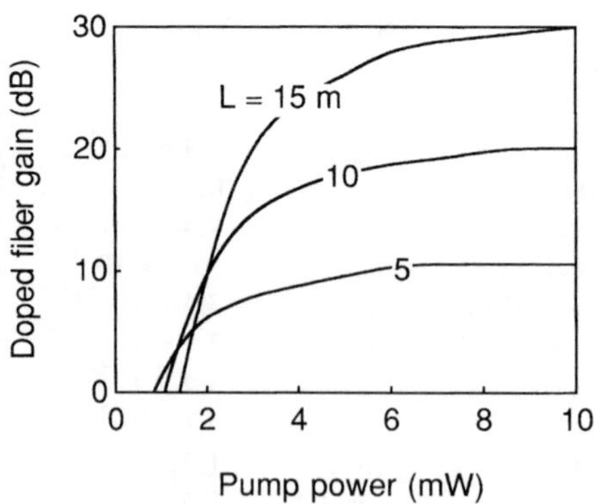

Fig. 5.16 Doped fiber gain versus pump power for different doped fiber lengths and in the absence of excited-state absorption. The fiber parameters are reported in Table 5.2.

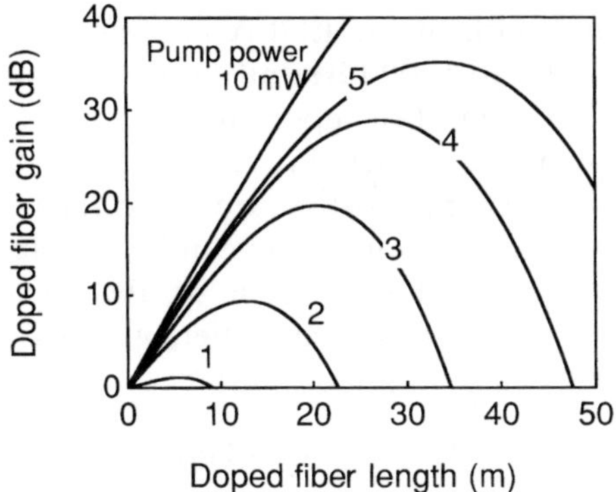

Fig. 5.17 Doped fiber gain versus fiber length for different pump powers in the absence of excited state absorption. The fiber parameter are reported in Table 5.2.

fiber length: the optimum length is a few tens of meters and increases with the pump power. Finally, the saturated gain versus the output power is reported in Fig. 5.18, at the signal wavelength. Gain saturation induced by the signal power is evident in the figure.

For $G > 20$ dB, gain saturation due to ASE cannot be completely neglected, so the curves start to become inaccurate, and for $G > 30$ dB they must be treated as qualitative only.

To analyse the spectral behavior of an EDFA, it is possible to use the same set of equations (5.40 to 5.43), inserting the frequency dependence of the different fiber parameters: both pump and signal waves are still assumed as perfectly monochromatic fields. Depending on the fiber para-

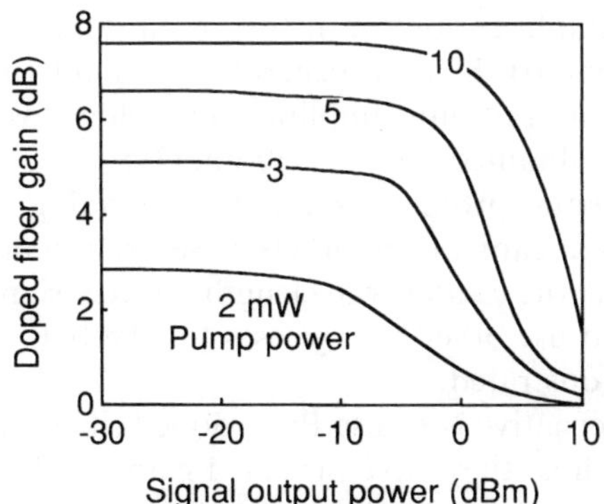

Fig. 5.18 Doped fiber saturated gain versus signal output power for different pump powers in the absence of excited state absorption. The fiber parameters are reported in Table 5.2.

meters, the 3 dB bandwidth of the EDFA in a small-signal amplification regime ranges from 3000 to 5300 GHz at the wavelengths of interest [37]. Such a wide bandwidth is another important feature of the EDFA, since in most applications it can be considered practically infinite.

As far as the noise characteristics are concerned, the basic theory reported for the semiconductor amplifiers can be adopted. For the sake of simplicity, here is a practical expression of the spontaneous emission power density at the output of the amplifier, which is generally used to perform calculations:

$$P_{sp} = N_{sp}(G - 1)h v_s \qquad (5.47)$$

where N_{sp} is the amplifier inversion factor. It an be defined as

$$N_{sp} = \frac{\int_0^L N_{me}(z)dz}{\int_0^L [N_{me}(z) - N_{gr}(z)]dz} \qquad (5.48)$$

5.3 1300 nm FIBER AMPLIFIERS: PRASEODYMIUM-DOPED FIBER AMPLIFIERS

Operations in the second window are advantageous because of the minimum dispersion in standard single-mode fibers. As a matter of fact, even if attenuation is larger than for the third window, overcoming fiber dispersion is one of the main tasks in contemporary fiber-optic communication systems. Unfortunately, state of the art technology has not yet provided practical amplifiers in the second window.

Various materials have been investigated to obtain amplification in the 1300 nm region. Amplification in neodymium-doped ZBLAN fibers is limited by two factors: (i) the ASE generation around 1050 nm, (ii) excited state absorption (ESA) prevents amplification below 1310 nm and realistic gain values are only obtained above 1330 nm [38].

This is a fundamental problem, since the upgrading of already existing 1300 nm systems by means of amplifiers is severely hampered if it is also necessary to change the center wavelength of the signal sources. Other solutions might be to use other host glasses for the Nd^{3+} ions, but this has not so far been demonstrated.

An interesting alternative is to use Pr^{3+}-doped fiber amplifiers (PDFAs); in a fluorozirconate host they have provided gain in a broad band centered within the 1300 nm window [39, 40]. Even if net gains around 30 dB have been achieved, a low quantum efficiency of the upper level in the stimulated transition leads to very high pump power requirements [41, 42]. Due to the large impact of a possible 1300 nm amplifier, a huge effort has recently been put into reducing the pump power requirements.

A model for PDFA amplifiers has been reported [38]. Here we briefly discuss some aspects relating to the optimum design.

5.3.1 Design aspects of a praseodymium-doped fiber amplifier

The amplifier performance is influenced by waveguide design and by the excited state (G_4) lifetime [43]. The G_4 lifetime is related to the question of how to choose a host glass composition, which may allow the pump power problems to be overcome in Pr^{3+} amplifiers. The efficiency of these 1300 nm amplifiers is limited by no-radiative decay, so there is strong motivation for finding a means to reduce them. How to tailor a new glass host for the maximum increase in lifetime is currently unknown.

Only codirectionally pumped fibers have been considered, pumping at 1017 nm. It has been reported that the small-signal gain has an optimum for a cutoff wavelength of approximately 800 nm, which is almost independent of both numerical aperture and pump power [43]. The scattering loss of the fiber can be ignored, since, according to the authors [43], it is typically around 0.1 dB/m, with the potential of further improvement, due to the predicted low-loss properties of ZBLAN glass.

As an example, Fig. 5.19 shows the small-signal gain as a function of the numerical aperture NA for codirectional pumping with the pump power as a parameter. Each point of the curve corresponds to the fiber length at which the small-signal gain is maximum. The gain increases with the increasing of NA due to the improved overlap between the fiber core, in

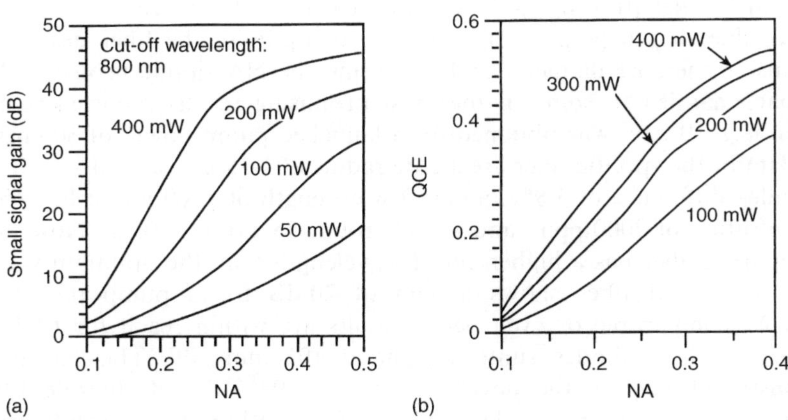

Fig. 5.19 (a) Small signal-gain versus NA for step index praseodymium-doped ZBLAN fibers pumped codirectionally. The signal wavelength is 1310 nm and the pump wavelength is 1017 nm.

which Pr^{3+} ions are homogeneously distributed, and the pump and signal modes. For a modest pump power of 50 mW, the NA requirement is very high, since a gain of only 10 dB requires NA > 0.4. For higher pump powers, more reasonable NA requirements can be set, but it is also true that for a pump power of 200 mW the largest improvement with NA is for NA < 0.35. A similar effect is seen for 400 mW pump power, where the gain curve already deviates for an NA of around 0.25. The reason is partly that the higher NA also increases the pump absorption due to the better overlap mentioned earlier. On the other hand, when the pump absorption rate becomes comparable to the total relaxation rate of the G_4 level $(1/\tau)$, an additional increase in pump absorption rate will not lead to as strong an improvement, and the gain will start to roll off. Another effect which contributes to the gain roll-off is that the amplifier starts to saturate due to ASE buildup.

5.3.2 Present results for a praseodymium-doped fiber amplifier

PDFAs are still far from the development level of EDFAs for 1550 nm. The primary difficulties are the fabrication properties and the fundamental problem of a very low quantum efficiency of about 1.5% for working amplifiers [44]. An attractive solution would be to find a host glass with a lower phonon energy; this would lower the phonon decay rate and prolong the effective relaxation lifetime. However, before such a host glass is found, other methods have to be adopted, and in rest of this section we will review them.

Among these approaches, one has been the effort of improving pumping efficiency by codoping with ytterbium, which has absorption in the 850–1050 nm wavelength range [45, 46]. Another method has been to employ a special fiber design [47]; in other words, to optimise the fiber design by minimising the core diameter and increasing the NA simultaneously. This approach has led to some of the most promising results reported so far [48]: a 38.2 dB gain was obtained for a launched pump power of 300 mW. The data of the specific fiber are a core radius of 1.15 μm, a relative refractive index difference of 3.8%, a cutoff wavelength of 1260 nm, a Pr^{3+} ion concentration of 2000 ppm, and a background loss of 0.1 dB/m. Note that the reported fiber has a higher cutoff wavelength than the optimum value. Since the actual fiber shows a gain of 20 dB for a pump power of 100 mW, comparable or even better results are within reach for PDFAs pumped by laser diodes emitting light at 1017 nm [49]. This has been demonstrated [50] by the development of a Pr^{3+}-doped fluoride fiber amplifier module pumped by laser diodes. The configuration was a bidirectional pumped amplifier in which the two 1017 nm diodes each emitted 160 mW, and an overall signal gain of 11.0 dB was obtained at 1300 nm. These results are also interesting because they demonstrate a low loss

(<0.5dB) and stable splice technique between the praseodymium-doped fluoride fiber and a silica fiber. The method is to use a V-groove connection in which each fiber is aligned and their end faces are mechanically polished.

Furthermore, the large-signal performance of PDFAs has recently shown good progress, since an output power of 212 mW has been obtained [51] in a counterpumped configuration where a 900 mW pump at 1010 nm was launched into the praseodymium-doped fiber. The power slope efficiency was reported to be 30%, and the fiber was specified by a refractive index difference of 0.04, a cutoff wavelength of 650 nm, a praseodymium concentration of 500 ppm, and a background loss of 0.14 dB/m.

Large-signal performance has recently been demonstrated with a PDFA by using the high-power output from an Nd:YLF laser as a pump source emitting at 1047 nm [52]. With a saturating input of +1 dBm, a maximum output power of 250 mW was obtained for 1.8 W pump power with a power slope efficiency of 18%. The advantage of this approach is that Nd:YLF pump lasers, because they are compact and reliable at high output powers, represent a realistic alternative to semiconductor laser pumping. Henceforth, this result can be considered relevant a step towards the realisation of practical 1300 nm amplifiers.

PDFAs exhibit good noise performance; noise figures between 3 and 4 dB have been obtained in the wavelength interval from 1290 to 1320 nm [53]. At wavelengths above 1320 nm, PDFA noise degrades because of the increased ground state absorption. In relation to these results, a PDFA has been used in conjunction with a pin-FET receiver, leading to a sensitivity of 37.3 dBm in a 2.5 Gb/s system [54].

REFERENCES

1. T. Saitoh and T. Mukai, Travellingwave semiconductor laser amplifiers, in Coherence, Amplification, and Quantum Effects in Semiconductor Lasers, Ed. Y. Yamamoto, John Wiley, New York (1991).
2. M. Gustavsson et al., Monolithically integrated 4x4 InGaAsP/InP laser amplifier gate switch arrays, *Electron. Lett.*, **28**, 2223–2224 (1992).
3. G. Metivier and M.A. Ali, Simulation of multi-channel AM-VSB CATV optical link employing semiconductor optical amplifier as external modulator, *IEEE Photon. Technol. Lett.*, **8**, 122–124 (1996).
4. R. Schnabel et al., Polarization insensitive frequency conversion of a 10-channel OFDM signal using four-wave mixing in a semiconductor laser amplifier, *IEEE Photon. Technol. Lett.*, **6**, 56–58 (1994).
5. N.A. Olsson, G.P. Agrawal and K.W. Wecht, 16 Gb/s, 70 km, pulse transmission by simultaneous dispersion and loss compensation with 1.5 μm optical amplifiers, *Electron. Lett.*, **25**, 603–605 (1989).

6. B.W. Hakki and T.L. Paoli, *J. Appl. Phys.*, **46**, 1299–1306 (1975).
7. Y. Yamamoto, *IEEE J. Quantum Electron.*, **QE16**, 1047–1052 (1980).
8. R.M. Jopson *et al.*, *Electron. Lett.*, **22**, 1105–1107 (1986).
9. T. Saitoh and T. Mukai, *Opt. Quantum Electron.*, **21**, S47–S58 (1989).
10. G. Grosskopf, R. Ludwig, R.G. Waarts, and H.G. Weber, *Electron. Lett.*, **23**, 1387–1388 (1987).
11. N.A. Olsson, *Electron. Lett.*, **24**, 1075–1076 (1988).
12. Y. Yamamoto, *IEEE J. Quantum Electron.*, **QE19**, 36–36 (1983).
13. T. Mukai and Y. Yamamoto, *IEEE J. Quantum Electron.*, **QE17**, 1028–1034 (1981).
14. A.E. Siegman, *Lasers*, Oxford University Press, Oxford (1986).
15. T. Saitoh and T. Mukai, *IEEE J. Quantum Electron.*, **QE23**, 1010–1020 (1987).
16. G. Eisenstein *et al.*, *Appl. Phys. Lett.*, **54**, 454–456 (1989).
17. I.D. Henning, M.J. Adams, and J.V. Collins, *IEEE J. Quantum Electron.*, **QE21**, 609–613 (1985); errata, *ibid.* **QE21** 1973 (1985).
18. T. Mukai, K. Inoue and T. Saitoh, *Appl. Phys. Lett.*, **51**, 381–383 (1987).
19. K. Shimoda, H. Takahashi, and C.H. Townes, Fluctuation in amplification of quanta with application to maser amplifiers, *J. Phys. Soc. Jpn.*, **12**, 686–700 (1957).
20. D. Marcuse, *Engineering Quantum Electrodynamics*, Harcourt Brace Jovanovich, New York (1970).
21. G.P. Agrawal and N.A. Olsson, Self-phase modulation and spectral broadening of optical pulses in semiconductor laser amplifiers, *IEEE J. Quantum Electron.*, **QE25**, 2297–2306 (1989).
22. G.P. Agrawal and N.K. Dutta, *Long-Wavelength Semiconductor Lasers*, Ch. 2, Van Nostrand Reinold, New York (1986).
23. L.M. Frantz and J.S. Nodvik, Theory of pulse propagation in a laser amplifier, *J. Appl. Phys.*, **34**, 2346–2349 (1963).
24. A. Iesevgi and W.E. Lamb, Jr., Propagation of light pulses in a laser amplifier, *Phys. Rev.*, **185**, 517–545 (1969).
25. G.P. Agrawal, *Nonlinear Fiber Optics*, Academic Press, Boston MA (1989).
26. G.P. Agrawal and M.J. Potasek, Effect of frequency chirping on the performance of optical communication systems, *Opt. Lett.*, **11**, 318–320 (1986).
27. M. Montecchi *et al.*, Gain and noise in rare-earth-doped optical fibers, J. Opt. Soc. Amer. B, **8**, 134–141 (1991).
28. S. Betti, G. De Marchis, and E. Iannone, *Coherent Optical Communications Systems*, John Wiley, New York (1995).
29. E. Desurvire and R.J. Simpson, Amplification of spontaneous emission in erbium-doped single-mode fiber amplifiers, *IEEE/OSA J. Lightwave Technol.*, **LT7**, 835–845 (1989).
30. P.R. Morkel and L.I. Laming, Theoretical modeling of erbium doped fiber amplifiers with excited state absorption, *Opt. Lett.*, **14**, 1062–1064 (1989).
31. C.R. Giles, E. Desurvire, and J.R. Simpson, Transient gain and crosstalk in erbium doped fibre amplifiers, *Optics Lett.*, **14**, 880–882 (1989).
32. K. Kikuchi, Generalized formula for optical amplifiers noise and its application to Erbium doped fibre amplifiers, *Electron. Lett.*, **26**, 1851–1853 (1990).

33. C.R. Giles and E. Desurvire, Modeling erbium doped fiber amplifiers, *IEEE/ OSA J. Lightwave Technol.*, **LT9**, 271–283 (1991).
34. M. Peroni and M. Tamburrini, Gain in erbium doped fiber amplifiers: a simple analytical solution for the rate equations, *Opt. Lett.*, **15**, 842–844 (1990).
35. F. Matera, M. Romagnoli, M. Settembre, and M. Tamburrini, Evaluation of chromatic dispersion in Ermium doped fiber amplifiers, *Electron. Lett.*, **27**, 1867–1969 (1991).
36. A. Yariv, *Quantum Electronics*, 2nd ed., Ch. 9, Wiley, New York (1989).
37. P.C. Becker, A. Lidgard, J.R. Simpson, and N.A. Olsson, Erbium doped fiber amplifier pumped in the 950–1000 nm region, *IEEE Photon. Technol. Lett.*, **PTL2**, 35–37 (1990).
38. A. Bjarklev, *Optical Fiber Amplifier: Design and System Application*, Artech House, Norwood MA (1993).
39. Y. Durteste, M. Monerie, J.Y. Allain, and H. Poignant, Amplification and lasing at 1.3 μm in Praseodymium-doped fluorozirconate fibres, *Electron. Lett.*, **27**, 626–628 (1991).
40. S.F. Carter, D. Szebesta *et al.*, Amplification at 1.3 μm in a Pr^{3+}-doped single-mode fluorozirconate fibres, *Electron. Lett.*, **27**, 628–629 (1991).
41. Y. Ohishi, T. Kanamori, T. Nishi and S. Takahashi, A high gain, high output saturation power Pr^{3+}-doped fluoride fiber amplifier operating at 1.3 μm, *IEEE Photon. Technol. Lett.*, **3**, 715–717 (1991).
42. Y. Miyajima, T. Sugawa and Y. Fukasaka, 38.2 dB amplification at 1.31 μm and the possibility of 0.98 μm pumping in Pr^{3+}-doped fluoride fiber, *Technical Digest Optical Amplifiers and their Applications OAA '91*, Snowmass, CO, post-deadline paper PD1 (1991).
43. B. Pedersen, W.J. Miniscalco, and R.S. Quimby, Optimization of Pr^{3+}-ZBLAN fiber amplifiers, *IEEE Photon. Technol. Lett.*, **4**, 446–448 (1992).
44. S.F. Carter, R. Wyatt, D. Szebesta, and S.T. Davey, Quantum efficiency and amplification at 1.3 μm in a Pr^{3+}-doped fluorozirconate single-mode fiber, *in Proceding of the 17th European Conference on Optical Communication ECOC '91*, paper MOA2-3, pp. 21–24 (1991).
45. J.Y. Allain, M. Monerie, and H. Poignant, Energy transfer in Pr^{3+}/Yb^{3+}-doped fluorozirconate fibres, *Electron. Lett.*, **27**, 1012–1014 (1991).
46. Y. Ohishi *et al.*, Gain characteristics of Pr^{3+}/Yb^{3+} codoped fluoride fibre for 1.3 μm amplification, *IEEE Photon. Technol. Lett.*, **3**, 990–992 (1991).
47. Y. Miyajima, Progress in 1.3 μm fluoride fiber amplifiers, *Technical Digest of Optical Amplifiers and their Applications OAA '92*, Santa Fe NM, invited paper WB1, pp. 4–7 (1992).
48. Y. Miyajima, T. Sugawa, and T. Fukasaku, 38.2 dB amplification at 1.31 μm and the possibility of 0.98 μm pumping in Pr^{3+}-doped fluoride fiber, *Electron. Lett.*, **27**, 1706–1707 (1991).
49. H. Zarem *et al.*, High-power single-mode InGaAs lasers for pumping praseodymium-doped fiber amplifiers, *Proc. OFC '92*, San Jose CA, poster paper WL7, p. 160 (1992).
50. M. Shimuzu *et al.*, 1.3 μm band Pr-doped fluoride fiber amplifier module pumped by laser diodes, *Technical Digest of Optical Amplifiers and their Applications OAA '92*, Santa Fe NM, postdeadline paper PD3 (1992).

51. T. Whitley, R. Wyatt, D. Szebesta, and S. Davey, High output power from an efficient Praseodymium doped fibre amplifier, *Technical Digest of Optical Amplifiers and their Applications OAA '92*, Santa Fe NM, paper WB2 (1992).
52. T. Whitley *et al.*, Quarter Watt output at 1.3 μm from a praseodymium doped fluoride fibre amplifier pumped with a diode-pumped Nd:YLF laser, *Technical Digest of Optical Amplifiers and their Applications OAA '92*, Santa Fe NM, postdeadline paper PD4, pp. 8–11 (1992).
53. T. Sugawa, and Y. Miyajima, Noise characteristics of Pr^{3+}-doped fluoride fibre amplifier, *Electron. Lett.*, **28**, 246–247 (1992).
54. R. Lobbett *et al.*, System characterization of high gain and highsaturated output power, Pr^{3+}-doped fluorozirconate fibre amplifier at 1.3 μm, *Electron. Lett.*, **27**, 1472–1474 (1991).

6

Optical Networking Devices

The technology discussed so far relates to devices useful for optical communication links. The realisation of optical networking elements requires other devices such as optical couplers, multiplexers and demultiplexers, optical filters, and key elements for optical switching such as optical space switching matrices. They are considered here with particular reference to wavelength division multiplexing (WDM) elements.

6.1 GENERALITIES ABOUT OPTICAL NETWORKING

Wavelength division multiplexing (WDM) represents the main technique used in the realisation of all-optical networks. In facts any optical wavelength can be used to carry an optical signal or, in general, a communication channel such as time division multiplexed telecommunication traffic or a set of video channels. The channel can be routed and switched directly in the optical domain operating at different optical wavelengths. This concept, usually known as wavelength routing, represents a key issue in the achievement of optical networking.

A key question for WDM systems is *how many wavelengths* can be used simultaneously and what is the optimal *wavelength separation*. The separation determines the class of WDM and has a significant relation on the ease of implementation. Currently the 1550 nm window represents the most convenient region to be used for fiber optical communications, mainly because of the low losses (Chapter 3) and the availability of EDFAs (Capter 5). An observation of the fiber attenuation curve reveals that approximately 100 nm bandwidth is available in the 1550 nm region. However, under the assumption that most advanced systems will use optical amplification, the maximum bandwidth accessible is reduced to approximately 30 nm, the bandwidth of a single EDFA. The equivalent frequency range is then

$$\Delta f = -\frac{c}{\lambda^2}\Delta\lambda \tag{6.1}$$

where λ is the operating wavelength and c the speed of light. Thus at 1550 nm with $\Delta\lambda = 30$ *nm*, $\Delta f = 3.710^{12}$ *Hz*. The maximum of 2.5 Gb/s channels that can be accommodated within this bandwidth is $N_{\max} = \Delta f /$ $2B = 740$, assuming that the baseband spectrum of the signal is twice the Nyquist bandwidth $B = 2.5$ GHz, allowing sideband modulation, and that a guard band equal to twice the Nyquist frequency is used to allow for imprecise filtering.[1]

This number is a theoretical maximum, however, as such a high packing density demands very high performances from the optical components. The avoidance of crosstalk between channels would in fact require a laser source having extreme wavelength stability, and a very high selectivity at the receiver to separate out the individual channels. Any chirping of the laser source would also spread spectral components into adjacent channels. So the above bound is highly ideal.

WDM usually specifies a wavelength spacing greater than 1 nm, with a number of channels between 2 and 8; whereas HD-WDM (high-density WDM) specifies spacing less than 1 nm. A further approach, even if there are not many investigations on it, is represented by coherent multichannel (CMC), in which coherent heterodyne detection techniques are used to provide very high selectivity, besides good detection sensitivity. In CMC systems the channel spacing could be very small, such as 0.1 nm.

6.2 OPTICAL FILTERS

The optical filter is the basic building block of WDM systems, because it must provide the necessary wavelength selection to isolate independent channels. The transmission characteristics of the filter are most important as they have a significant effect on the performance of the system as a whole. In particular, the passband characteristics of a filter and its roll-off slope are directly related to the system crosstalk characteristics. Furthermore, for a system using more than one filter, the overall passband characteristic will be different than for an individual filter and will be determined by the cascade of the individual characteristics. There are different types of optical filters. Here we consider fixed interference filters, tunable Fabry-Perot (FP) filters, grating filters, acousto-optic tunable filters and active filters.

6.2.1 Fixed interference filters

An interference filter is constructed from multiple thin-film layers. The dielectric constant and the thickness of each layer as well as the number of layers determine the filter transmission characteristic. A typical component is a three-stage double half-wave device, which has a passband character-

---

[1]Assuming that the baseband spectrum of the signal is twice the Nyquist bandwith $B = 2.5$ GHz, that single sideband modulation is possible and that a guard band equal to twice the Nyquist frequency is used to allow for imprecise filtering.

istic of this form [1]:

$$T(\Lambda) = \frac{1}{1 + (2\Lambda)^6} \qquad (6.2)$$

where λ is the normalised wavelength defined as

$$\Lambda = \frac{\lambda - \lambda_0}{\delta\lambda} \qquad (6.3)$$

with λ_0 the center wavelength and $\delta\lambda$ the 3dB bandwidth of the filter. The characteristic of this type of filter is that its response falls off very rapidly outside the passband, thus providing a very good choice for WDM. However, its tuning capability is limited, although it may be tuned slightly by tilting the filter with respect to the incident signal. The insertion loss is generally more than 2 dB and bandwidths are generally greater than 1 nm.

6.2.2 Tunable Fabry-Perot filters

This type of filter consists of an optical Fabry-Perot cavity which is formed between the ends of two fiber tails, coated to assure a very high reflectivity. There are two kinds of FP filter, cavity in air and cavity in glass, as shown in Fig. 6.1. To tune the device one of the fibers may be moved to alter the

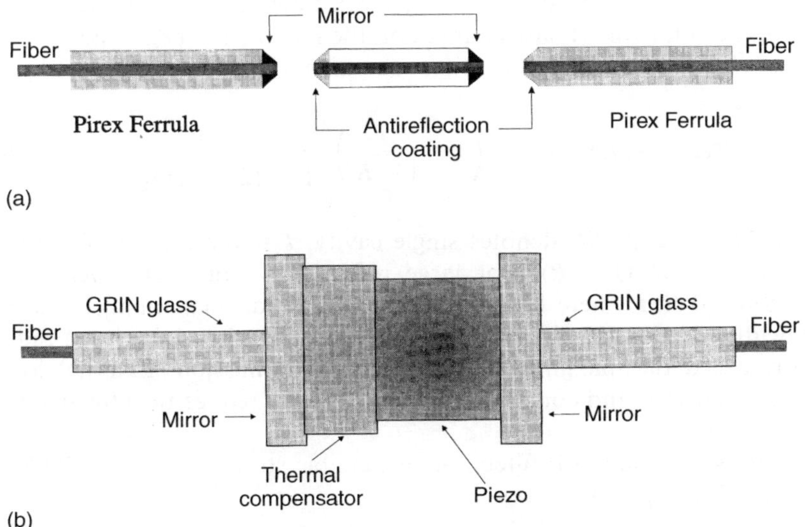

Fig. 6.1 With cavity FP filters: (a) fiber filter in air, (b) micro-optic filter with cavity in glass.

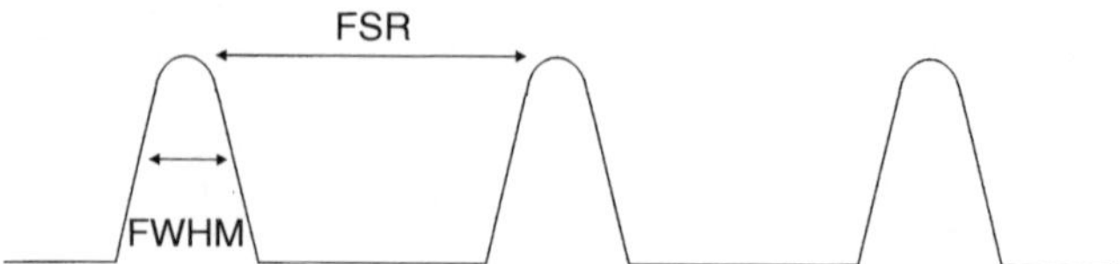

Fig. 6.2 The periodic transmission characteristic of FP filters.

cavity length. A piezoelectric device, allows slight modifications of the cavity length L, changing the resonance frequency of the filter. The electrical tunability of the filter, arising from its piezoelectric properties, make this device particularly suitable for applications in optical networks.

The transmission characteristic (Fig. 6.2) is periodic. The wavelength difference between the transmission peaks is called the free spectral range (FSR) and is a function of the cavity length.

If R and T represent the reflectivity and the transmittance of both mirrors, and the mirrors are assumed to be lossy with loss parameter $A = 1 - R - T$, the field transfer function of an FP filter is given as follows [2, 3]:

$$t(\delta) = (1 - R - A) \sum_{i=0}^{\infty} (Re^{j2\pi\delta})^i \tag{6.4}$$

where δ is the normalised optical frequency equal to $2fnL/c$, so that $t(\delta)$ is periodic in δ with period 1, n being the refractive index of the medium between both mirrors. Phase changes on the mirrors are neglected.

The power transfer function, or transmission of the filter, is given by

$$T_{SC}(\delta) = t(\delta)t^*(\delta) = \left(1 - \frac{A}{1 - R}\right)^2 \frac{1}{1 + [(2F/\pi)\sin(\pi\delta)]^2} \tag{6.5}$$

where the subscript SC denotes single cavity. F is the finesse of the filter defined as $\pi\sqrt{R}/(1 - R)$. For large reflectivity values, $(1 - R) \ll 1$ so $F \approx FSR/FWHM$ where FWHM (full width at half maximum) is the 3dB bandwidth of the FP filter. The first factor on the right hand side is usually called the maximum intrinsic filter transmission. External losses, such as connector and coupling losses, necessarily reduce the effective filter transmission.

Besides single-stage FP filters, other possible structures are available [2], as depicted in Fig. 6.3:

(1) Double-pass FP filter (Fig. 6.3b): in this case the cavity is somewhat 'reused', by resubmitting the output of the filter through the cavity.

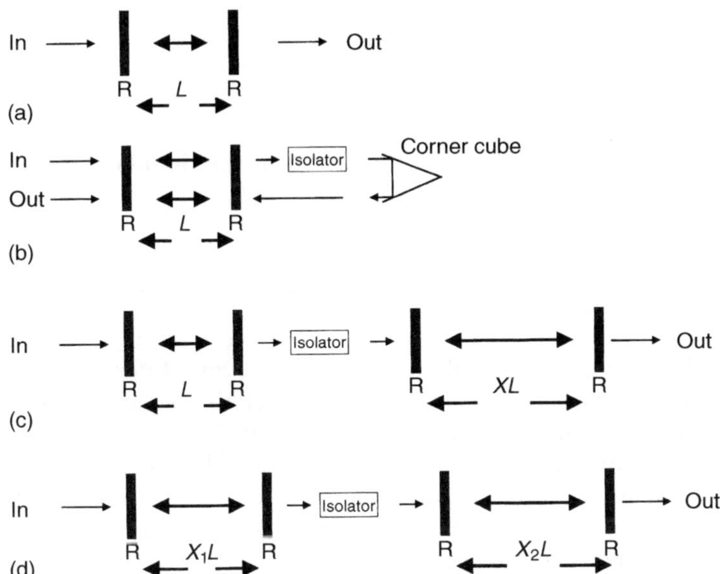

Fig. 6.3 Different FP structures: (a) single cavity, (b) double pass, (c) two stage and (d) vernier.

   (2) Two-stage double-cavity FP filter (6.3 c): the filter consists of two cavities with equal or comparable F-values, but widely different free spectral ranges (and thus widely different FWHM values). This is usually called two-stage, since the first filter may be considered as the first, or rough, stage filter with an FSR equal to the system bandwidth that suppresses all distant channels, and the second filter as the fine stage filter that provides the required selectivity by suppressing adjacent channels.
   (3) Vernier double-cavity FP filter (Fig. 6.3d): here two cavities are also used, but with FSR values such that K FSRs of filter 1 cover $K + 1$ FSRs of filter 2, thereby enchanging the effective FSR and reducing the FWHM.

Clearly there is a trade-off between transmission (or attenuation) and selectivity (or FWHM) for the different structures. Actually a double-stage filter shows a better roll-off than a single-stage filter, but a worse attenuation.

6.2.3 Grating-based filters

Optical filters based on diffraction grating technology can be also used. The transmission characteristics can be approximated by the Gaussian function [1]:

$$T(\Delta) = \exp[-(4ln2)\Lambda^2].$$ (6.6)

An example of a grating filter is illustrated in the discussion on demultiplexers below.

6.2.4 Acousto-optic tunable filters

Integrated optical, acoustically tunable wavelength filters (AOTFs) are attractive due to their narrowband filter characteristics, large tuning range, low driving power and simultaneous filtering capability.

An AOTF normally employs the large-angle codirectional acousto-optic interaction [4–6]. Strong acousto-optic interaction occurs only when the Bragg condition (conservation of momentum) is satisfied. If the incident light beam contains many spectral components, only one will satisfy the Bragg condition at a given acoustic frequency. In other words, only one spectral component will be diffracted at a given acoustic frequency. Therefore, by varying the acoustic frequency, the frequency (or wavelength) of the diffracted beam can also be varied. The fraction of power transferred from the incident beam to the diffracted beam in an interaction length L can be derived by applying the coupled-mode analysis of Bragg diffraction [4]:

$$T = \frac{sin^2\left[\kappa_{12}L\sqrt{1 + (\Delta\beta/2\kappa_{12})^2}\right]}{1 + (\Delta\beta/2\kappa_{12})^2}$$ (6.7)

where κ_{12} is the magnitude of the coupling constant, and $\Delta\beta$ is the momentum mismatch:

$$\Delta\beta = \beta_1 - \beta_2 \pm K$$ (6.8)

β_1 is the component of the wave vectors of the incident beam, along the direction of propagation of the acoustic wave; β_2 is the same component for the diffracted beam.

An ideal AOTF is normally operated at the condition when

$$\kappa_{12}L = \pi/2$$ (6.9)

so that the power conversion is 100% when the phase matching condition $\Delta\beta = 0$ is satisfied. Substituting condition (6.9) into equation (6.7) we obtain

$$T = \frac{sin^2\left[\pi/2\sqrt{1 + (\Delta\beta L/\pi)^2}\right]}{1 + (\Delta\beta L/\pi)^2}.$$

(6.10)

Hence the conversion efficiency drops to 50% when $\Delta\beta L = \pm 0.80\pi$.

A practical AOTF is depicted in Fig. 6.4 [6, 7]. It employs a surface acoustic wave (SAW) device to generate a birefringence grating seen by an incoming polarised beam of light. In a collinear AOTF, birefringence causes a polarisation transformation between TE and TM states. If the interaction strength, integrated over the length of the interaction, is sufficiently strong, a perfect polarisation flip (100% TE-TM conversion) is achieved. The required phase matching conditions ensures a narrow spectral passband. The resonant polarisation conversion bandwidth is approximately $\Delta\lambda/\lambda = L_b/L$, where $L_b = \lambda/\Delta n$ is the polarisation beat length, and Δn is the waveguide effective index birefringence. Since the single-stage structure has some drawbacks because of the high sidelobes of the filter characteristic and the frequency shift imposed on the filtered wave, a double-stage configuration emerged as a favorable solution. The filter characteristics of such a device is just the squared response of a single-stage filter. This yields a narrowing of the filter characteristics, a strong suppression of the sidelobes and no net frequency shift. Double-stage acousto-optical filters, in Fig. 6.4, have been reported in the literature [6, 7].

The sinc^2 optical transmission function of the AOTF has high sidelobes, which create severe intensity crosstalk in WDM systems. However, by tailoring the interaction profile, the corresponding passband slope can be

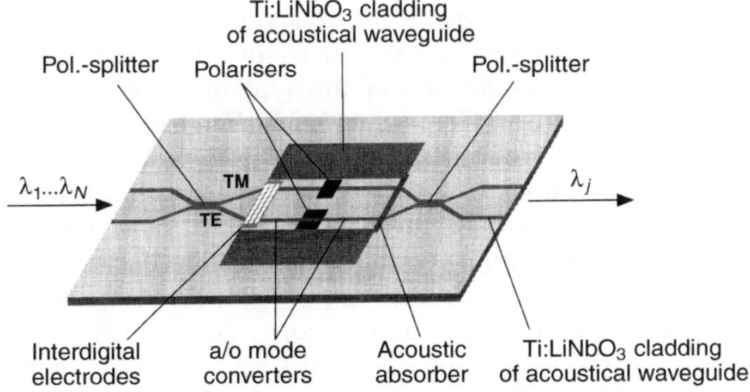

Fig. 6.4 Structure of a double-stage polarisation independent acousto-optic tunable filter.

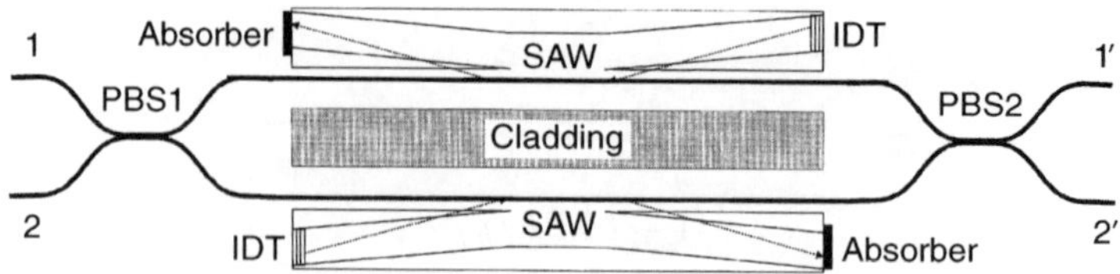

Fig. 6.5 Design of a polarisation-diversity AOTF with engineered passbands and unidirectional optical frequency shift. TM (passing the top waveguide) and TE (passing the bottom waveguide) modes experience the same frequency shift due to the opposite directions of SAW propagation. (After [9] with permission of IEEE).

altered [8]. For instance, a half-sinusoidal interaction strength apodisation produces a −10 dB sidelobe reduction. The gradual onset and cutoff of the acousto-optic interaction can be achieved by imbedding the active optical waveguide in one arm of a SAW directional coupler. Further tapering of the interaction onset and cut-off, Fig. 6.5 reduces the sidelobes more than the sinusoidal interaction [9].

Passband engineering becomes yet more elaborate than mere sidelobe suppression. Although apodisation significantly reduces interchannel crosstalk [10], channel misalignment tolerance is still severe because any slight displacement of the input wavelength from resonance leads to incomplete polarisation conversion. The chosen channel is incompletely switched, producing interport crosstalk. Waveguide selective routers for WDM systems ideally should possess a rectangular spectral profile, to allow cascadability and wavelength misalignment tolerance. Different techniques have been proposed for flattening. Continued improvements in passband engineering in the near future are expected.

The AOTF has already been used in WDM systems for wavelength add/ drop multiplexing as well as for larger WDM cross-connect elements. Add/ drop multiplexers are a key element for optical networks. They are used to add or drop optical channels into or out of an optical network, at given wavelengths. An integrated version of this type of device was realised by the University of Paderborn, Germany [11, 12] (Fig. 6.6a).

The performance of device is illustrated in Fig. 6.6b, for the different input-output configurations (i_1 and i_2, and o_1 and o_3, are the input and output ports, respectively). The transmission spectra have a 3dB width of 2 nm. At the cross-state output ports, passband filter curves are obtained ($i1 \rightarrow o2$ and $i2 \rightarrow o1$); the sidelobes are more than 18.5 dB below the maximum level. At the bar-state output ports, notch filter curves appear ($i1 \rightarrow o1$ and $i2 \rightarrow o2$); a rejection of −20 dB was achieved for the worst-case polarisation at $\lambda = 1556$ nm. However, the devices can be locked at any wavelength within the EDFA window without significant change of the transmission spectra.

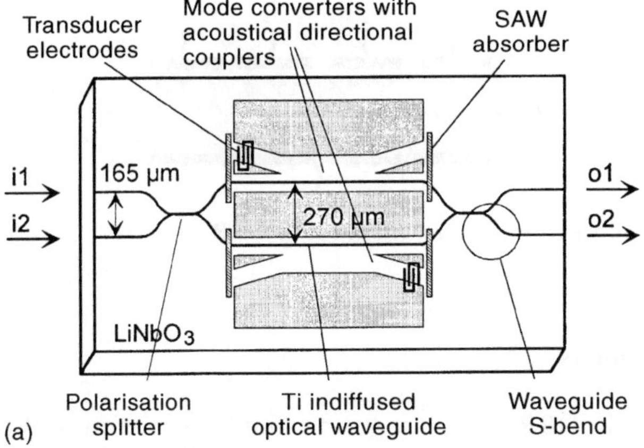

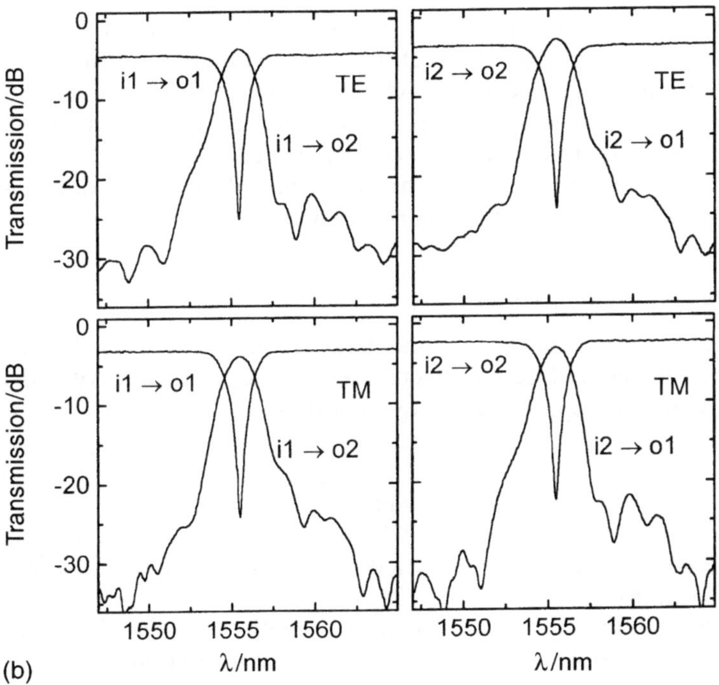

Fig. 6.6 The integrated optical 2×2 switch-add/drop multiplexer (a) and its switching performance (b). The diagrams in (b) show the transmission for TE and TM input polarisation versus optical wavelength into the cross and bar states for both input ports. (After [12] with permission of IEEE).

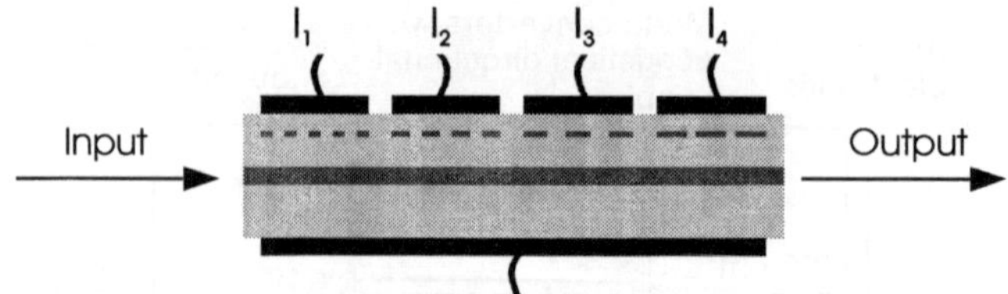

Fig. 6.7 Typical structure of a resonant active DBR filter.

6.2.5 Active filters

Resonant active DBR filters are valid candidates for future WDM systems [13]. A typical structure is depicted in Fig. 6.7. In the form of laser diodes biased below threshold, they have attractive features such as narrow linewidth, net gain, electronic tuning and compact size. On the other hand, they are affected by several problems: spontaneous emission degrades the signal-to-noise ratio (SNR); the optical linewidth is typically only a few gigahertz, causing problems at high bit rates with large information bandwidth; gain saturation occurs at high input power. A key feature is a wide dynamic range, that is a large difference between the maximum allowed input signal to the filter (before strong gain saturation effects set in) and the minimum required input signal (determined by SNR degradation). Some design aspects for active filters are briefly treated here.

Linewidth and linear crosstalk

If we neglect saturation effects and treat the gratings as lossless discrete mirrors, each one adding an effective length $L_{B,eff}$ to the cavity, the linewidth of high-finesse filter can be expressed as:

$$\Delta \nu = \frac{c}{\pi n L_{tot} \sqrt{\Theta}}. \tag{6.11}$$

where we have neglected the fact that the refractive index n is different in different sections.

The total length L_{tot} t is given by the lengths of the active and phase sections and the effective lengths of the two grating sections:

$$L_{tot} = L_a + L_\phi + 2L_{B,eff} = L_a + L_\phi + \frac{\tanh(\kappa \cdot L_B)}{\kappa}, \tag{6.12}$$

where κ is the grating coupling coefficient. The contrast factor θ is given by

$$\Theta = \frac{4RG_0}{(1 - RG_0)^2} \qquad (6.13)$$

where the single-pass gain G_0 is $G_0 = \exp(g_a\, L_a\, \Gamma_a)$ (g_a being the material gain coefficient in the active section, Γ_a the optical power confinement factor) and $R = \sqrt{R_1 R_2}$ is the effective Bragg reflectance of the two Bragg reflectors, whose reflectance can be expressed as $R_1 = R_2 = \tanh^2(\kappa\, L_B)$. θ is related to the finesse by the relation $F = (\pi/2)\,\sqrt{\Theta}$. In transmission the resonant chip (cavity) G_{rc} is

$$G_{rc} = |\,T\,|^2 = \frac{(1 - R)^2 \cdot G_0}{(1 - R \cdot G_0)^2} \qquad (6.14)$$

where T is the complex field transmission. Thus the linewidth can be expressed as

$$\Delta v = \frac{c}{2\pi n L_B \sqrt{G_{rc}}} \cdot \frac{1 - \tanh^2(\kappa L_B)}{\kappa(L_a + L_\phi) + \tanh(\kappa L_B)} \cdot \frac{\kappa L_B}{\tanh(\kappa L_b)} \qquad (6.15)$$

A trade-off exists between gain and bit rate. One should always keep the phase and active sections as short as possible. For large values of κL_B the linewidth is small (for a given chip gain). Hence the internal cavity irradiance becomes high and saturation power low.

The sidemode rejection ratio (SMRR) is a key parameter, which determines the worst-case channel-to-channel crosstalk figure. The FWHM of a Bragg reflector can be approximated by

$$\Delta\lambda_B \approx \Delta\lambda_{B,\infty} \cdot \left(1 + \frac{1}{\kappa L_B}\right) = \frac{\kappa\lambda^2}{\pi n} \cdot \left(1 + \frac{1}{\kappa L_B}\right), \qquad (6.16)$$

where $\Delta\lambda_{B,\infty}$ is the stopband of a lossless and infinitely long Bragg section. To maximise SMRR, the longitudinal mode distance $\delta\lambda_m$ should satisfy the condition

$$\delta\lambda_m = \frac{\lambda^2}{2 n_g L_{tot}} \geq \Delta\lambda_B. \qquad (6.17)$$

which can be written as

$$L_\alpha + L_\phi < L_{B,\text{eff}} \approx \left\{ \begin{array}{l} 1/2\kappa, \ \kappa L_B \gg 1 \\ L_B/2, \ \kappa L_B \ll 1. \end{array} \right. \qquad (6.18)$$

The maximum number of channels depends on what penalty level is tolerable.

Much better crosstalk performances are obtained in reflection than in transmission. With a well-designed filter, the crosstalk comes entirely from single-pass amplification of channels outside the stopband $\Delta\lambda_B$. Since G_0 is typically of the order of 3–5 dB, the number of channels will always be limited in transmission. However, there is no such crosstalk in reflection. If very many channels are required, reflection operation has to be accepted. With a sufficient channel separation (e.g. around 30 GHz), the number of channels is limited by tuning range, not by crosstalk. There are other advantages with smaller G_{rc}, such as reduced sensitivity to optical feedback and more stable operation in general.

Typically, the best design [13] is a tapered structure, with a 200–300 μm long active plus phase section, with a small κ (< 25 cm^{-1}) and with a thin active layer.

Noise properties

Since the active DBR structure introduces spontaneous emission noise, the SNR degrades when a signal passes though it. As a matter of fact, amplifier noise theory can be applied in this case. BER simulations were accomplished to evaluate the noise performance of a DBR active filter for three different values of the spontaneous factor n_{sp}: 0 (no filter), 3 and 10 [13]. The calculated penalty was always less than 1 dB.

Saturation and dynamic range

A crucial problem with resonant amplifiers is the low saturation power. Also, the filter peak wavelength shifts to longer wavelengths, due to the refractive index change associated with the optically induced decrease of the gain coefficient g. The input saturation power $P_{3\text{dB}}$ is defined as the optical power level at which the filter gain G_{rc} decreases by 3 dB. For brevity we directly report the expression derived in [13]:

$$P_{3dB} = \rho I_{sat} A_{\text{mode}} \frac{1}{\eta_i \cdot Q}, \rho \approx \frac{(\sqrt{2} - 1) \cdot (1 - RG_0)}{RG_0} \cdot \frac{1}{\log(G_0)} \tag{6.19}$$

where Q is the cavity power enhancement factor (mean power in cavity/ input power), I_{sat} is the irradiance at which gain would decrease by 3 dB in a traveling wave amplifier, A_{mode} is the mode area, η_i the input coupling coefficient. The peak values reported in the literature [13] are $P_{3dB} = -23$ dBm for $\kappa L_B = 0.50$ and $P_{3dB} = -32$ dBm for $\kappa L_B = 1.50$.

The spectral shift of the resonant peak (now disregarding peak suppression effects) is

$$\delta\nu = \frac{c\log(G_0)}{\lambda n L_{tot}} \cdot \frac{\tau_{rec}}{h\omega} \cdot \frac{\delta n}{\delta N} \cdot \frac{\eta_i P_{in} Q}{A_{\mathrm{mode}}} \qquad (6.20)$$

where τ_{rec} is the carrier spontaneous lifetime. With $\delta n / \delta N \approx \varnothing 10^{-20}\ cm^3$, and using the same values as before, the following shifts are obtained: $|\delta\nu / P_{in}| = 2.7\ GHz/\mu W$ for $\kappa L_B = 1.5$, and $|\delta\nu / P_{in}| = 1.6\ GHz/\mu W$ for $\kappa L_B = 0.5$. The FWHM of the resonance peaks are about 1 GHz and 6 GHz for the respective devices. As an estimate of the maximum allowable input peak power P_{in} we take

$$\delta\nu < \Delta\nu/2. \qquad (6.21)$$

The induced shift $\delta\nu$ should not move the resonance peak by more than half its FWHM value. The criteria expressed in the last expression also guarantee, with a twofoldmargin, that the transfer function is single valued. The influence of saturation will be strongest if very long sequences of identical bits are allowed.

Polarisation effects

Strong polarisation dependence is a severe problem. Because of different effective indices and confinement factors, TM components have weaker transmission maxima, usually separated many nanometers from the TE contributions. It is impracticable to reproducibly fabricate filters whose TE

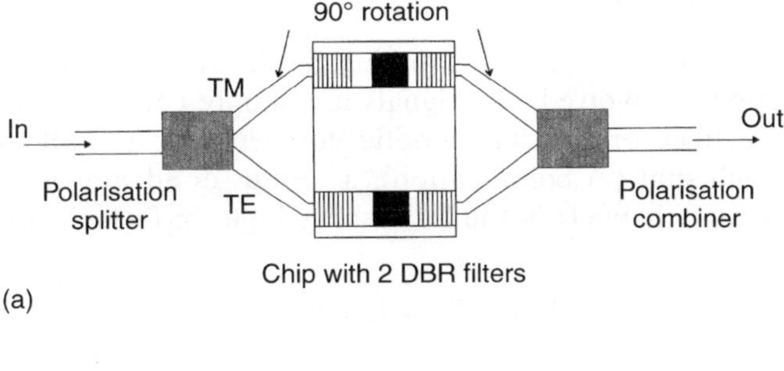

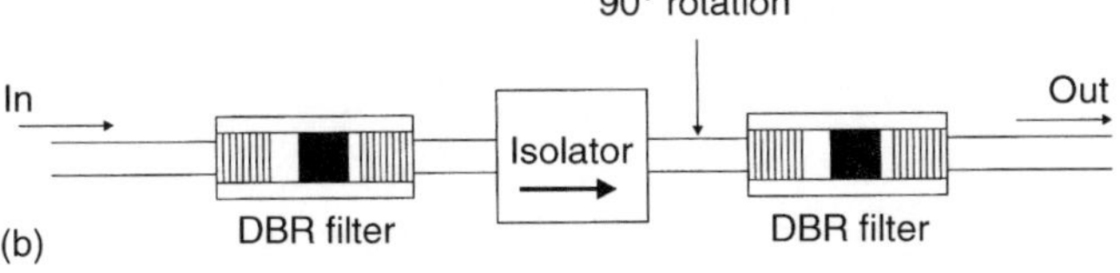

Fig. 6.8 Two methods for overcoming the polarisation problem. (a) One-chip polarisation diversity; the splitter and the combiner could be fiber based or monolithically integrated. (b) Two-chip cascading. (After [13] with permission of IEEE).

and TM modes coincide within about 1 GHz. Rather, one should make sure that the difference between TE and TM resonance wavelengths is larger than the tuning range, and solve the polarisation problem by scrambling, diversity or cascading [13]. Alternatively polarisation control or polarisation-maintaining fibers could be employed.

Scrambling works well in single-stage distributive switching systems, since it could be performed once and for all after the emitter. It is less attractive in multiple-stage, long-distance and wavelength routing systems, where scrambling has to be performed before each individual filter.

Diversity, shown in Fig. 6.8a, ideally leads to a 3 dB penalty (the ASE will double), apart from splitting and combining losses. In cascading (Fig. 6.8b), the first filter operates on the TE part of the signal, which then passes unaffected through the second filter. The second filter operates only on the TM part. A problem is that the SNR is degraded for TM signal, which becomes attenuated by a factor $\eta_i + \eta_o - G_0(TM)$, where η_i, and η_o the input and output coupling efficiencies, before it reaches its filter.

6.3 MULTIPLEXERS AND DEMULTIPLEXERS

Multiplexing can be accomplished in two ways: a simple passive combiner or a wavelength selective combiner. Demultiplexing can only be achieved by using wavelength selective components.

6.3.1 Passive combiner

The basic principle of a passive combiner is shown in Fig. 6.9. No wavelength selection is involved, the signals are simply combined through the use of either fiber or integrated optic couplers. As a 3 dB power loss occurs for each split (or combination), a serious disadvantages is that for large numbers of channels N the loss is very high. In fact, the loss is given by

$$Loss(dB) = 10log_{10}(N).$$
(6.22)

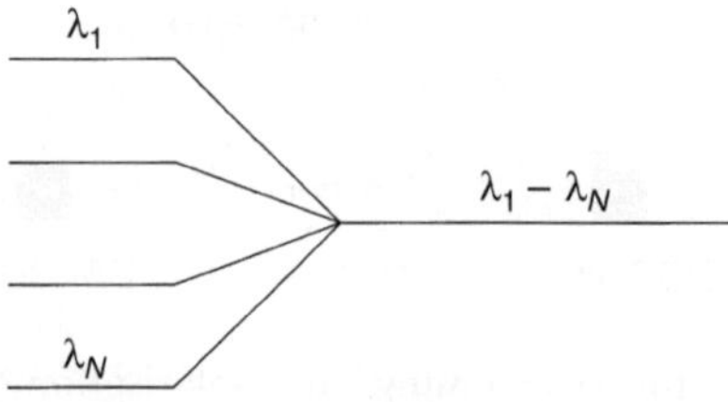

Fig. 6.9 The basic structure of a passive combiner.

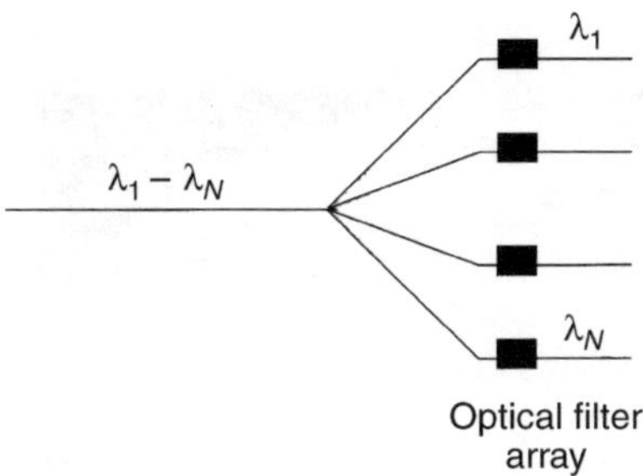

Fig. 6.10 Passive splitter used in conjunction with an optical filter.

In practice an excess loss would also be associated with the passive combiner/splitter due to non-ideal realisation. Nonetheless, this type of combiner is particularly useful in broadcast star network topologies where all signals are sent to all nodes.

6.3.2 Wavelength-selective mu-demultiplexer

Because of the need to select a wavelength at the receiver, a passive splitter must be used in conjunction with an optical filter (Fig. 6.10) or other wavelength selective component. The same splitting loss occurs as with the simple combiner; however, the filter introduces a further insertion loss. Hence a WDM system using four wavelengths and simple lossless passive combiner multiplexer together with a lossless passive splitter and filters would have its power budget per channel reduced by 15 dB. This clearly shows the reduction in power budget associated with the extra components necessary to support the multiplexing and demultiplexing of additional wavelength channels.

A promising technology is represented by micro-optic devices employing interference filters. The advantage in such more complicated devices is that the response of the device can be tailored by the selection of an appropriate filter. It is common for these devices to have a characteristic where the loss associated with a signal wavelength is maintained over a significant range (usually more than 10 nm). Cascades of these devices can be used to combine more than two wavelengths.

6.3.3 Grating mu-demultiplexer

This type of component can be used as a multiplexer or demultiplexer. A typical device is shown in Fig. 6.11, comprising an array of fibers, a lens and an optical grating. In operation the multiplexed signal enters on one

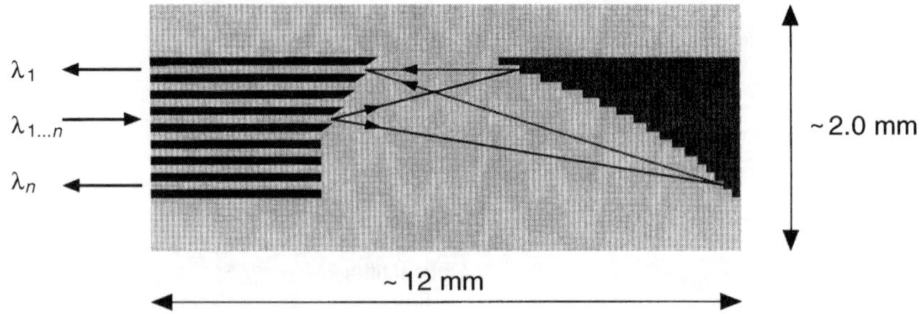

Fig. 6.11 A grating-based device for multiplexing or demultiplexing operations.

fiber, where it is focused by the lens onto the grating. The grating returns the incident light in a beam where the different wavelength components are spatially-separated. The geometry of the device is such that separate components of the beam illuminate a particular fiber; thus the output fiber array comprises signals corresponding to the individual wavelengths. The advantage is that the loss is fairly independent of the number of wavelengths. Commercial devices with a wavelength separation of 4 nm have an insertion loss of approximately 5 dB. Thus, for large numbers of wavelengths (e.g. more than five), grating-based multiplexers are advantageous; however, they are complicated components requiring careful positioning of the fiber array.

Similar components can be achieved using integrated optics and different technologies. Even InP-based grating devices have been reported in the literature, which allow integration of several components in the same chip.

6.4 INTEGRATED OPTICS FOR NETWORKING DEVICES

Integrated optics are of great interest in optical signal processing and telecommunications [14, 15]. Several factors are crucial for its successful implementation. Beyond just producing waveguide structures with low propagation losses and low in-plan scattering, any technology must be compatible with microelectronic device processing, involve a reasonable number of steps and masking levels, and provide high reproducibility. Especially important is the coupling between an optical integrated circuit and the other elements of an optical link, such as optical fibers, light sources and detectors. On all these points, silicon-based technologies offer significant advantages over competing technologies.

As a matter of fact, silica-on-silicon technology is a derivation of the planar technology widely used for electronic manufacturing. Its main advantages are compatibility with microelectronic processes, and the possibility of using the silicon substrate to hold and align the fibers to the device. Indeed, silicon crystals oriented along the (100) direction present preferential etching, which allows the creation of V-grooves bounded by (111) planes. The groove's depth and width can be controlled very accurately. Therefore, fibers can be coupled to component waveguides by simply positioning the fiber in the V-groove, producing a very low cost, efficient and reliable fiber-waveguide coupling process. This ultimately ends in a major impact on technoeconomic aspects of single-mode fiber-optic systems. A further advantage is represented by the possible hybridisation on the same substrate of optical components and ICs using the waveguides as optical interconnections between different parts of the circuit. A very compact optoelectronic module can then be realised.

Silica layers can be deposited in controlled steps on a silicon substrate. Single mode planar optical waveguides can be obtained by depositing three layers of SiO_2 with different doping. Channel waveguides can be patterned on the deposited layer by reactive ion etching. The physical characteristics of the waveguide core can be chosen so they are fully compatible with the fibers, both being made of the same material. Confined waveguides have been demonstrated with very low losses.

6.5 OPTICAL SPACE SWITCHING ARRAYS

Optical space switches are also fundamental for the achievement of an all-optical network. In fact, an all-optical network needs to perform switching and routing functions directly in the optical domain, so avoiding optoelectronic or electro-optic conversion, and electronic switching. In this way the device could act independently of the modulation format and speed.

Different devices have been demonstrated and reported in the literature, based on different switch points, such as directional couplers, digital optical switches or laser amplifier gates; employing technologies based on $LiNbO_3$ [16], InP [17] and silica waveguides [18]. Attractive advantages of using laser amplifier gates are the possibility of zero-loss operation; their compact physical size which, used in combination with tightly confining passive waveguides, allows switch matrices of relatively small substrate area; nanosecond switch time; low crosstalk; and detection possibility provided by the laser amplifiers.

One of the most promising devices, based on InP to be substantially lossless, has been realised by Ericsson Components. Illustrated in Fig. 6.12 it is a 4 × 4 switch matrix of a strictly non-blocking tree structure based on InGaAsP/InP laser amplifiers [19]. The key element is the traveling wave semiconductor optical amplifier (SOA). In fact, 16 SOAs are used in the interconnection region to control the switch state. The SOA is a logical

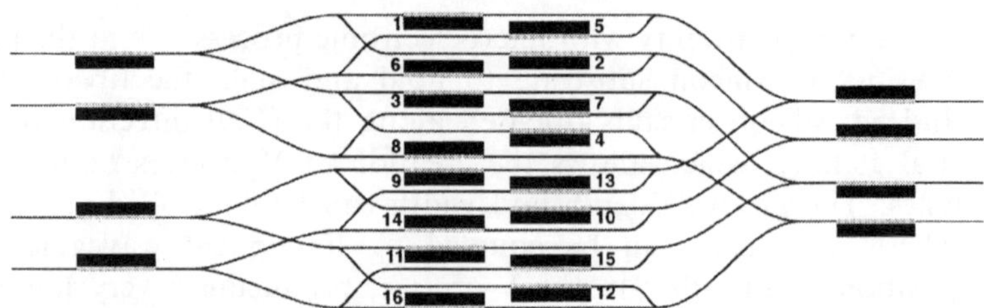

Fig. 6.12 Structure of a 4 × 4 InP-based optical space switching matrix. The key element is the semiconductor optical amplifier, which is used either as a logic gate or as a booster amplifier. (After [19] with permission of IEE).

gate that block or passes the light. For any possible path, an SOA acts as a gate. When a bias current is lower than a given value, the SOA absorbs the radiation so that light does not exit the port associated with that path. Conversely, if the current biasing the SOA is sufficient, the radiation passes through and reaches the corresponding output port. The ON/OFF ratio depends on the current values. A critical issue of this type of device is that the SOA end facets must be coated by an antireflection coating in order to strongly avoid any Fabry-Perot effect, otherwise lasing could occur and this would prevent the device from functioning properly.

The four input SOAs act as booster amplifiers, partially compensating for losses, properly controlled by electronic circuitry, the four output SOAs are used to equalise the different light output powers, due to the different lightwave paths.

Two main parameters are used to characterise this type of device: losses and ON/OFF ratio. The losses have to be compensated by amplification, and they have to be as low as possible to avoid severe limitations in the signal to noise ratio. On the other hand, the ON/OFF ratio of the individual output has to be the highest possible, in order to limit channels crosstalk, which represents one of the main drawback of all-optical networks (Chapter 12).

As an example, Fig. 6.13 depicts the performance of the 4 × 4 InGaAsP/ InP optical switch matrix described above, in terms of the fiber-to-fiber gain versus the current of gate switch section, for one of the transmission paths. Crosstalk can be defined as the optical power in a fiber at the considered output crosstalk port divided by the optical power in a fiber at the output signal port. The crosstalk measures [19] provided a mean value of approximately –40 dB. To analyse the contributions to crosstalk levels, it is practical to distinguish between two types of state:

(1) those which involve waveguide crossing as a possible source of crosstalk;

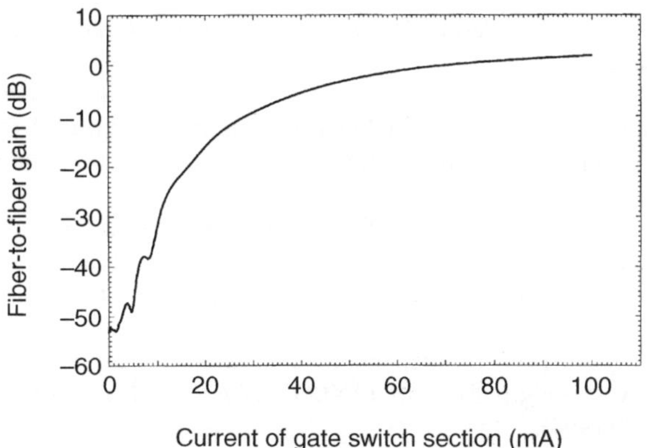

Fig. 6.13 Switching characteristics of one of the transmission paths of a 4 × 4 semiconductor optical amplifier gate switch matrix. The TE mode of the switch is excited by the input optical signal, wavelength 1.543 \mum at a power level of around −20 dBm. (After [19] with permission of IEE).

(2) those which do not.

For the considered switch, crosstalk levels less than −35 dB have been obtained for states of type 1, and less than −50 dB for states of type 2. However, examples of high crosstalk levels have been observed for states of type 1; this was probably due to defects in some of the waveguide crossing introduced in the semi-insulating regrowth and/or due to excessively deep reactive ion etching when the contact layers outside the contacts were removed to increase the electrical isolation between the active elements.

Moreover, digital transmission experiments, including transmission over 50 km, at 2.5 Gb/s involving such a monolithic 4 × 4 switch array, operating at a net positive fiber-to-fiber gain, have been reported in the literature [20]. It was found that this device introduces merely a small power penalty (<1 dB) at a FET receiver sensitivity of −27 dBm for a wide range (>10 dB) of switch input powers, demonstrating its suitability for transport network applications.

REFERENCES

1. M.J. O'Mahony, Wavelength/optical frequency division multiplexing, in *High Capacity optical transmissions explained,* Ed. by D.M. Spirit and M.J. O'Mahony, John Wiley, New York (1995).

2. P.A. Humblet, and W.M. Hamdy, Crosstalk analysis and filter optimization of single- and double cavity Fabry-Perot filters, *IEEE J. on Select. Areas on Comm.,* **8**, 1095–1107 (1990).

3. G. Hernandez, *Fabry-Perot Interferometers*. Cambridge University Press, Cambridge (1986).

4. A. Yariv and P. Yeh, *Optical Waves in Crystals*, John Wiley, New York (1984).

5. K.W. Cheung, Acoustooptic tunable filters in narrowband WDM networks: system issues and network applications, *J. Select. Areas in Commun.*, **8**, 1015–1025 (1990).

6. H. Hermann, D.A. Smith, and W. Sohler, Integrated optical, acoustically tunable wavelength filters and switches and their network applications, in *Proc. Eur. Conf. Integrated Optics* (*ECIO '93*), Neuchâtel, Switzerland, pp.10.1–10.3 (1993).

7. F. Tian *et al.*, Polarization-independent integrated optical, acoustically tunable double-stage wavelength filter in LiNbO$_3$, *IEEE/OSA J. Lightwave Technol.*, **12**, 1192–1197 (1994).

8. J.L. Jackel, J.E. Baran, A. D'Alessandro, and D.A. Smith, A passband-flattened acousto-optic filter, *IEEE Photon. Technol. Lett.*, **7**, 318–320 (1995).

9. D.A. Smith and Z. Bao, Technology and applications of the integrated acousto-optic filter, *MELECON'96*, Bari, Italy, pp. 100–107 (1996).

10. D.A. Smith et al., 'Evolution of the acousto-optic wavelength routing switch', IEEE J. Lightwave Technol., vol. 14, no. 6, pp. 1005–1019, 1996.

11. F. Wehrmann *et al.*, Fully packaged, integrated optical, acoustically tunable add-drop multiplexers in LiNbO$_3$, in *Proceding of the European Conference on Integrated Optics* (*ECIO '95*), Delft, The Netherlands, pp. 487–490 (1995).

12. F. Wehrmann et al., Integrated optical, wavelength selective, acoustically tunable 2 × 2 switches (add-drop multiplexers) in LiNbO$_3$, *IEEE J. Select. Topics in Quantum Electron.*, **2**, 263–269 (1996).

13. O. Sahlén, Active DBR filters for 2.5-Gb/s operation: linewidth, crosstalk, noise, and saturation properties, *IEEE/OSA J. Lightwave Technol.*, **10** 1631–1643 (1992).

14. N. Takato *et al.*, Silica-based single-mode waveguides on Silicon and their application to guided-wave optical interferometers, *IEEE/OSA J. Lightwave Technol.*, **6**, 1003–1009 (1988).

15. S. Valette *et al.*, Si-based integrated optics technologies, *Solid State Technol.*, 69–75 (1989).

16. P. Granestrand *et al.*, Strictly nonblocking 8 × 8 integrated optical switch matrix, *Electron. Lett.*, **22**, 816–818 (1986).

17. M. Janson *et al.*, Monolithically integrated 2 × 2 InGaAsP/InP laser amplifier gate switch arrays, *Electron. Lett.*, **28**, 776–777 (1992).

18. Y. Yamada *et al.*, Hybrid-integrated 4 × 4 optical gate matrix switch using silica-based optical waveguides, *IEEE/OSA J. Lightwave Technol.*, **LT10**, 383–390 (1992).

19. M. Gustavsson *et al.*, Monolithically integrated 4 × 4 InGaAsP/InP laser amplifier gate switch arrays, *Electron. Lett.*, **28**, 2223–2224 (1992).

20. M. Gustavsson, M. Janson, and L. Lundgren, Digital transmission experiments with monolithic 4 × 4 InGaAsP/InP laser amplifier gate switch array, *Electron. Lett.*, **29**, 1083–1084 (1993).

7
Wavelength Translators

7.1 INTRODUCTION

Wavelength division multiplexing (WDM) techniques offer a very effective utilisation of the fiber bandwidth directly in the wavelength domain, rather than in the time domain. In addition, wavelength can be used to perform functions such as routing and switching [1], allowing the realisation of an all-optical transparent layer in the network [2]. The number of allowed wavelengths in WDM networks determines the number of independent wavelength addresses, or paths. Although this number may be large enough to fulfil the required information capacity, it is often insufficient to support a large number of nodes. Then the blocking probability rises due to possible wavelength contention when two channels, at the same wavelength, are to be routed to the same output. One method to overcome this limitation is to convert signals from one wavelength to another.

The benefit of wavelength translation varies with network architectures and traffic patterns. In general, a 'small' network with a fixed traffic pattern may not require wavelength conversion and will maintain a low blocking probability. A 'large' network with dynamic traffic patterns will instead greatly benefit from wavelength conversion. It is generally accepted that the benefit of wavelength conversion increases with increased traffic.

Wavelength translation also allows the distribution of network control and management into smaller sub-networks and as flexible wavelength assignments within the sub-network. Figure 7.1 illustrates this accomplishment when three subnetworks are employed. Network operators 1, 2 and 3 are required to manage only their own sub-networks, and wavelength conversion may be needed for communications between the sub-networks.

For WDM networks to serve a vast number of users and evolve further in the future, the maximum degree of transparency and interoperability are desired. Transparent networks in turn facilitate implementations of complicated security architectures. The first problem in achieving transparency in WDM networks lies in wavelength translation, as the majority of wavelength conversion methods offer only limited transparency.

Wavelength conversion can be obtained in different ways, using semiconductor lasers, passive fibers, semiconductor optical amplifiers and semiconductor waveguides. However, their application in high-speed optical communication systems demands a high speed of operation. For this reason the chapter will treat only wavelength converters which can operate at high speed. In this scenario there are currently five possible conversion techniques:

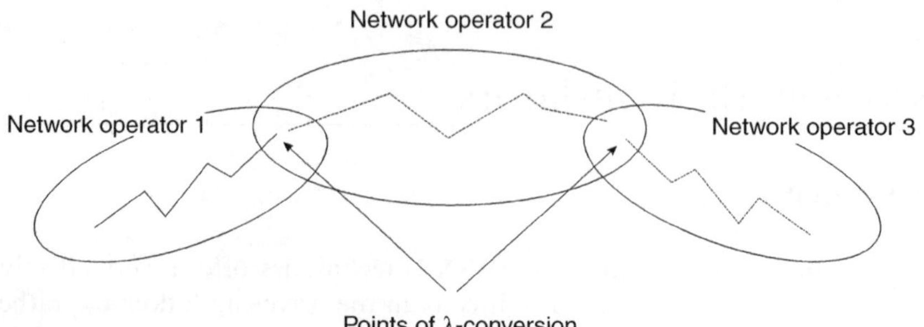

Fig. 7.1 Wavelength translation in optical networks. Using wavelength translation, the network control and management can be distributed into smaller subnetworks, and the wavelength assignments within the subnetwork can be achieved more flexibly.

(1) optoelectronic conversion,
(2) cross-gain modulation (XGM) in semiconductor optical amplifiers (SOAs),
(3) cross-phase modulation (XPM) in SOAs,
(4) four-wave mixing (FWM) in SOAs,
(5) difference frequency-generation in semiconductor waveguides.

The different types of wavelength converter (W-C) are discussed in the following sections.

7.2 OPTOELECTRONIC CONVERSION

Optoelectronic wavelength conversion can be achieved either via regenerative or non-regenerative WCs.

A regenerative WC simply consists of a cascade of receiver and transmitter modules that regenerate the digital signal (Chapter 8). The transmitter can either emit at a fixed optical wavelength or be tunable at different wavelengths. Transmitter tunability, offers a greater flexibility of the WC. Such device prevents signal transparency, since it is only operate for a precise signal modulation format and speed. Regenerative WCs have been used as key elements for optical cross-connects [2].

The non-regenerative WC [4] allows a certain degree of transparency [3]. It basically consists of a photodetector (e.g. a p-i-n diode), followed by an electronic front end (the amplifying stage, possibly including a filtering stage), and an optical transmitter which, as in the other case, can be tunable or not. The transmitter itself may consist of a directly or externally modulated laser source. In the last case it also contains an optical modulator. The non-regenerative WC introduces some trasmission penalty through two mechanisms [5]: the noise introduced by the electronic front end and the chirp introduced by the optical transmitter. The chirp depends on the transmitter that is used in the device itself: it is larger if the laser is directly modulated, and smaller if the laser is externally modulated (even zero if a Mach-Zehnder (MZ) modulator is used and properly designed). The chirping/dispersion induced distortions accumulate along the signal path.

It is meaningful to compare the simulated performance of different types of non-regenerative WC, considering for instance directly modulated lasers for 622 Mb/s systems, and externally modulated lasers for 2.5 Gb/s systems, and assuming either MZ or Franz-Keldysh (FK) modulators. In order to do that, we simulate the transmission performance of a generic path through an optical network containing these devices (see later for the performance evaluation criteria). For details on the modeling see Chapter 3. The main systems parameters are reported in Table 7.1 and the para meters of the optoelectronic wavelength converter are reported in Table 7.2. In the simulations, dispersion-shifted fibers (λ = 1550 nm, D = 4 ps/km/nm) were assumed. The space switch matrices were passive with a loss of 8 dB, and the optical filters inside the OXC were double-stage acousto-optic filters whose single-stage bandwidth was 1.7 nm (the overall bandwidth is 1.2 nm).

As a result, Figs. 7.2 and 7.3 show the performance of a transmission path through the network for the optoelectronic converter with directly

Table 7.1 System parameters used in the simulation

WDM comb in the transport network	4
Input/output fibers in the OXCs	4
Channel spacing	4 nm
OXC spacing	200 km
Number of fiber trunks	8
Inline amplifiers spacing	50 km
Optical amplifier noise factor	2.6

Table 7.2 Parameters for the optoelectronic wavelength converter. EM = External modulator; DML = directly modulated laser.

Laser currents: threshold, bias, modulation	33.5 mA, 38.8 mA, 40 mA
Rise time of the driving current	0.3 T
Transmitter extinction ratio	10 dB (E.M), 6 dB (D.M.L)
Reicever sensitivity	–29 dBm (E.M), –33 dBm (D.M.L)

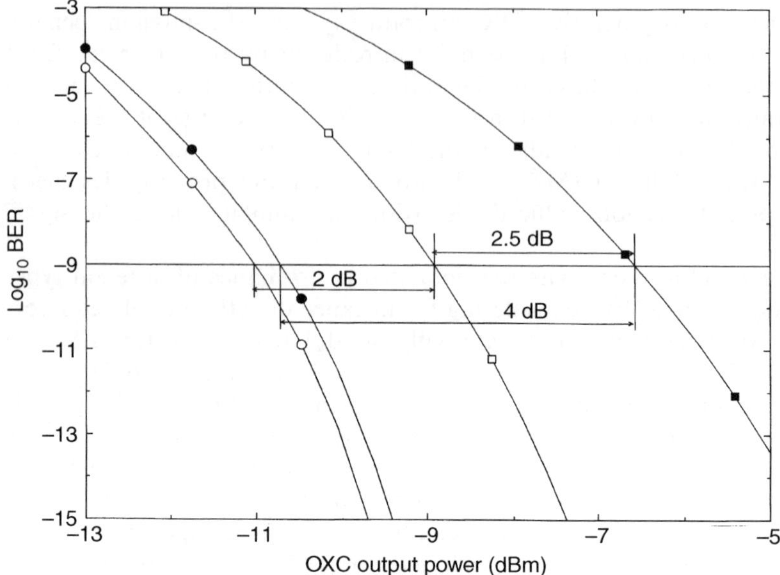

Fig. 7.2 Simulated transmission performance at 622 Mb/s. Squares and circles relate to the presence or the absence of wavelength converters inside the optical cross connect, respectively. Open symbols indicate dispersion compensation; filled simbols indicate no dispersion compensation.

modulated lasers at 622 Mb/s, and with externally modulated lasers at 2.5 Gb/s. These curves are compared with the case in which no wavelength conversion is adopted in the OXC node.

Concerning dispersion, consider the possibility of compensation by passive fibers ($D = -80$ ps/km/nm) after each inline optical amplifier. It can notice that, as expected, the introduction of wavelength converters introduces a power penalty related to the noise factor of the converter itself and by the accumulation of the chirping/dispersion induced distortions. Dispersion compensation always improves the performance. The effect of chromatic dispersion is quite irrelevant for 622 Mb/s systems in which the wavelength conversion is absent; however, it is significant if optoelectronic converters are introduced in the OXC node (about 2.5 dB penalty in Fig. 7.2). In fact, small variations of the driving current of the laser cause appreciable variations of the output optical power with a consequent degradation of the transmission performance. The introduction of the considered wavelength converters produces a penalty of 2 dB with compensation and 4 dB without.

For the 2.5 Gb/s systems, introduction of WCs, introduces a penalty of at least 2.5 dB. Moreover, when a converter using an MZ modulator with a chirping parameter $\theta = 0$ is used, there is a measurable difference

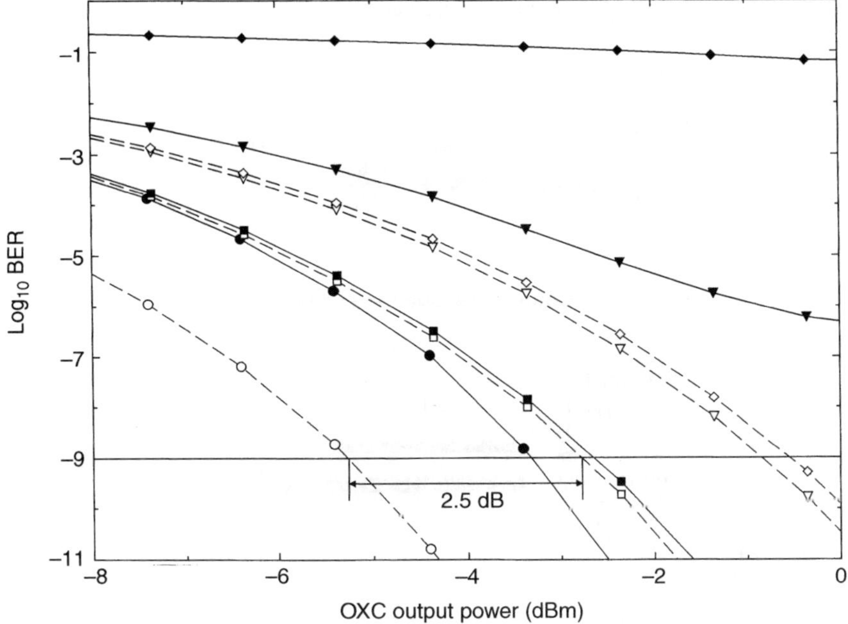

Fig. 7.3 Simulated transmission performance at 2.5 Gb/s. Circles indicate the absence of wavelength converters inside the optical cross-connect. Squares, triangles and diamonds indicate the use, in the converter, of external modulators with chirping parameters of 0, 0.8 and 1, respectively. Open symbols indicate dispersion compensation; filled symbols indicate no dispersion compensation.

between the compensated and non-compensated cases (Fig. 7.3). That is, when chirping is suppressed, dispersion compensation is not critical.but if modulators with $\theta > 0$ are used in the converters (e.g. FK modulators), dispersion compensation could be needed.

The reported results show that non-regenerative optoelectronic wavelength converters can be favorably employed in optical networks, provided the chirping introduced by the optical transmitters inside the converters is quite low, otherwise dispersion compensation might be required.

7.3 CROSS-GAIN MODULATION IN SOA DEVICES

7.3.1 Basic characteristics

The principle of XGM is shown in Fig. 7.4 [6]. An intensity-modulated input signal modulates the gain in the SOA, via gain saturation. A continuous wave (CW) signal, at the desired output wavelength, is modulated by

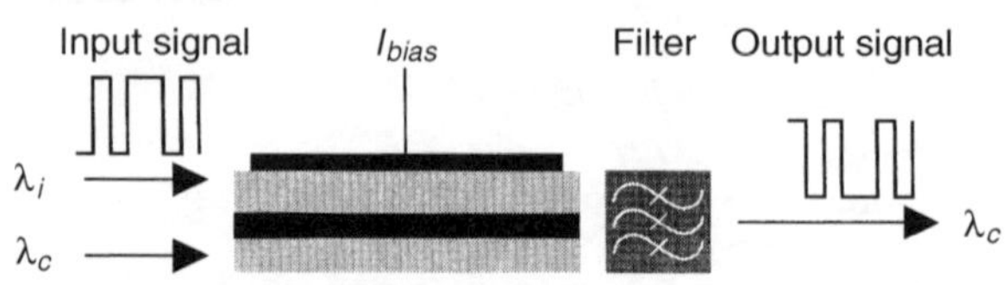

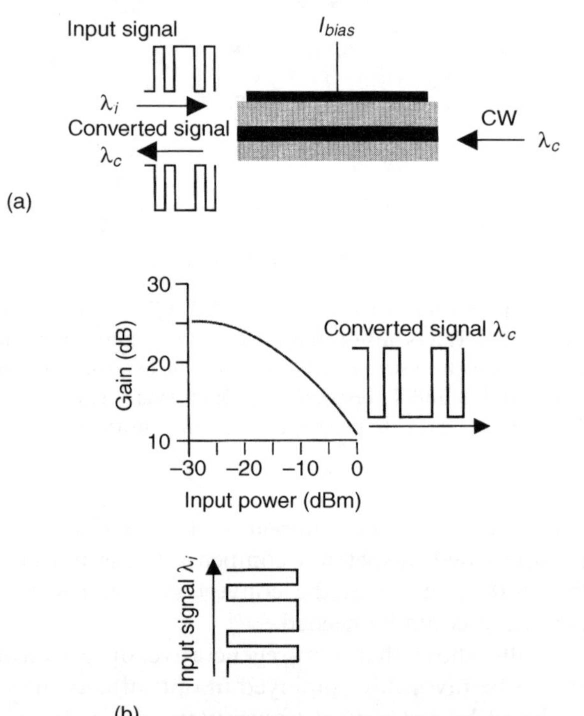

Fig. 7.4 Wavelength conversion based on XGM in SOAs: (a) converter configuration schemes; (b) principle of conversion by gain saturation.

the gain variation; so after the SOA it carries the same information as the intensity-modulated input signal. As shown in the figure, the input signal and the CW signal can be launched either co- or counterdirectionally into the SOA. In the counterdirectional case, the output filter needed for the copropagation scheme can be avoided and it is possible to convert the same wavelength.

The XGM SOA converter is polarisation independent if SOAs with a polarization-independent gain are employed. These amplifiers are now fabricated in many laboratories [7].

7.3.2 XGM converter modeling

Accurate modeling of all the converters based on SOAs, operating at high speed, generally requires dynamic models for the SOA [8, 9], similar to those reported in Chapter 5. However, in many practical cases, even at high speed such as 10 Gb/s, simpler approaches can be employed, as described below.

A fundamental hypothesis, called 'adiabatic approximation', has to be made for all the considered converters: the device responds instantaneously to the modulating signal. This assumption holds up to very high modulation speeds (as high as 100 G/s) for the FWM converter if a high enough detuning is adopted (see later). For XPM converters, this hypothesis holds up to 10 G/s if SOAs are suitably designed [6], but it is more critical for XGM converters. As a matter of fact, even if good transmission performance has been obtained up to 20 Gbit/s adopting one XGM converter, the pulse distortion introduced by the converter cannot be neglected when a large number of converters are cascaded [8]. However, if a large number of XGM converters are cascaded, the link performances are also degraded by the reduction of the signal extinction ratio occurring in each conversion. Due to this last effect, the number of XGM converters that can be cascaded is quite limited, so the adiabatic approximation becomes feasible up to a transmission speed of approximately 10 G/s.

Under the adiabatic approximation, XGM and XPM converters have been simulated by an amplifier model developed following a reported procedure [9]. Three equations are used. The first gives the saturated gain as a function of the overall input signal power:

$$\frac{P_0}{P_\sigma} = \frac{g_0 - \gamma}{\gamma} \cdot \frac{1 - (G/G_0)^{\gamma/g_0}}{G - (G/G_0)^{\gamma/g_0}} \tag{7.1}$$

where P_0 and P_σ are the input power and the saturation power respectively, G is the saturated gain and G_0 is the linear device gain. The scattering losses and the local effective gain are indicated with γ and g_0 respectively.

The second equation gives the phase change $\Delta\phi$ experienced by the field during propagation in the SOA:

$$\Delta\phi = 2\pi r_0 \frac{I}{\lambda} + \alpha[\ln(G) - \ln(G_0)] \tag{7.2}$$

where α is the SOA linewidth enhancement factor and r_0 the refractive index in the absence of optical power.

Finally the third equation gives the power spectral density S_n of the ASE noise:

$$
\begin{aligned}
S_n = \hbar\omega\frac{\overline{N}}{\overline{N} - N_0}\Bigg\{ &\frac{g_0}{g_0 - \gamma}(G - 1) \\
&+ \frac{\gamma g_0}{(g_0 - \gamma)^2}\frac{P_0}{P_\sigma}\cdot G\ln\left[\frac{(g_0 - \gamma)P_\sigma G - \gamma P_0 G}{(g_0 - \gamma)P_\sigma - \gamma P_0 G}\right]\Bigg\} \\
&+ \hbar\omega\frac{N_0}{\overline{N} - N_0}\frac{P_0}{P_\sigma}[\ln(G) + \gamma L]
\end{aligned}
\tag{7.3}
$$

where $\overline{N}$ and N_0 are the carrier densities in the linear regime and at transparency, respectively. Note that equation (7.3) no longer holds when the amplifier is too close to transparency (as a rule of thumb when $G < 4$–5 dB) since in this case a more accurate expression of the SOA noise factor has to be taken into account.

By using (7.1) and (7.3), XGM wavelength converters can be easily simulated in the time domain.

7.3.3 XGM converter performance

The gain curve in Fig. 7.4b shows that the extinction ratio for the signal at the output of the converter is generally smaller than for the signal at the input. This leads to an excess penalty in the transmission performance. In general, a larger extinction ratio for the output signal can be obtained by strongly saturating the SOA via a high dynamic of the input signal. On the other hand, higher signal levels, driving the SOA deeper into saturation, produce excess penalty due to larger turn-on delay for the converted signal and increase the noise contribution of the device itself. Moreover, gain and refraction index are linked in SOAs so that gain modulation induces a spurious index modulation. As a consequence, a spurious chirp effect is also linked to XGM wavelength conversion.

The cascade of XGM converters causes a progressive degradation of the signal extinction ratio. In fact, once the signal has passed through an XGM converter, it experiences a lower extinction ratio. This signal then drives the second converter with a lower dynamic, hence the output signal extinction ratio is even smaller, and so forth. The use of XGM converters in optical transport networks is mainly limited by such progressive degradation of the performance. The error probability of a signal crossing several optical cross-connects adopting this type of device is reported in literature [9]. The results are shown in Fig. 7.5 (the parameters used in the

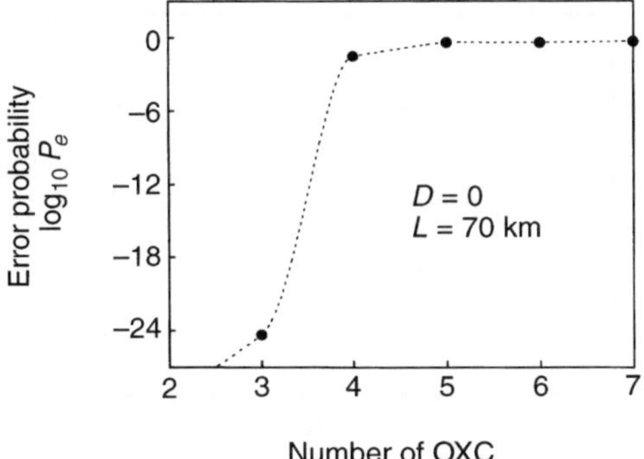

Fig. 7.5 Error probability versus the optical cross-connect number (that coincides with the number of crossed converters).

simulation are not reported here for the sake of brevity, but can be found in the article). In this system, an optical cross-connect spacing of 70 km and a bit rate of 2.5 Gb/s were assumed. The performance degraded quickly.

7.4 CROSS-PHASE MODULATION IN SOA DEVICES

7.4.1 Basic characteristics

The XGM scheme allows a simple realisation of wavelength converters. Unfortunately, the strong extinction ratio degradation prevents their employment in optical networks. To overcome this problem, the SOA converter can be used in a cross-phase modulation (XPM) mode. The XPM scheme relies on the dependency of the refractive index on the carrier density in the active region of the SOA [10–12]. An incoming signal that decreases the carrier density will modulate the refractive index and thereby result in phase modulation of a CW signal (at a different wavelength) coupled into the converter. The phase-modulated CW signal can be demultiplexed after the converter, or even better the SOA can be integrated into an interferometer so that an intensity-modulated signal format results at the output of the converter. Non-linear loop mirrors [11], Mach-Zehnder interferometers (MZIs) [12] and Michelson interferometers (MIs) [13] have been proposed.

The XPM scheme has the distinct feature that the converted signal can be either inverted or not, compared to the input signal, depending on the slope of the demultiplexer. Normally, it is advantegeous for the converted signal to be non-inverted. The power efficiencies for the XPM scheme are much large than for the XGM scheme. This is illustrated in Fig. 7.6, where

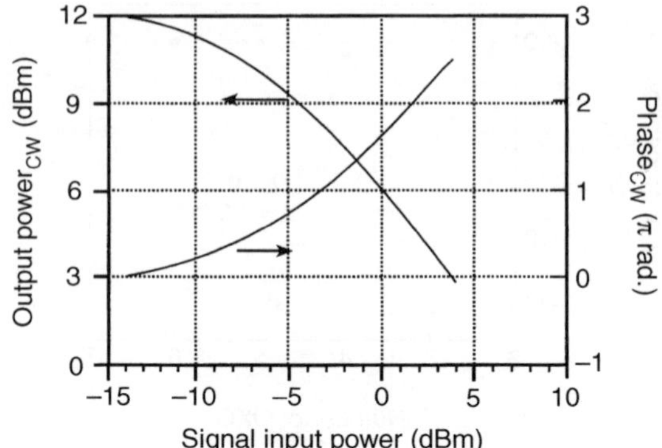

Fig. 7.6 Output power and phase of the amplified CW signal versus the signal input power for a 1200 \mum long SOA. A phase change of \pi radians, which is needed to operate an interferometric converter, is attained for a gain variation of only 4 dB. (After [15] with permission of IEEE).

both the output power and the phase of the amplified CW signal are shown versus the signal input power in SOA 1200 μm long. A phase change of π radians, which is needed to operate an interferometric converter, is attained for a gain variation of only 4 dB. A gain variation of at least 10 dB is required for the XGM converter.

As an example, Fig. 7.7 shows the principle of operation for interferometric WCs based on XPM converters in SOAs. The SOAs are placed in asymmetric configurations so that the phase change in the two amplifiers is different. As a result, the CW light is modulated according to this phase difference. In the first configuration (Fig. 7.7a), asymmetric splitters ensure that an intensity-dependent phase difference is achieved between the interferometer arms, due to the different saturation of the two amplifiers. In the second configuration (Fig. 7.7b), the MZI is formed by symmetric splitters and the input signal is fed into only one of the SOAs through an additional coupler. The saturation is asymmetric since the other SOA is not affected by the input signal power. The converter could also be realised with only one amplifier in one of the interferometer arms [14], but this scheme gives less output power and will be sensitive to changes in the polarisation of the CW signal.

7.4.2 XPM converter modeling

To analyse the behavior of XPM converters, it is possible to exploit the modeling reported in the case of an XGM device, adapted in the new

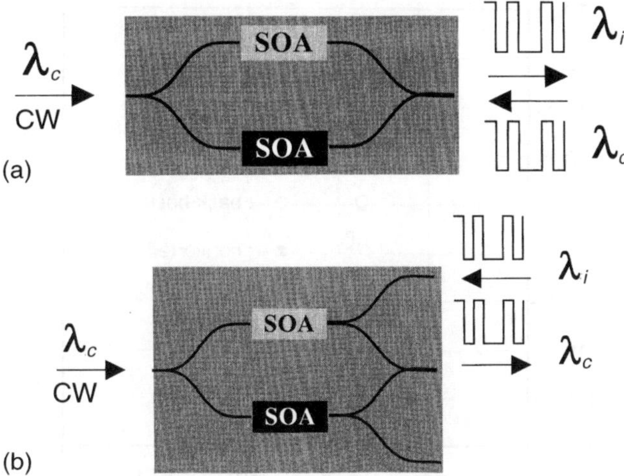

Fig. 7.7 Typical configurations of XPM converters: (a) asymmetric splitters ensure that an intensity-dependent phase difference is achieved between the interferometer arms, due to the different saturation of SOA1 and SOA2; (b) the MZI is formed by symmetric splitters and the input signal is fed to only one of the SOAs through an additional coupler.

conversion scheme. All the ideas for modeling SOAs remain valid for XPM devices.

7.4.3 XPM converter performance

Typical conversion performance in a system experiment [15] at 5 Gb/s is illustrated in Fig. 7.8. The BER curves for the converted signal (1543 nm) together with back-to-back measurements of the input signal (1531 nm) indicate that penalty-free conversion is obtained.

The wavelength dependence of the input signal and the CW signal is also important. An ideal converter would be expected to operate with equal performance at all wavelengths within the EDFA window (about 30 nm), where it is likely that the future WDM system will operate. Figure 7.9 gives the measured penalty for conversion from 1543 nm, as a function of wavelength for the converted signal. The MZI converters show good performance for both up- and downconversion. Furthermore, they can regenerate signals both with respect to extinction ratio and spectral quality.

This type of converter can be favorably employed in an optical network [9]. A performance evaluation of networks using this type of converter is given in Chapter 12 (12.15 to 12.20).

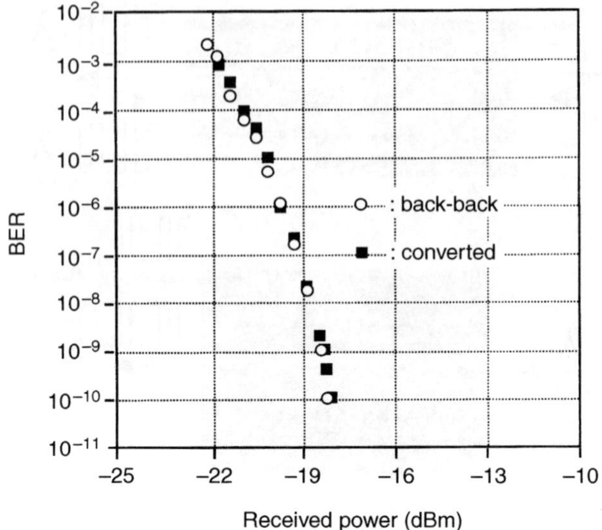

Fig. 7.8 Conversion performance for a typical XPM converter in a system experiment at 5 Gb/s: BER curves for the converted signal (1543 nm) together with back-to-back measurements of the input signal (1531 nm). Penalty-free conversion is obtained. (After [15] with permission of IEEE).

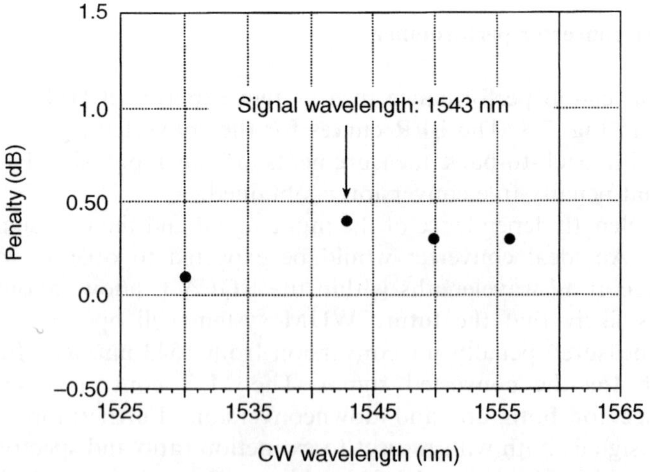

Fig. 7.9 Measured penalty for conversion from 1543 nm, as a function of wavelength for the converted signal. (After [15] with permission of IEEE).

7.5 FOUR -WAVE MIXING IN SOA DEVICES

7.5.1 Basic characteristics

Four-wave mixing is a non-linear effect that takes place when two waves (signal and pump at different wavelengths are injected into an SOA. In this situation, a third optical field is generated at the device output, whose angular frequency ω_c is given by $\omega_c = 2\omega_p - \omega_i = \omega_p - \Omega$, where ω_i and ω_p are the angular frequencies of the signal of the pump field respectively, and $\Omega = \omega_i - \omega_p$ is the detuning between signal and pump. The basic sketch of this type of converter is shown in Fig. 7.10.

Different physical phenomena can generate FWM in an SOA. At low detunings (Ω within a few tens of gigahertz) the main mechanism is the carrier density pulsation induced by the pump-signal beating inside the active region [16]. For higher values of the detuning, carrier pulsation is no more effective and FWM is mainly caused by non-linear gain saturation due to intraband carrier dynamics. In particular, two intraband mechanisms have been proposed in the literature as causes of FWM: spectral hole burning [16, 17] and carrier heating [18, 19]. The characteristic times of these phenomena are of the order of hundreds of femtoseconds. Because of such small times, FWM can be observed at very high detunings and experimental observations have been carried out for detuning levels above 1 THz [20–22].

The field at frequency ω_c has a spectrum equal to that of the signal except for spectral inversion, so that signal modulation is preserved. Spectral inversion is quite an important feature of the FWM converter. If

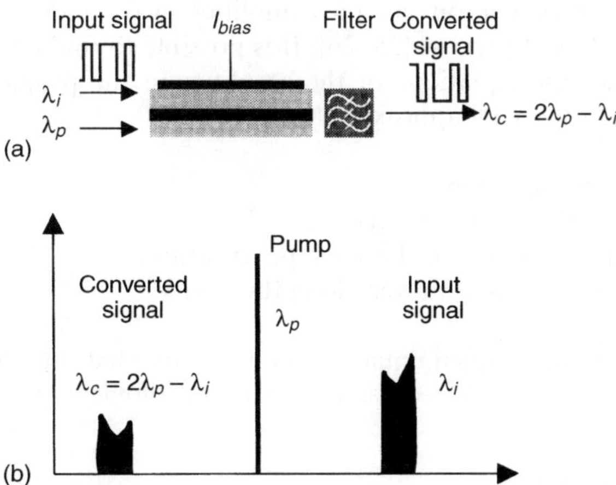

Fig. 7.10 Typical configuration (a) of a wavelength converter based on FWM in an SOA, and the resulting spectra (b).

this kind of converter is placed in the middle of a dispersive optical link, dispersion compensation can be attained [23]. Moreover, in the nonlinear regime, even the Kerr effect in a fiber can be partially compensated by midway spectral inversion [24].

Frequency conversion takes place independently of the modulation format and the bit rate; it is not sensitive to the signal power level, with the only conditions that the converter response is almost constant in the signal bandwidth and that the signal power is somewhat smaller than the pump power. This allows the realisation of a frequency converter transparent to both transmission format and bit rate.

The main disadvantage of the FWM converter is its low conversion efficiency: although XGM and XPM converters are active devices, providing signal amplification besides wavelength conversion, the FWM converter attenuates the signal. Moreover, the FWM converter introduces phase noise in the signal at each conversion, due to the phase noise that is unavoidably present in the optical pump from a semiconductor laser.

7.5.2 FWM converter modeling

To simulate the FWM converter, equations (7.1) and (7.3) are not sufficient since an accurate model must be carried out for the propagation in the amplifier of all three waves in the presence of fast non-linearity. A proper analysis requires the evaluation of the converter efficiency and the noise spectral density as a function of the amplifier parameters and the injected optical fields.

The converter can be described by a set of coupled-mode equations describing the propagation in the amplifier waveguide of the fields involved in the FWM process [25, 26]. It is possible to find an approximate solution for the rate equations of the amplifier in the propagation equations within the following approximations [27]:

(i)   slowly varying fields
(ii)   negligible dispersion effects
(iii) pump and signal with the same polarisations
(iv)   constant saturation power along the cavity.

As a result, a set of coupled equations can be obtained, in which the nonlinear processes are introduced using a phenomenological approach [27].

Conversion efficiency

A numerical solution of the coupled equations mentioned above leads to an estimate of the filed power at the device output and therefore of the conversion efficiency [26]. The results are in very good agreement with those of more detailed models [28, 29].

However, it has been shown in [30] that it is possible to solve the set of coupled wave equations in an analytical manner by assuming that the conjugate power is lower than the signal power at any position along the cavity. This hypothesis is quite realistic if we consider the application of these converters in optical systems, in which the conversion interval is higher than a few tens of gigahertz. Adopting this conjecture, a simple formula for the efficiency can be obtained [27]:

$$\eta = \frac{|E_2(L)|^2}{|E_1(0)|^2} = \left[\frac{|E_0(0)|^2}{S(0)}\right]^2 \cdot P_s^2 G \left(\ln \frac{G_0}{G}\right)^2 \left|\frac{F(\Omega)}{2}\right|^2 \tag{7.4}$$

where E_0, E_1, and E_2 represent the pump, the input signal and the conjugate (output) signal, respectively; S is the total power, P_s is the saturation power, G_0 and G are the unsaturated and the saturated gain, respectively. The function $F(\Omega)$ accounts for the non-linear effects and is a function of the detuning Ω.

Several non-linear effects determine the behavior of the converter and affect $F(\Omega)$. The most important ones are the *carrier pulsation*, in the high-detuning regime, the *carrier heating* and the *spectral hole burning*. In fact, they can be physically explained as follows. When the optical fields interact with the medium, they give rise to stimulated emission processes, which decrease the number of carriers in the conduction band of the inverted semiconductor. The electrical injection tends to restore the dynamical equilibrium in a characteristic time of the order of 200 ps. Moreover, the optical pump field digs a hole in the intraband carrier distribution, which fills up within the carrier-carrier scattering time, of the order of 100 fs (spectral hole burning). Meantime, stimulated emission and free carrier absorption give rise to a heated carrier density distribution. Stimulated emission subtracts cold carriers from the lowest states in the band, thus heating the carrier distribution. Free carrier absorption contributes to carrier transitions toward higher states in the conduction band and also increases the average temperature of the carriers (carrier heating). Two characteristic times are associated with this effect: one is the time required to establish a heated carrier distribution (carrier-carrier scattering time), which is around 100 fs; the other is the time required by the phonon-carrier scattering to cool the carrier temperature down to lattice temperature, between 0.7 and 1.3 ps.

Other processes that perturb the distribution equilibrium, faster than those described above, are *two-photon absorption* and the *Kerr effect*. Their dynamics are related to the polarisation dephasing time, which is faster than 100 fs.

An expression for $F(\omega)$ is reported in [27]:

$$F(\Omega) = -\left\{ \frac{1}{P_s}(1 - i\alpha)H_c(\Omega) + \varepsilon_{sh}H_{sh} + (1 - i\beta)\varepsilon'_{ch}H_{ch} \right.$$
$$\left. + \left[1 + \frac{G+1}{2(G-1)} \ln \frac{G_0}{G} \right](1 - i\beta)\varepsilon^\circ_{ch}H_{ch} \right\},$$

(7.5)

where H_c, H_{sh}, and H_{ch} are the linear responses of the gain relating to the carrier pulsation, spectral hole burning (SH) and carrier heating (CH), respectively; ε_{sh} and ε_{ch} represent the strength of SH and CH, respectively. In particular, ε'_{ch} and ε°_{ch} represent the strengths of CH due to stimulated emission, and due to free carrier absorption, respectively. α and β are the linear and the non-linear parts of the Henry constant, respectively.

Phase noise

If the pump laser were perfectly monochromatic, the converted signal would be a spectral inverted replica [31], scaled by the efficiency, of the process of the input signal. However, since the pump laser consists of a semiconductor laser, it is affected by phase noise. Assuming a perfectly Lorentzian lineshape, the pump phase can be represented by a Wiener process of variance $\sigma_\phi^2 = 2\pi\Delta\nu_p T$ [32], where T is the observation time (the bit interval in the case of a digital transmission), and $\Delta\nu_p$ the pump FWHM linewidth. Under this condition, even the phase noise added to the output signal is a Wiener process whose variance is given by $4\sigma_\phi^2 = 8\pi\Delta\nu_p T$ [33].

These consideration, lead to the following equation relating the baseband power spectral density $S_i(\omega)$ of the input signal to that of the frequency-translated signal at the SOA output, $C_o(\omega)$:

$$C_0(\omega) = \chi(\Omega)S_i^{inv}(\omega) * L(\omega, 8\pi\Delta\nu_p) + S_c,$$

(7.6)

where $\chi(\Omega)$ is the conversion efficiency, $S_i^{inv}(\omega)$ represents the spectral inverted input signal, $L(\omega,8\pi\Delta\nu_p)$ is a Lorentzian function with FWHM equal to $4\Delta\nu_p$, and S_c is the power spectral density of the ASE noise introduced by the SOA. The convolution operation is indicated by asterisk.

Equation (7.6) allows the power spectral density to be directly evaluated. However, in many practical simulations, it could be necessary to evaluate the Fourier transform of the propagating signal, at each point of a fiber-optic communication link, instead of the power spectral density. This type of simulation can be carried out as follows.

The wavelength conversion is simulated in the time domain by three steps: the incoming signal is phase conjugated and multiplied by the conversion efficiency, a random phase is introduced to take into account the phase noise and the obtained signal is filtered by the filter at the converter output. The random phase is a sample of a Wiener process simulating the phase noise of a semiconductor laser with a linewidth four times that of the pump laser.

7.5.3 FWM converter performance

The measured performance of an FWM converter was reported by HHI researchers [34]. Error-free frequency conversion over 500 GHz at bit rates from 1 to 10 Gb/s has also been reported [35]. This experiment revealed that the frequency-converted signals are unaffected by the ultrafast conversion process. In fact, Fig. 7.11 shows the performance, in terms of BER, at 1, 2.5, 5 and 10 Gb/s. The same receiver was used for all measurements, with a sensitivity (BER $= 10^{-9}$) of -38.4 dBm at 1 Gb/s, -37.2 dBm at 2.5 Gb/s, -32.6 dBm at 5 Gb/s, -32.7 dBm at 10 Gb/s. Compared to these baselines, a penalty of 2.2 dB at 1 Gb/s, 0.7 dB at 2.5 Gb/s, 0.9 dB at 5 Gb/s, and 0.4 dB at 10 Gb/s occurs for the frequency-converted signals. Surprisingly, the largest penalty is measured for the lowest bit rate.

If FWM converters are used in an optical network, the conversion efficiency depends critically on the conversion interval. In a real network, the conversion interval depends on the optical node architecture. Performances are reported in Chapter 12 along with XPM devices. An interesting feature of FWM devices is their ability to compensate dispersion-induced distortions. As a matter of fact, the FWM device can be seen as an optical phase

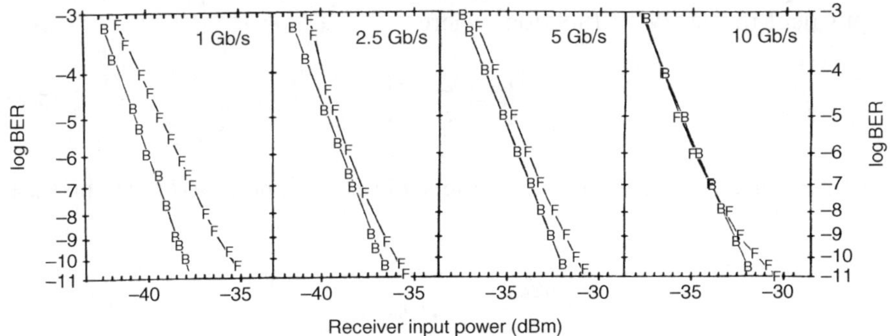

Fig. 7.11 Measured performances of wavelength converters based on FWM in SOAs. The BER is reported versus the receiver input power, at bit rates of 1, 2.5, 5 and 10 Gb/s, respectively, and compared with the respective receiver sensitivity curves. B = baseline measurement, F = 500 GHz frequency-converted measurement. (After [35] with permission of IEE).

conjugator. Used in the middle of an optical link, it allows fiber dispersion to be compensated.

The efficiency limit, which depends on the wavelength detuning, will be progressively reduced by proper design of the SOA in which the FWM effect takes place [36].

7.6 DIFFERENCE FREQUENCY GENERATION

7.6.1 Basic characteristics

Difference frequency generation (DFG) is a consequence of non-linear interaction of the material with two optical waves: a pump wave and a signal wave. Similarly to FWM, DFG offers a transparent wavelength conversion with a quantum bnoise limited operation. It is also capable of chirp reversal (due to spectral inversion) and multiwavelength conversion.

DFG wavelength conversion uses second-order optical non-linearities to produce transparent interchange of wavelengths. The mapping function is a frequency mixing relationship:

$$\frac{1}{\lambda_o} = \frac{1}{\lambda_p} - \frac{1}{\lambda_i},\tag{7.7}$$

where λ_o, λ_p, and λ_i are the output, input and pump wavelengths, respectively. Figure 7.12 shows this mapping function. For a limited bandwidth of WDM channels, this is analogous to taking a mirror image of the input wavelength about the mirror plane at twice λ_p. This mapping function is equivalent to the FWM function, where

$$\frac{1}{\lambda_o} = \frac{2}{\lambda_p} - \frac{1}{\lambda_i}.\tag{7.8}$$

The key difference is the fact that the symmetry plane (mirror) in this case

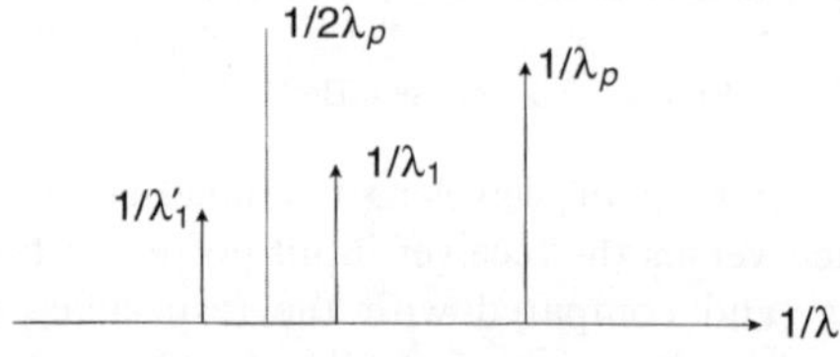

Fig. 7.12 Mapping function for a DFG wavelength converter.

is the actual pump wavelength itself. This causes difficulties in filtering the unwanted waves in the case of FWM.

At present, DFG wavelength translators have been realised in research laboratories, based either on an $LiNbO_3$ waveguide [37] or an AlGaAs waveguide [38]. DFG in semiconductor waveguides is of particular interest to realise an efficient wavelength conversion device monolithically integrated with a semiconductor laser.

The main difficulty in accomplishing semiconductor-based wavelength translators lies in the phase matching of interacting waves [37]. The fabrication issues relating to this type of device are discussed in the literature [38]. Note that the design of the AlGaAs waveguide takes account of polarisation sensitivity. The zincblende semiconductor possesses non-zero off-diagonal second-order susceptibility tensor elements $\chi_{ijk}^{(2)}$, with $i \neq j$, $j \neq k$, $k \neq i$. If type II phase matching [39] is used in a waveguide formed on an xy plane, the TE signal wave interacting with the TE pump wave yields a TM converted wave. Correspondingly, the TM signal wave interacting with the same pump wave yields a TE converted wave. The two processes are symmetric, follow an identical phase matching condition, and produce the same conversion efficiency. In the case where an input signal wave with an arbitrary polarisation state is injected, the TE and TM components of the wave will simultaneously sustain the two conversion processes to produce the TM and TE components of the converted wave. Althoughthe conversion process changes the polarisation state, it produces a polarisation-independent wavelength conversion efficiency. As a result, the designed waveguide can operate as a polarisation diversified as well as a polarisation-independent wavelength converter.

7.6.2 DFG converter performance

Unlike the othertypes converter types, modeling a DFG converter is quite simple, since its efficiency is almost constant along the optical bandwidth of interest in optical communications networks. Figure 7.13a shows the measured output spectra of the wavelength converters and Fig. 13b shows the measured [38, 40] conversion efficiency as a function of the input wavelength for two input signal polarisation states. Both sets of data points are curve fitted according to theory [38]. When Bellcore was making the measurements, the efficiency was limited to −17 dB, being far below theoretically predicted −4 dB (see the theory of a DFG [38]). The main cause of reduction in conversion efficiency was attributed to scattering loss at the pump wavelength (45 dB/cm) due to waveguide corrugation. A significant improvement of the conversion efficiency can be obtained by reducing the corrugation height (so reducing losses), which can be obtained through the improvement of organometallic chemical vapor deposition (OMCVD)

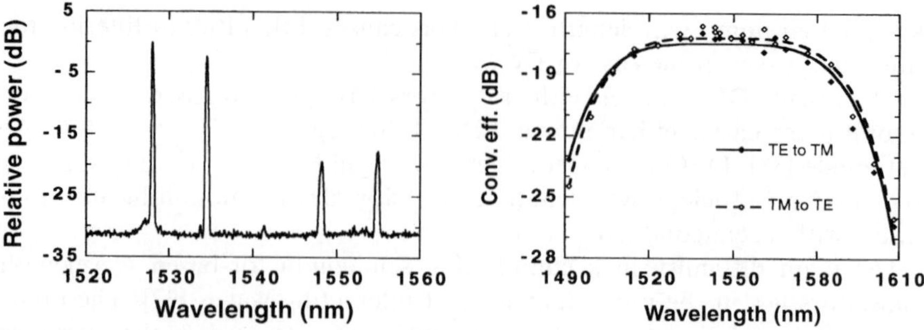

Fig. 7.13 (a) Measured output spectra of the wavelength converters [38]. The two input waves are at 1534 nm and 1528 nm; the resulting converted waves are at 1549 nm and 1555. (b) Measured conversion efficiency as a function of the input wavelength for two input signal polarisation states. Both sets of data points are curve fitted according to theory. (After [37–39] with permission of the author and of IEEE).

growth and fabrication.

The conversion efficiency varied by less than 3 dB through the wavelength span of 90 nm, and the variation was less than 1 dB in the 1520–1560 nm region (the window of major interest). Notice that the efficiency independence on the optical bandwidth, is a very important feature in WDM networks, since it avoid the use of channels equalisation. Furthermore, it allows the channels to be separated as much as possible, minimising the crosstalk contribution (Chapter 11).

The two conversion processes, TE input wave conversion to TM output wave and vice versa, were confirmed by testing the polarisation states of the signal and the converted waves. Note that the two conversion processes had nearly identical efficiency, as expected. Thispolarisation sensitivity was

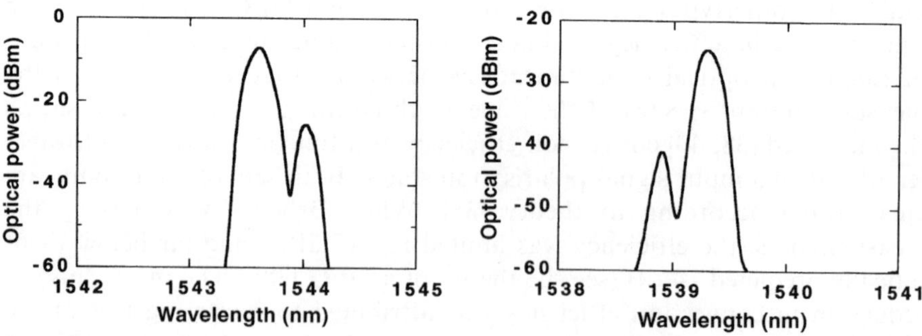

Fig. 7.14 Optical spectra for input (a) and output (b) waves of a DFG converter [39]. These spectra are inverted. (After [37–39] with permission of the author and IEEE).

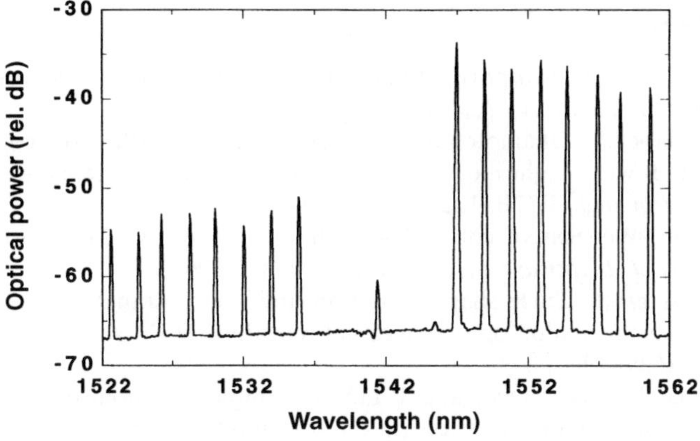

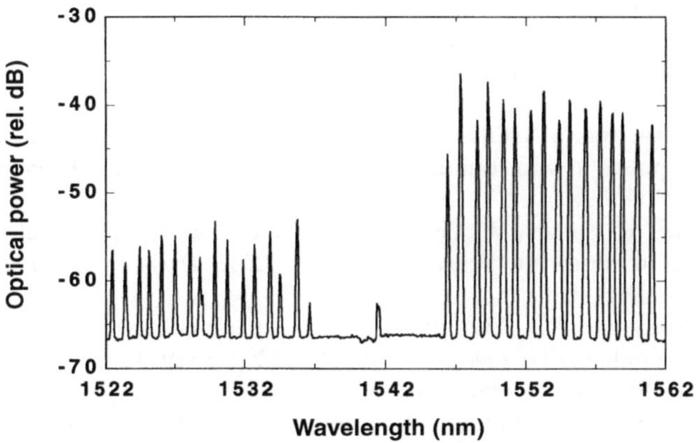

Fig. 7.15 Wavelength translation of a set of WDM channels [39]. (After [37–39] with permission of the author and of IEEE).

also tested by monitoring the conversion efficiency for arbitrary input polarisations, and the measured conversion efficiency deviated by no more than 0.5 dB for arbitrary alterations of a polarisation controller.

The spectral inversion property of this type of conversion process is shown by Fig. 7.14, where the optical spectra of input and output waves are plotted [39]. This device does not include any noise contribution since the waveguide is passive and it can simultaneously convert a set of WDM signals. This is shown in Fig. 7.15, where a conversion of eight channels is depicted [39].

REFERENCES

1. G.R. Hill *et al.*, A transport network layer based on optical network elements, IEEE/OSA *J. Lightwave Technol.* **11**, 667–679 (1993).
2. A. Watanabe, S. Okamoto, and K. Sato, Optical path cross-connect node architecture with high modularity for photonic transport networks, *IEEE Trans. on Commun.* **E77B**, 1220–1229 (1994).
3. S.J.B. Yoo, Wavelength conversion technologies for WDM network applications, *IEEE J. Lightwave Technol.* **14**, 955–966 (1996).
4. M. Schiess *et al.*, Pulse shape evolution and noise estimates in concatenated fiber links using analog optoelectronical repeaters, *IEEE J. Lightwave Technol.* **14**, 1621–1629 (1996).
5. R. Sabella, M. Avattaneo, and E.Iannone, Impact of non-regenerative optoelectronic wavelength converters on the transmission performance of all-optical networks, *Microwave Opt. Technol. Lett.*, November (1997).
6. K.E. Stubkiaer *et al.*, Optical wavelength converters, in *Proc. ECOC'94*, Florence, Italy, pp. 635–642 (1994).
7. P. Doussiere *et al.*, 1.55 μm polarization independent semiconductor optical amplifier 25 dB fiber to fiber gain, *IEEE Photon. Technol. Lett.* **6**, 170–172 (1994).
8. G.P. Agrawal and N.A. Olsson, Self-phase modulation and spectral broadening of optical pulses in semiconductor laser amplifiers, *IEEE J. Quantum Electron.* **25**, 2297–2306 (1989).
9. E. Iannone, R. Sabella, L. de Stefano, and F. Valeri, All-optical wavelength conversion in optical multicarrier networks, *IEEE Trans. on Commun.* **44**, 716–724 (1996).
10. D.M. Patrick and R.J. Manning, 20 Gb/s wavelength conversion using semiconductor nonlinearity, *Electron. Lett.* **30**, 252–253 (1994).
11. M. Eiselt, W. Pieper, and H.G. Weber, Decision gate for all-optical data retiming using a semiconductor laser amplifier in a loop mirror configuration, *Electron. Lett.* **29**, 107–109 (1993).
12. T. Durhuus *et al.*, All optical wavelength conversion by SOA's in a Mach-Zehnder configuration, *IEEE Photon. Technol. Lett.* **6**, 53–55 (1994).
13. B. Mikkelsen *et al.*, Polarization insensitive wavelength conversion of 10 Gb/s signals with SOA's in a Michelson interferometer, *Electron. Lett.* **30**, 260–261 (1994).
14. T. Durhuus *et al.*, 2.5 Gb/s optical gating with high on/off ratio by use of SOA's in Mach-Zehnder interferometers, in *Proc. CLEO '93*, Baltimore, MD, paper CThH3 (1993).
15. T. Durhuus *et al.*, All-optical wavelength conversion by semiconductor optical amplifiers, *IEEE/OSA J. Lightwave Technol.* **14**, 942–954 (1996).
16. G.P. Agrawal, Population pulsations and non-degenerate four-wave mixing in semiconductor lasers and amplifiers, *J. Opt. Soc. Amer. B* **5**, no. 1, pp.147–158, 1988.
17. K. Kikuchi, M. Kakui, C. Zah, and T.P. Lee, Observation of highly nondegenerate four-wave mixing in 1.5 mm travelling wave semiconductor optical amplifiers and estimation of nonlinear gain coefficient, *IEEE J. Quantum Electron.* **28**, 151–156 (1992).

18. M. Willatzen *et al.*, Nonlinear gain suppression in semiconductor laser due to carrier heating, *IEEE Photon. Technol. Lett.* **3**, (1991).

19. L.F. Tiemeijer, Effects of nonlinear gain on four-wave mixing and asymmetric gain saturation in a semiconductor laser amplifier, *Appl. Phys. Lett.* **59**, 499–501 (1991).

20. J. Zhou, N. Park, J.W. Dawson, and K.J. Vahala, Highly nondegenerate four-wave mixing and gain nonlinearity in a strained multiple-quantum-well optical amplifier, *Appl. Phys. Lett.* **62**, 2301–2303 (1993).

21. S. Murata *et al.*, Observation of highly nondegenerate four-wave mixing (> 1 THz)in an InGaAsP multiple quantum-well laser, *Appl. Phys. Lett.* **58**, 1458–1460 (1991).

22. A. D'Ottavi *et al.*, Four wave mixing investigation of carrier heating and spectral hole burning in semiconductor amplifiers, *Appl. Phys. Lett.* **64**, 2492–2494 (1994).

23. M.C. Tatham, G. Sherlock, and L.D. Westbrook, Compensation of fibre chromatic dispersion by mid-way spectral inversion in a semiconductor laser amplifier, *Proc. ECOC '93*, Montreaux, paper Th12.3 (1993).

24. K. Kikuchi and C. Lorattanasane, Compensation for pulse waveform distortion in ultra-long distance optical communication systems by using midway optical phase conjugator, *IEEE Photon. Technol. Lett.* **6**, 104–105 (1994).

25. A. D'Ottavi *et al.*, Efficiency and Noise Performances of Wavelength Converters Based on FWM in Semiconductor Optical Amplifiers, *IEEE-Photon. Technol. Lett.* **7**, 357–359 (1995).

26. A. Mecozzi *et al.*, Four wave mixing in travelling wave semiconductor amplifiers, *IEEE J. Quantum Electron.* **31**, 689–699 (1995).

27. S. Scotti and A. Mecozzi, Frequency converters based on FWM in travelling-wave optical amplifiers: theoretical aspects, *Fiber and Integrated Optics* **15**, 243–256 (1996).

28. G.P. Bava, P. Debernardi, and G. Osella, Frequency conversion in travelling wave semiconductor laser amplifiers with bulk and quantum-well structures, *IEE Proc. Optoelectron.* **141**: 119–125 (1994).

29. M. Uskov, J. Mørk, and Mark, Wave mixing in semiconductor laser amplifiers due to carrier heating and spectral hole burning, *IEEE J. Quantum Electron.* **30**, 1769–1781 (1994).

30. A. Mecozzi, Analytical theory of four-wave mixing in semiconductor amplifiers, *Opt. Lett,* **19**, 892–894 (1994).

31. A. Yariv, *Quantum Electronics*, 3rd ed., John Wiley, New York (1989).

32. G.P. Agrawal and N.K. Dutta, *Long-Wavelength Semiconductor Lasers*, Ch. 6,Van Nostrand Reinold, New York (1986).

33. R. Hui and A. Mecozzi, Phase noise of four wave mixing of semiconductor lasers, *Appl. Phys. Lett.* **60**, 2454–2456 (1992).

34. R. Ludwig *et al.*, Four-wave mixing in semiconductor laser amplifiers: applications for opticl communication systems, *Fiber and Integrated Optics* **15**, 211–223 (1996).

35. R. Ludwig and G. Raybon, BER measurements of frequency converted signals using four-wave mixing in a semiconductor laser amplifier at 1, 2.5, 5 and 10 Gb/s, *Electron. Lett.* **30**, 338–339 (1994).

36. P. Spano (ed.), Frequency conversion in WDM optical networks: an overview of the european research, *Fiber and Integrated Optics* **15**, no. 3 (1996).
37. S.J.B. Yoo, Wavelength conversion technologies for WDM network applications, *IEEE/OSA J. Lightwave Technol.* **14**, 955–966 (1996).
38. S.J.B. Yoo *et. al.*, Wavelength conversion by difference frequency generation in AlGaAs waveguides with periodic domain inversion achieved by wafer bonding, *Appl. Phys. Lett.* **68**, 2601–2611 (1996).
39. S.J.B. Yoo and K. Bala, Parametric wavelength conversion and cross-connect architecture for transparent all-optical networks, in Proc. *SPIE '96*, Boston, pp. 2919–20 (1196)

8

Optical Receivers

8.1 INTRODUCTION

This chapter concentrates on receivers intended for the direct detection of digital data consisting of a stream of light pulses, where the presence and the absence of a pulse correspond to the transmission of a binary mark and space, respectively. The block diagram of a basic telecommunications receiver is depicted in Fig. 8.1.

The receiver consists of a photodetector, which converts the optical power signal into an electrical current that reproduces the envelope of the received optical signal. The electrical current is then amplified by a low-noise preamplifier. After the front-end (photodiode and preamplifier), there is a linear section that consists of an equaliser, to correct for roll-off in the front end (not always needed), a high-gain postamplifier (usually with some kind of automatic gain control), and a low-pass filter, generally to reject the outband noise and to shape the signal to for minimum the intersymbol interference if necessary. After the linear section there is a clock recovery circuit, which provides the synchronisation signal to the decision circuit. With a return-to-zero signal (RZ) data format, a spectral component at the baud rate is present in the received signal, and the clock extraction can be accomplished with a narrow bandpass filter such as a phase-locked loop (PLL) or surface-acoustic-wave (SAW) filter. With non-return-to-zero (NRZ) data format, a non-linear operation such as squaring must be performed on the received signal to generate a spectral line at the baud rate.

The decision circuit regenerates the signal. It consists of a comparison circuit whose threshold level is set to the center of the received eye pattern to give an equal error probability for decisions on both marks and spaces.

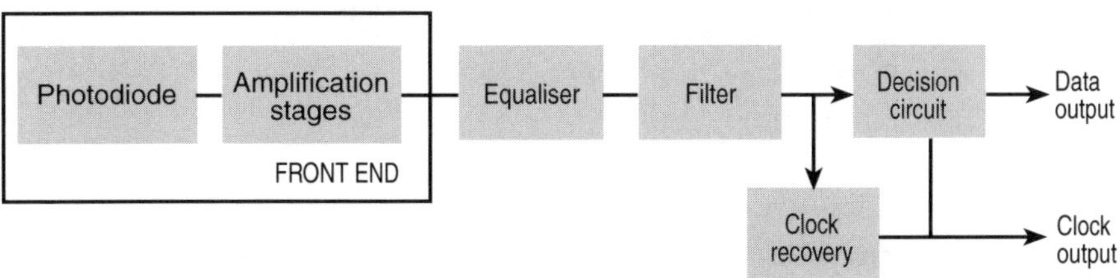

Fig. 8.1 A basic receiver for fiber-optic communications.

After the comparator there is a D flip-flop clocked by the recovered timing signal. The clock phase is timed to store the best estimate of the transmitted data by triggering the flip-flop at the center of each bit when the eye opening is the greatest.

This chapter analyses the key devices used in the receiver are analysed and discusses the different technologies.

8.2 THE PHOTODETECTOR

Several types of photodetectors have been proposed in the literature. In high-speed optical receivers, normal p-i-n photodiodes are usable up to approximately 30 GHz. Side-illuminated p-i-n diodes are somewhat faster. At higher speed the metal-semiconductor-metal (MSM) photodetector is a strong candidate. In the windows of interest (1300 and 1550 nm light wavelengths), photodetectors are made using a variety of compounds (ternary and quaternary) semiconductors based on InP technology.

8.2.1 The p-i-n photodiode

The basic structure of a p-i-n diode is illustrated in Fig. 8.2. It consists of three regions made of different materials:

p^+ region: usually heavily doped p-type InP;
i region: it is made of the absorbing material (InGaAs), nominally undoped;
n region: n-type InP.

The material for the p and n regions is usually chosen to be InP, because it presents a band gap larger than that of InGaAs, so as to obtain a lower dark current.

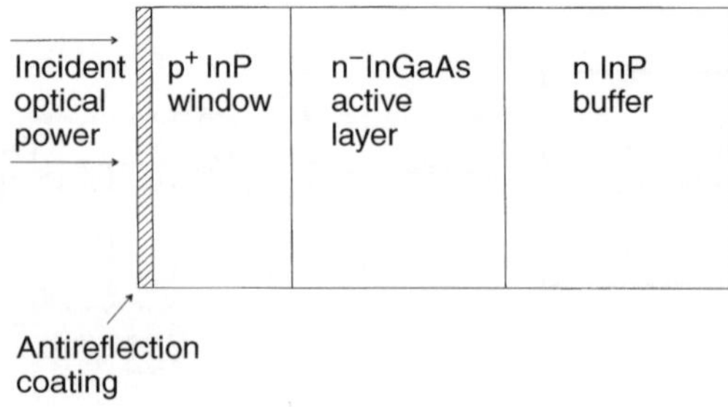

Fig. 8.2 Basic structure of a p-i-n diode: p + InP = i InGaAs = n InP.

This diode operates in a reverse-biased mode which creates a depletion of the i region (and consequently a strong electric field). The presence of the electric field assures that electron-hole pairs, created by photon absorption in the InGaAs layer, are quickly separated and collected at the photodiode terminals before recombination can occur. Such electron-hole pairs form a current, whose average value can be calculated straightforwardly:

$$i = \eta q \frac{P}{h\nu} \qquad (8.1)$$

where P is the incident power, η the detector quantum efficiency (defining the percentage of incident photons which generate electron-hole pairs), q the electron charge, h Planck's constant and ν is the optical frequency.

Photodiodes can be divided into three classes, depending on how light impinges on the device: top illuminated, back illuminated and side illuminated. The first two types are collectively known as surface-illuminated photodiodes; they constitute the conventional p-i-n structure. The difference between them is mainly a question of mounting technique. With back-illuminated diodes it is possible to flip-chip mount the diode, which gives lower series inductance. On the other hand, it is often preferable to have a matching inductance on the amplifier input to get large bandwidth. In the surface-illuminated structures, a trade-off exists between speed and quantum efficiency, because a thin absorber region is required for short carrier transit time and a thick absorber section for high quantum efficiency. Side-illuminated p-i-n diodes overcome this trade-off between speed and quantum efficiency due to a longer interaction length of light in the absorbing layer obtained without suffering from long carrier transit times. These photodiodes are also called waveguide p-i-n or edge-coupled p-i-n diodes.

Modeling a p-i-n diode modeling

For high-speed operations, the frequency response of the photodiode is one of the main issues. The three main mechanisms limiting the photodiode frequency response are as follows [1]:

(1) the finite transit time of the carriers drifting across the depletion layer;
(2) the effect of the diode capacitance plus any parasitic capacitance;
(3) the finite diffusion time of carriers produced in the p and n regions.

To evaluate the effect of the last two limiting factors, it is possible to refer to a diode equivalent AC circuit, as shown in Fig. 8.3. Here R_d is the diode incremental (AC) resistance, C_d the junction capacitance, R_s the contact

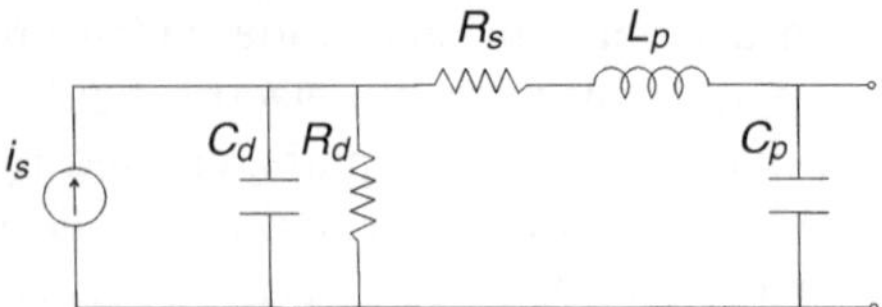

Fig. 8.3 Equivalent circuit of a pin diode. R_d is the diode incremental (AC) resistance and C_d the junction capacitance; R_s represents the contact and series resistance, L_p the parasitic inductance associated mostly with the contact leads, and C_p the parasitic capacitance due to contact leads and contact pads.

and series resistance, L_p the parasitic inductance associated mostly with the contact leads, and C_p the parasitic capacitance due to contact leads and contact pads.

On the other hand, to evaluate the effect of transit time, it is necessary to look into the physics of the device, as describerd by Poisson's equation and by the continuity equations. Different numerical and analytical models have been reported in the literature. Here we describe an analytical model which allows the frequency response of the p-i-n diode to be accurately evaluated, taking into account both the influence of the electric field on the carrier velocities and their non-uniform generation.

The reference p-i-n diode structure is illustrated in Fig. 8.2 In order to lead to a simplified yet realistic description of the device, some physical aspects are considered:

(1) The applied electric field is expected to be great enough to completely deplete the absorbing layer, in this case only fixed charges are present.
(2) It is assumed that, as far as charge is concerned, the device can be divided into different regions, as shown in Fig. 8.4. This hypothesis is usually known as 'abrupt heterojunction' [2, 3].
(3) The light-generated carriers do not actually perturb the field produced by external bias, so the field distribution in the intrinsic region can be treated as time independent.
(4) All donors and acceptors are ionised.

Under these assumptions, Poisson's equation can be written as

$$divE(x) = \frac{dE}{dx} = \begin{cases} -qN_A/\varepsilon & -L_p < x, 0 \\ qN_1/\varepsilon_i & 0 < x < d \\ qN_D/\varepsilon & x < L_n + d \end{cases} \qquad (8.2)$$

where L_p, L_n, d are the lengths of the p$^+$ depleted region, of the n depleted region and of the intrinsic layer, respectively; N_A, N_I, N_D are the doping

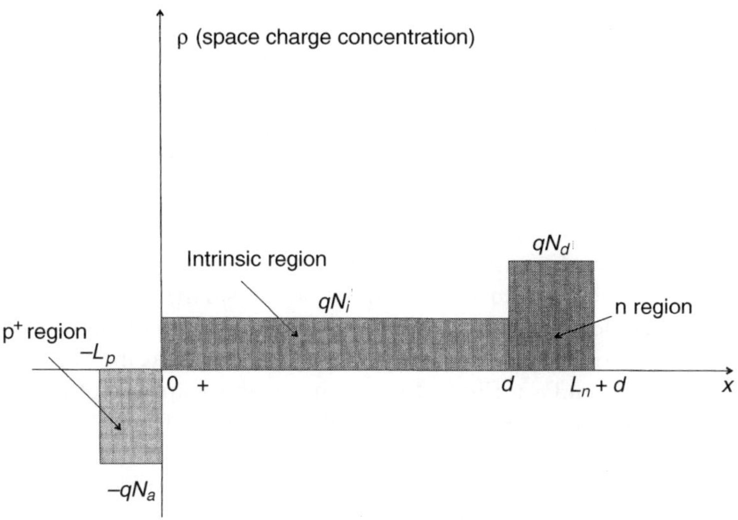

Fig. 8.4 Partition of the depleted region, assuming an abrupt heterojunction.

densities of the three layers ε is the InP dielectric constant, and ε_i is the InGaAs dielectric constant.

In each zone of the depleted region, the potential and electric field are defined. Then it is necessary to impose the continuity conditions of the potential and the displacement field ($D = \varepsilon E$) at the interfaces $x = 0$ and $x = d$. The following expressions for the potential are obtained:

$$\phi(x) = \begin{cases} -\Psi - qN_A(x + L_p)^2/(2\varepsilon) \\ -\Psi - qN_A L_p^2/(2\varepsilon) + qN_1 x^2/(2\varepsilon_i) - q(N_D L_n/\varepsilon + N_I d/\varepsilon_i)x \\ qN_D(x - d - L_n)^2/(2\varepsilon) \end{cases}$$

(8.3)

where ψ is the difference between the built-in voltage and the applied voltage [4]. The values of L_n and L_p are calculated by the solution of the system of equations, obtained by imposing the matching conditions at the two heterointerfaces. The minimum external bias voltage to obtain complete depletion (so verifying the first hypothesis) can be calculated by imposing the condition $L_n = 0$ (or $L_n = 0$ if $N_D > N_A$).

The analytical expression of the field is obtained from equation (8.3) quite simply. In the intrinsic region, the electric field is given by:

$$E(x) = -qN_I x/\varepsilon_i + qN_D L_n/\varepsilon + qN_I d/\varepsilon_i.$$

(8.4)

In order to determine the carrier distribution in the intrinsic region, the continuity equations must be solved for electrons and holes:

$$\frac{\partial p(x,t)}{\partial t} = G(x,t) - R(x,t) - \frac{1}{q}div[J_p(x,t)] \qquad (8.5)$$

$$\frac{\partial n(x,t)}{\partial t} = G(x,t) - R(x,t) - \frac{1}{q}div[J_n(x,t)] \qquad (8.6)$$

where $G(x,t)$ and $R(x,t)$ are the generation and the recombination functions, respectively. In practice the recombination term $R(x,t)$ can be neglected, since the product pn is quite a bit smaller than the squared intrinsic carrier density, as long as the illumination is not high (as it is in normal transmission systems, where the received light power does not exceed values around -2 dBm). Furthermore, the trapping effects can be neglected because they can be made irrelevant technologically in different ways; for example, by inserting a thin quaternary layer between the p^+ and the intrinsic layers, or by a graded bandgap structure, or by locating the electrical junction in the ternary layer, very close to the window layer [5].

The current densities can be written in the drift-diffusion form:

$$J_p(x,t) = q \cdot v_p(E) \cdot p(x,t) - qD_p\frac{\partial p(x,t)}{\partial x} \qquad (8.7)$$

$$J_n(x,t) = q \cdot v_n(E) \cdot n(x,t) - qD_n\frac{\partial n(x,t)}{\partial x} \qquad (8.8)$$

where v_n and v_p are the field-dependent electron and hole drift velecities, respectively; D_n and D_p are the diffusion coefficients.

In such equations the diffusion terms can be omitted, since they are considered negligible in the depleted region as soon as the potential across the device exceeds kT/q.

Therefore equations (8.5) and (8.6) reduce to the following forms:

$$\frac{\partial p(x,t)}{\partial t} = G(x,t) + p(x,t)\frac{\partial v_p(E)}{\partial x} + v_p(E)\frac{\partial p(x,t)}{\partial x)} \qquad (8.9)$$

$$\frac{\partial n(x,t)}{\partial t} = G(x,t) - n(x,t)\frac{\partial v_p(E)}{\partial x} - v_n(E)\frac{\partial n(x,t)}{\partial x)} \qquad (8.10)$$

where $V_n(E)$ and $V_p(E)$ are given by the following empirical expressions:

$$v_n(E) = \frac{\mu_n E(x) + \beta V_{nl} E(x)^\gamma}{1 + \beta E(x)\gamma} \qquad \text{(for the electrons)} \qquad (8.11)$$

$$\nu_p(E) = V_{pl} \tanh\left[\mu_p \frac{E(x)}{V_{pl}}\right] \quad \text{(for the holes)} \tag{8.12}$$

being μ_n and μ_p the mobilities of the electrons and holes, respectively [6], V_{nl} and V_{pl} the high field velocities of the electrons and holes, respectively [7, 8]; β and γ are two fitting parameters.

As far as the generation term $G(x, t)$ is concerned, it takes account of the thermal and the radiative generation:

$$G(x, t) = g_{th} + g_0 e^{-\alpha x} G(t), \tag{8.13}$$

where $G(t)$ is the time variation of the generation function, $g_{th} = n_i/2\tau_0\,\tau_0$ being and the average carrier lifetime n_i the intrinsic carrier density, and a is the absorption coefficient; with the assumptions made above, the two continuity equations become independent of each other and can be solved separately.

The term g_0 is given by $g_0 = \alpha\eta P_o/h\nu$ where η is the quantum efficiency, P_o is the optical input power per unit area, $h\nu$ is the photon energy.

Evaluation of the photoresponse

The photoresponse is evaluated by considering CW incident light. By setting the time derivatives in (8.5) and (8.10) to zero, we have

$$g_{th} + g_0 e^{-ax} + p(x)\frac{\partial \nu_p(x)}{\partial x} + \nu_p(x)\frac{\partial p(x)}{\partial x} = 0 \tag{8.14}$$

$$g_{th} + g_0 e^{-\alpha x} - n(x)\frac{\partial \nu_n(x)}{\partial x} - \nu_n(x)\frac{\partial n(x)}{\partial x} = 0 \tag{8.15}$$

where n^* and p^* are the space-dependent stationary electron and hole concentrations.

The equations above are two ordinary first-order differential equations. Their particular solutions are obtained by imposing the boundary conditions: $p(d) = 0$ and $n(0) = 0$, hence

$$p*(x) = \frac{1}{\nu_p(x)}\left[-g_{th}(x - d)g_0(e^{-\alpha x} - e^{-\alpha d})/\alpha\right] \tag{8.16}$$

$$n*(x) = \frac{1}{\nu_n(x)}\left[g_{th}x - g_0(e^{-\alpha x} - 1)/\alpha\right]. \tag{8.17}$$

Evaluation of the frequency response

To evaluate the frequency response it is sufficient to solve equations (8.9) and (8.10) for a sinusoidal variation of the incident light. In particular, the time variation of the generation function can be chosen to have the form: $G(t) = \Re\,(1 - Ae^{j\omega t})$ where the modulation index A is less than 1 so that $G(t) > 0$ for all t. $\Re$ represents the real part operation.

The time-dependent solution may be written in terms of a frequency response function which, by using transform techniques, allows the calculation of the response to an arbitrary function of time.

The partial differential equations to be solved are[1]

$$\frac{\partial p(x,t)}{\partial t} = g_{th} + g_0 e^{-\alpha x}(1 + Ae^{j\omega t}) + p(x,t)\frac{\partial v_p(x)}{\partial x} + v_p(x)\frac{\partial p(x,t)}{\partial x} \quad (8.18)$$

$$\frac{\partial n(x,t)}{\partial t} = g_{th} + g_0 e^{-\alpha x}(1 + Ae^{j\omega t}) - n(x,t)\frac{\partial v_n(x)}{\partial x} - v_n(x)\frac{\partial n(x,t)}{\partial x} \quad (8.19)$$

The approach to solve the two equations is the same, so we will refer to just the electrons. We search for a general solution of the following form:

$$n(x,t) = n*(x) + N(x,t), \quad (8.20)$$

where $n^*(x)$ is the electron steady state-solution found in the previous section. Then (8.19) reduces to an equation in the variable $N(x,t)$, which can be solved with the hypothesis of linearity, after the stabilisation of the temporal response. The detailed steps to solve the equations are reported in the literature [5]. The solutions for electrons and the holes are:

$$n(x,t) = n*(x) + \frac{1}{v_n(x)}f(x)e^{-\alpha x + j\omega t} \quad (8.21)$$

$$p(x,t) = p*(x) + \frac{1}{v_p(x)}g(x)e^{-\alpha x + j\omega t} \quad (8.22)$$

The expressions of the functions $f(x)$ and $g(x)$, and the analytical steps of their derivations are reported in the literature [5].

In this case the current density must also contain the displacement contribution

$$J(x,t) = q[v_p(x)p(x,t) + v_n(x)n(x,t)] + \varepsilon\frac{\partial E}{\partial t} \quad (8.23)$$

By assuming that the diode is in series with a battery of potential V_0 and

---

[1] The velocities are expressed as a function of x, since the electric field is stationary and not perturbed, as mentioned in the earlier assumptions.

an impedance Z (load + parasitic) as shown in Fig. 8.4, the following expression can be obtained:

$$J(x, t) = \frac{q}{d} \int_0^d [\nu_p(x)p(x, t) + \nu_n(x)n(x, t)]dx \cdot \left[\frac{1}{1 + j\omega ZC}\right] \tag{8.24}$$

where C is the diode capacitance. Note that the numerator is the average conduction current in the intrinsic region. The real part of the current density can be derived from (8.24) by omitting the electric pole ($C = 0$), that is

$$J(x, t) = \frac{q}{d} \int_0^d [\nu_p(x)\Re[p(x, t)] + \nu_n(x)\Re[n(x, t)]]dx \tag{8.25}$$

Finally, equation (8.25) can be put in a more practical form as

$$J(t) = J_0 + H(\omega) \cos[\omega t + \phi(\omega)] \tag{8.26}$$

where $H(w)$ and $\phi(\omega)$ are given in the literature [5]. The first term of (8.26) on the right-hand side is the DC component of the current density and the second one is the AC component. As previously anticipated, equation (8.26) gives the solution in terms of a frequency response function for the diode. Hence, by using the transform techniques, the response of the p-i-n diode to an arbitrary function of time can be calculated.

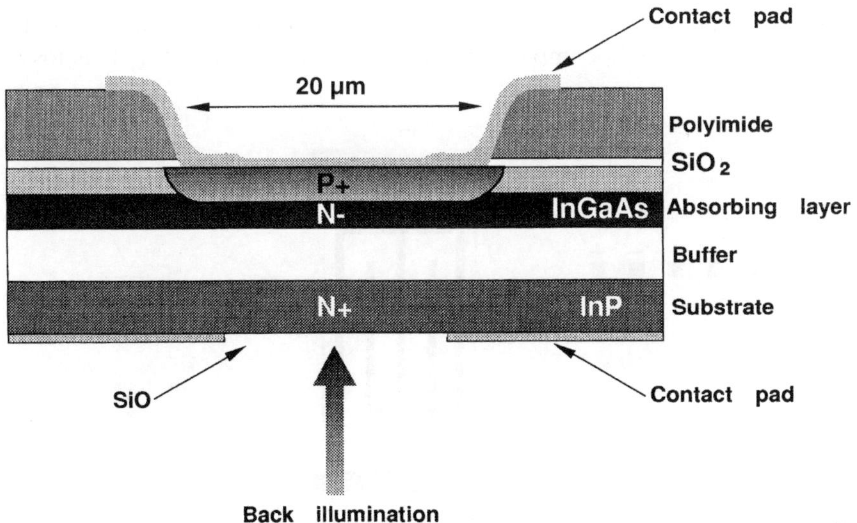

Fig. 8.5 Schematic of a planar p-i-n photodiode.

Note that if the carrier velocities are treated as constant, we obtain the same results reported in the literature [5]. This assumption is not always valid, as it depends on the structure of the diode itself.

A practical structure of a back-illuminated p-i-n photodiode, with planar configuration, is depicted in Fig. 8.5.

8.2.2 The MSM photodetector

The planar MSM photodetector is shown schematically in Fig. 8.6. It consists of a semiconductor absorbing layer on which interdigitated electrodes have been deposited to form back-to-back Schottky diodes. This type of detector is extremely easy to manufacture. The compatibility of the detector structure with FETs, sketched in Fig. 8.7, and the planarity of the contacts, results produces a minimal increase in the process complexity when an MSM is integrated with MESFET or HEMT circuitry.

Apart from its planarity and compatibility with FETs, the interdigitated electrode structure of the planar MSM gives the detector a very low capacitance per unit area. This means that a large-area MSM may be used in conjunction with a high-impedance integrating front-end configuration without necessarily compromising the bandwidth. Furthermore, as the input noise current in FETs is proportional to the square of the total front-end capacitance, i.e. $(C_{detector} + C_{gs} + C_{stray})^2$, a low detector capacitance benefits receiver sensitivity.

Bandwidths of around 40 GHz have already been demonstrated [9], and values as high as 105 GHz have been reported [10]. The fingers of the MSM are blocking out some of the light but with modern lithography this can be as low as 10% compared with planar p-i-n diodes [11]. This loss of

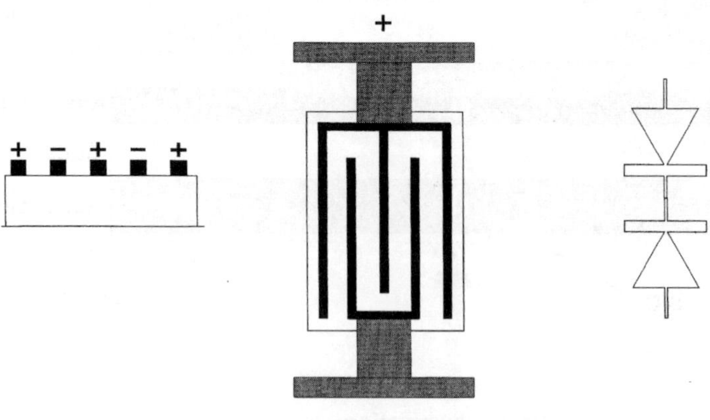

Fig. 8.6 Schematic of a planar MSM photodetector.

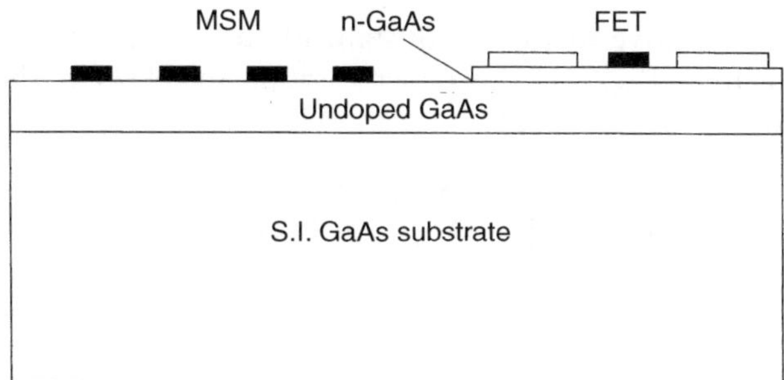

Fig. 8.7 Sketch of a structure including a detector with a FET.

light is well compensated by lower total noise due to the very low detector capacitance. The MSM design should be able to detect several hundreds of gigahertz.

Diode structure

The absorbing layer of an MSM diode consists of undoped InGaAs (n^-). Unfortunately, this material presents a low Schottky barrier height ($\sim$ 0.2 V [12]). This means that direct deposition of the electrode on the InGaAs gives detectors with unacceptably large leakage current, even at low biases. Various schemes have therefore been explored to enhance this Schottky barrier and so limit the dark current.

The most successful barrier enhancement technique reported so far has been the growth of a thin layer of undoped lattice-matched InAlAs on the InGaAs. InAlAs lattice-matched to InP (E_g = 1.47 V) has a high Schottky barrier ($\sim$ 0.8 V [13]) and a thin layer grown epitaxially on InGaAs has the effect of raising both the electron and holes barriers. An alternative approach is the growth of $In_xGa_{1-x}As$ on GaAs substrates [14, 15].

Diode capacitance

One significant advantage of interdigitated MSM detectors is their very low capacitance per unit area. Typical devices, photolitographically defined, have electrode fingers widths of $\sim$1 mm, and interdigitated spacings of 1–3 mm. This electrode structure has a capacitance of just a few femtofarads per 10 m sqare.

In general, for an undoped and infinitely thick semiconductor, the capacitance is given as follows [12, 16]:

$$C = \frac{K(k)}{K(K\prime)} \varepsilon_0 (1 + \varepsilon_r) \frac{A}{P_{finger}} \qquad (8.27)$$

where E_r is the relative dielectric constant of the semiconductor, A is the interdigitated area, P_{finger} is the finger period, and $K(k)$ is the complete elliptic integral of the first kind:

$$K(k) = \int_0^{\pi/2} \left[1 / \sqrt{(1 - k^2 sin^2\varphi)} \right] d\varphi \qquad (8.28)$$

with

$$k = \tan^2[\pi W_{finger} / 4 P_{finger}] \qquad (8.29)$$

and

$$k' = \sqrt{(1 - k^2)} \qquad (8.30)$$

where W_{finger} the finger width.

As an example, Fig. 8.8 plots the capacitance as a function of the squared interdigitated area for three electrode spacings: 1, 2 and 3 mm [12]. Solid and dotted lines are the capacitance for electrode finger widths of 1 mm and 0.5 mm, respectively. Notice that the difference is only

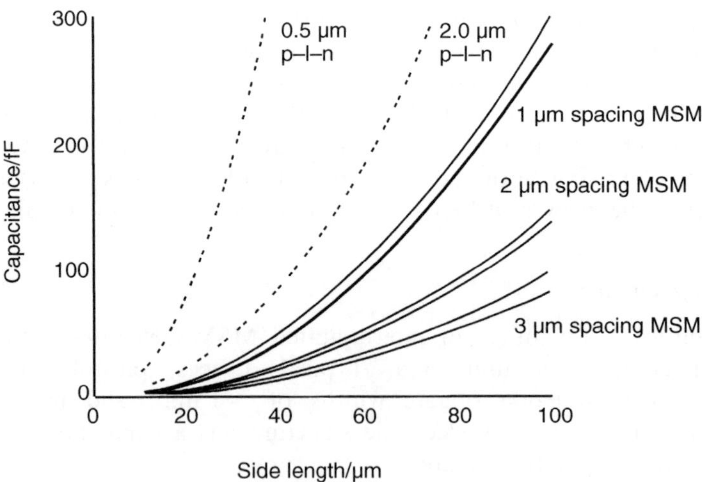

Fig. 8.8 Diode capacitance versus the squared interdigitated area for three electrode spacings: 1, 2 and 3 μm. The solid lines are the capacitance for an electrode finger width of 1 mm and the dotted lines are for an electrode width of 0.5 μm. The capacitance curves for p-i-n diodes with 0.5 and 2.0 μm InGaAs layers are also plotted. (After [12] with permission of IEEE).

marginal. For comparison, the capacitance of p-i-n diodes with 0.5 and 2 mm InGaAs layers are also plotted. The capacitance of the MSM is significantly lower than the p-i-n device, even when comparing the rather thick (2 mm) p-i-n and the narrow spacing (1 mm) MSM. This indicates that the bandwidth of a receiver incorporating an MSM-based front end will rarely be limited by the detector capacitance. For instance, consider a detector which has to be coupled with a single-mode fiber. Its active area should be in the region 50×50 mm. An MSM with 1 mm electrodes and a 2 mm spacing, typical of the detectors reported to date, has a capacitance of ~ 40 fF. This gives an RC charging time of just 1.9 ps into a 50 W load. In comparison, the transit times of photogenerated carriers would be about half the electrode separation divided by the saturation velocity, which is a few tens of picoseconds. The response into 50 W is therefore limited by the carrier transit times.

Diode photoresponse

The CW photoresponse of a typical device is reported in Fig. 8.9a [17], for different incident light powers. The shoulder at 1–1.5 V occurs because sufficient bias must be applied to obtain a flat band at the anode and separate the generated carriers. After the knee, a fairly flat photoresponse is obtained, rising gently as the bias is increased. At the lower biases, this is largely due to the increase of the depletion volume. At 7 V the internal quantum efficiency becomes unity. At higher biases, the photoresponse continues to rise slowly. Therefore, a net internal gain is present at these higher biases.

Diode quantum efficiency

The MSM internal quantum efficiency is close to unity over most of the operating range, with some small gain present at the higher biases. The external quantum efficiency is therefore determined by the electrode structure and the thickness of the InGaAs absorbing layer.

This external quantum efficiency is given as follows [12]:

$$\eta = (1 - r)\frac{s}{s + w}[1 - \exp(-d/L_{pen})] \tag{8.31}$$

where w is the finger width, s is the interdigital spacing, L_{pen} is the optical penetration length at the considered wavelength, and d is the thickness of the InGaAs layer. The Fresnel reflectivity r is included, although it may be reduced to zero by using an antireflection coating.

The dark I-V characteristic of a typical device is shown in Fig. 8.9b [17]. The full curve is shown in the inset and is symmetrical. There is a knee at

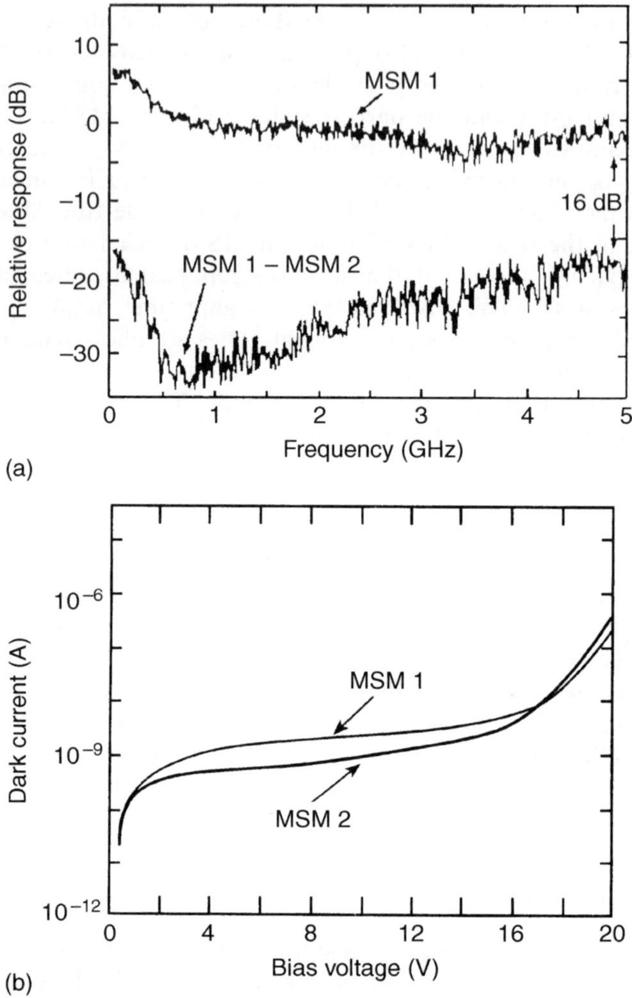

Fig. 8.9 (a) CW photoresponse and (b) dark *I-V* characteristics of a typical device. (After [17] with permission of IEEE)

low bias. This occurs as the anode contact becomes forward biased. For typical detectors, the voltage required is 1–5 V, depending on the electrode spacing and the residual doping level.

Although a two-dimensional analysis incorporating the details of the barriers is necessary for a full understanding of the *I-V* characteristics, it is still interesting to examine the extent to which a simple one-dimensional analysis [18] models the experimental situation. Using a simple thermoionic emission analysis, and neglecting the details of the InAlAs barriers and all the tunneling contributions, an expression for the current is obtained [18].

8.3 THE FRONT END

The design of the receiver front end (FE) has a strong impact on the receiver performance in terms of sensitivity and dynamic range. Receiver FEs can be classified into three broad categories [19, 20]: (1) the low-impedance FE, where the load resistor R_L, is small and the input RC time constant does not limit the FE bandwidth to less than the signal bandwidth; (2) the high-impedance FE, where R_L is large for improved sensitivity but the FE bandwidth is less than the signal bandwidth, resulting in a need for equalisation; (3) transimpedance FE, where the load resistor is replaced by a large feedback resistor, and negative feedback around a wideband amplifier is used to obtain increased bandwidth. The three types of FE circuits will be discussed below.

8.3.1 Low-impedance front end

The scheme of a low-impedance FE is shown in Fig. 8.10. It typically consists of a photodiode operating into a low-impedance (e.g. 50 W) amplifier, often through coaxial cable or other transmission line. The terminating resistor R_L is equal to the transmission line impedance in order to suppress standing waves for uniforn frequency response. Low-impedance FEs cannot provide high sensitivity since only a small signal voltage can be developed across the amplifier input impedance and resistor R_L. Therefore they can be used where a high sensitivity is not required. The variance of the input noise current for this FE can be expressed as

$$\langle i^2 \rangle_a = \frac{4KTF}{R_L} \Delta f \tag{8.32}$$

where F is the amplifier noise figure, Δ_f is bandwidth, k is the Boltzmann's

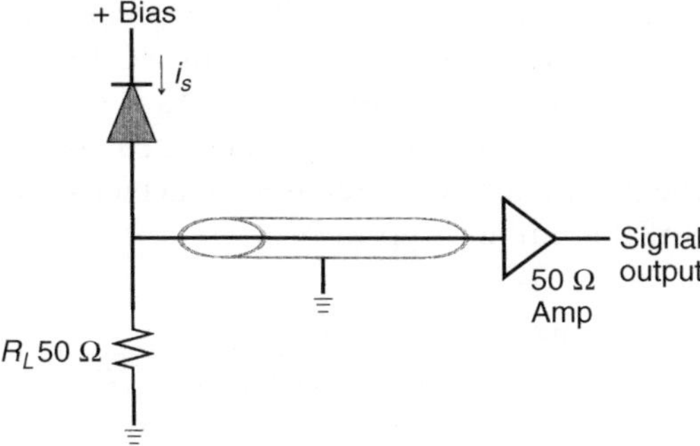

Fig. 8.10 Schematic of a low-impedance front end.

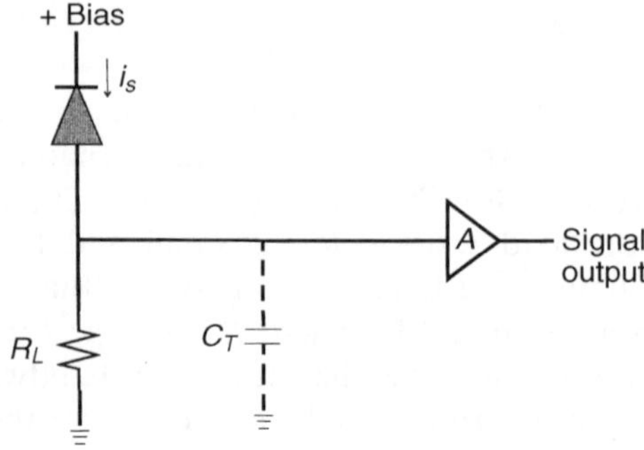

Fig. 8.11 Schematic of a high-impedance front end.

constant and T is the absolute temperature. Typical values for microwave amplifier noise figures at a few gigahertz lie between 3 and 6 dB.

8.3.2 High-impedance front end

The scheme of a high-impedance FE is sketched in Fig. 8.11. Since the variance of the input noise current is inversely proportional to the load resistor R_L, it is clear that a an increase in the load resistor causes a noise reduction, since a high impedance implies an integration the input. Moreover, the nature of the photodiodes as a current source makes it possible to increase the signal voltage for a given photocurrent by operating the photodiode into an amplifier with higher input resistance. The signal voltage is therefore increased before the addition of noise from the following amplifier, improving the sensitivity. In this FE the load resistor is very large and a high impedance field-effect transistor (FET) amplifier is employed. The total input capacitance C_T is the sum of the FET input capacitance, the detector capacitance and the stray capacitance.

On the other hand, because of the large input RC time constant, the FE bandwidth is less than the signal bandwidth. As a consequence, the pole placed by the RC constant has to be compensated, perhaps by equalisation. A basic scheme for an RC equaliser is depicted in Fig. 8.12. The transfer function of this circuit is given by

$$H(\omega) = \frac{V_0(\omega)}{V_i(\omega)} = \frac{R_2(1 + i\omega RC)}{R + R_1 + R_2 + i\omega RC(R_1 + R_2)}.$$ (8.33)

The numerator and the denominator of this transfer function specify a zero and a pole, respectively, with corner frequencies f_L and f_U given by

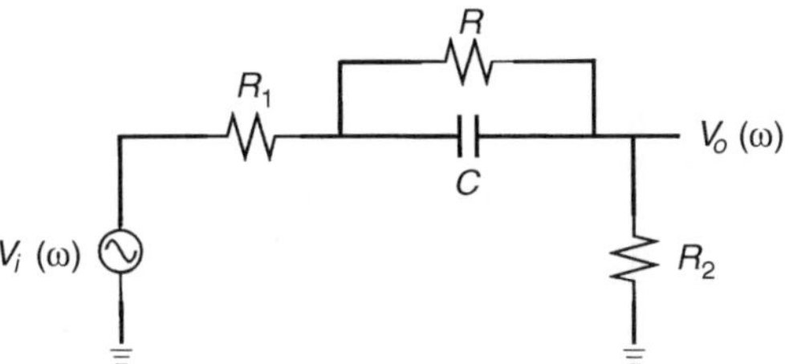

Fig. 8.12 Typical design of an equaliser for high-impedance front ends.

$$f_L = \frac{1}{2\pi RC}, \qquad f_U = f_L \frac{R + R_1 + R_2}{R_1 + R_2}. \tag{8.34}$$

The values of R and C are chosen so the zero cancels the pole of the high-impedance FE, and the overall receiver bandwidth is extended to f_U.

Now consider the position of the equaliser. The noise from stages following the equaliser can lower the sensitivity. On the other hand, the saturation of amplifiers preceding the equaliser reduce the receiver's dynamic range. A limited dynamic range is the major drawback of high-impedance FEs, as the high level of low-frequency signal components present in the integrated signal from the FE can easily saturate amplifier stages before equalisation. The equaliser should therefore be as close as possible to the FE in the amplifier chain, where signal levels are still small, but not so close that its attenuation causes noise from the following stage to be significant compared to the FE noise.

Thus high-impedance FEs can be used in applications where a wide dynamic range is not needed or where the low-frequency signal content is reduced using technique such as coding.

8.3.3 Transimpedance front end

The transimpedance FE is depicted in Fig. 8.13. The load resistor R_L is connected as a feedback resistor. As a consequence, the bandwidth is increased by a factor of $A + 1$ (A being the amplifier gain) over that of a high-impedance FE with the same R_L and C_T. Hence the need for further equalisation is reduced or eliminated. In addition, low-frequency gain is reduced by the same factor $A + 1$, decreasing the susceptibility to saturation and improving the dynamic range.

A critical factor for this FE is stability, due to the feedback mecchanism. In fact, the capacitive input impedance of the preamplifier causes a phase shift of 90° at high frequencies, and the inverting amplifier contributes a phase shift of 180°. In general, one would like to maintain a phase margin

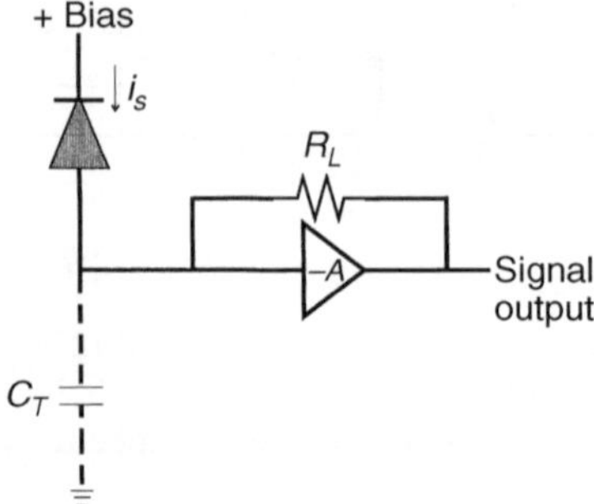

Fig. 8.13 Schematic of a transimpedance front end.

of about 45°, in order to allow for a maximum additional open-loop phase shift of only 45° up to the frequency at which the open-loop gain becomes smaller than unity. Such a small tolerable phase shift limits the gain that can be covered within the feedback loop, hence it is usually impossible to obtain the desired preamplifier bandwidth if a large feedback resistor is used. Moreover, R_C equalisers that could extend the bandwidth are often avoided in practice because of their need for adjustment and because of difficulty in maintaining pole-zero matching over a wide range of operating conditions.

The transimpedance FE presents greater simplicity and much wider dynamic range than the high-impedance FE, and it is very common in practical design.

8.3.4 Aspects of front end design for high-speed systems

In the early 1990s it was not clear which integrated circuit (IC) technology would be optimum at 10 Gb/s and beyond. Before 1990, silicon ICs were preferred; after 1990 GaAs technology became an attractive competitor. Transistors with cutoff frequencies beyond 30 GHz and digital ICs operating at speeds above 10Gb/s were thought necessary to provide sufficient performance margins. This requirement eliminate Si MOSFETs and greatly reduced the choice of Si processes. The majority of commercially avaliable GaAs MMIC processes were suitable for bit rates of 2.5 Gb/s and under, whereas GaAs HBT devices had demonstrated their promise in 10 Gb/s applications.

As far as the FE scheme is concerned, integrating designs generally offer slightly better noise performance than transimpedance designs, because the latter show a tendency to self-oscillation unless the value of the feedback resistor is sufficiently low. The use of inductor coupling between the photodetector and its bias resistor results in a four-terminal rather than a two-terminal interstage between photodiode and input stage, thus increasing the bandwidth for given values of bias resistor and shunt capacitance. Combined with a simple CR differentiator as used to equalise the response of integrating FEs, or with an active equaliser which can itself widen the bandwidth by a factor of 5 or 6, the integrator bandwidth can be extended very significantly.

For instance, in a 12 GHz pin-HEMT optical receiver FE was proposed [21]. A two-terminal (untuned) design with an active equaliser extending the bandwidth fivefold requires the input CR product to be $10/5 = 2$ GHz. Assuming a value for C of 0.4 pF, R must be 200 ω. Inductor coupling can readily give a threefold increase, thus reducing the 2 GHz to 667 MHz, and for the same value of C, increasing R to 600 ω. Equations for equivalent input noise current spectral density for simple models of both untuned and tuned input have been presented. By inserting the numerical values used earlier and others in literature [21], the results plotted in Fig. 8.14 are obtained. The improvement is substantial. The Nyquist bandwidth corresponding to 10 GHz is 5.6 GHz, and the total noise from 0 to 5.6 GHz, found by integrating the curves in Fig. 8.14, resulted in values of 847 and 519 nA corresponding to sensitivities of -22.9 to -25.1 dBm for the untuned and tuned cases, respectively. The noise is mostly due to the bias resistor and input transistor.

There are several reasons why hybrid receivers become increasingly impractical at bit rates beyond 10 Gb/s [21]. Some of the limitations could be overcome by using an optical preamplifier.

Another approach, for a simple and low-cost design, using commercial packaged transistors and a hybrid assembly on a soft substrate has been

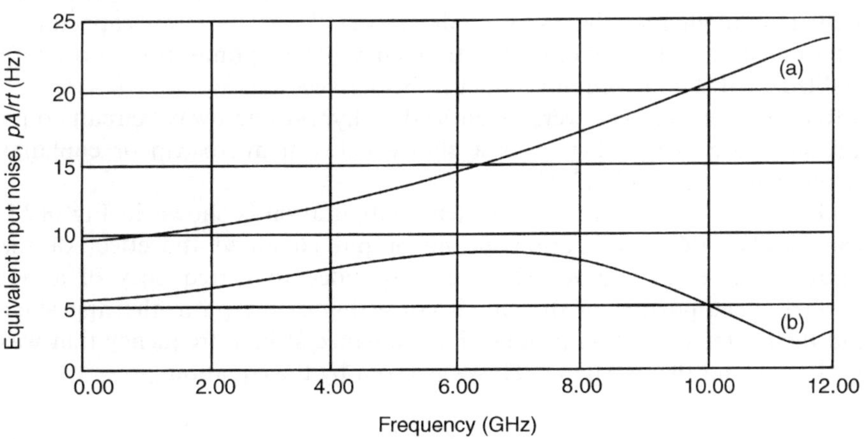

Fig. 8.14 Calculated equivalent input noise current spectral density versus frequency for interstages between the photodiode and input transistor: (a) two-terminal interstage, (b) four-terminal interstage. (After [21] with permission of IEE)

realised reproducibly [22]. The design ensured that the overall performance was largely insensitive to normal spreads in photodiode and transistor parameters, and to packaging and bonding parasitics.

The basic design considers the following. Discrete chip devices, individually characterised and assembled on a ceramic substrate, pose potential problems in terms of performance reproducibility and construction complexity, both of which increase costs in a manufacturing environment. The alternative approach is design a broadband receiver employing discrete, packaged, commercial devices with no individual characterisation, and a standard soft substrate.

In practice, a broadband receiver may be realised from a photodiode and a 50 ω amplifier, thereby taking advantage of readily available microwave amplifiers. However, as mentioned in section 8.3.1, the noise characteristics are far from optimal. Receiver sensitivity can be increased using high-impedance or transimpedance configurations. Cascaded or distributed circuit topologies can be used in either one. CAD studies showed that the present objectives could be realised using the high-impedance configuration with a cascaded amplifier. Simulation demonstrated that broader bandwidths could be achieved with this configuration because of transit time constraints that packaged transistors and a hybrid assembly imposed upon the feedback path of transimpedance designs.

Broader bandwidths could also be achieved with cascaded topologies rather than with distributed circuits, for the same packaged devices and substrate. They need to be disigned using CAD simulators that incorporate accurate models of all the important parasitics for the devices to be used. Three HEMT devices were modeled by the S-parameters given in the manufacturer's data sheets [22]. In that case considerably more care was needed in modelling the photodiode, especially as its stray capacitance determined the upper band edge frequency. The optimisation procedure enabled a satisfactory circuit model to be derived.

Receivers of this type were assembled as hybrid microwave circuits on a soft substrate, hence there was a choice between microstrip or coplanar waveguide technologies [23].

The circuit diagram for the chosen configuration is shown in Fig. 8.15. The parallel RC combination at the output equalised the effect of the input capacitances and R_1 which corresponded to a frequency of about 1.5 GHz. The portion of the circuit consisting of the p-i-n, the first stage and their interconnection produced a resonance at high frequency that was used to extend the passband. This was given by the equation

$$(2\pi f_p)^2 = \frac{(1 + C_D/C_{GS})}{C_D(L_S + L_G)} \tag{8.35}$$

The main characteristics were as follows. The 3 dB bandwidth was 10 GHz

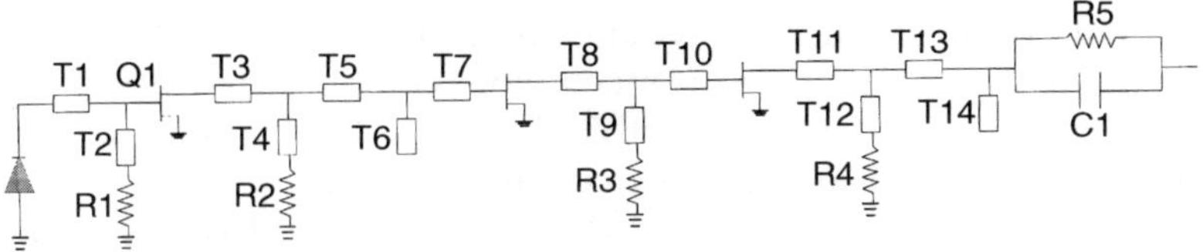

Fig. 8.15 Circuit diagram for a pin-HEMT broadband receiver.

and the gain ripple was less than ± 1 dB below 8 GHz. The predicted worst-case RMS noise current density was 12 $pA/(Hz)^{1/2}$, and the sensitivity was about –22.7 dBm at 2.4 Gb/s.

8.3.5 Front end amplifiers for high-speed systems

Research on systems operating at 10 Gb/s began in the late 1980s. Since then the receiver sensitivities have improved slowly. Values between –15 and –19.8 dBm were reported in 1989, whereas by 1993 the best reported measured value for a pin-FET receiver was –20.4 dBm (it is interesting to compare it with the best measured value of –20.7 dBm [24] using a pin-HEMT FE).

The principal options for hybrid broadband optical receiver implementations using ICs available in 1993 are summarised in the literature [24]. State-of-the-art high-speed silicon (f_T around 23 GHz) and GaAs HBT processes (f_T exceeding 40 GHz) had been used in the design of 10 Gb/s receiver FEs. The typical measured performance parameters of transimpedance FEs based on the GaAs HBT ICs were 3 dB bandwidths in the range 9.5–10.3 GHz, midband transfer impedances between 310 and 340 V/A and group delay variations of less than 30 ps within the passband. The measured results for the spectral density of the average equivalent input noise current was equal to 10 $pA/(Hz)^{1/2}$ and a sensitivity of about –22 dBm was inferred. As a result, the status of available IC technologies in late 1993 suggested that for the medium term the best choice appeared to be GaAs MMICs based on HBT technology, and this technology appeared to possess a considerable performance margin at 10 Gb/s.

8.4 RECEIVERS USING OPTICAL PREAMPLIFIERS

The sensitivity of an optical front end is limited by the noise generated within the photodiode, transistors and any dissipative elements present in the circuit. Coherent detection can provide better sensitivities than either p-i-n or avalanche photodiodes (APDs) and direct detection. However, besides polarisation control, it requires narrow-linewidth lasers for transmitter and local oscillator, and is much more complicated overall. The use of an optical preamplifier, e.g. an EDFA, before the photodiode provides better performance than direct detection using an APD, and is much simpler than coherent detection. On the other hand, an EDFA (or another optical amplifier) produces amplified spontaneous emission (ASE) noise and a standing output in the absence of an input signal.

Analysis of receiver performance with optical amplifiers in line (either with inline amplifiers or with preamplifier) is reported in Chapter 10. A narrowband optical filter can be inserted before the photodiode to limit the effect of ASE noise. As an example, the filter was inserted within the length of an erbium-doped fiber, so as to stop the buildup of both forward and backward travelling ASE power lying outside the signal band [25].

For use as areceiver preamplifier, a good low-noise performance is necessary, but without sacrifice in gain. These objectives are difficult to fulfil using a single EDFA. In fact, two-stage configurations can offer a better performance. A practical implementation of a transmitter and a receiver operating at 2×10 Gb/s and using a two-stage EDFA preamplifier is described in the literature [26]. A two-stage preamplifier pumped at 980 nm by two laser diodes provided 41 dB of small-signal gain and a noise figure of 3.8 dB. The FE bandwidth was limited to 14 GHz by the commercially available amplifier, which had been designed to give optimum pulse response.

Another two-stage EDFA configuration [27] is shown in Fig. 8.16. The main parameters were a combined noise figure of 4.1 dB with a high net gain of 20.7 dB at a signal wavelength of 1553 nm. The experiment achieved the highest reported 10 Gb/s receiver sensitivity of −31.1 dBm (corresponding to 606 photons per bit), which was 18.8 dB over the sensitivity of the non-preamplified receiver. The measured bit error rates (BERs) with and without the preamplifier are given in Fig. 8.17. They clearly show the improvements in BER obtainable with one or two EDFAs. The back-to-back sensitivity at 10^{-9} was −12.3 dBm Fig. 8.17c. Under these experimental conditions the gain was 16.8 dB for the first stage and 27.0 dB for both stages. For Fig. 8.17b the improvement in sensitivity was 15 dB, giving −27.3 dBm for a net gain of 16.5 dB. The increase in gain of 10.5 dB in going from one to two stages corresponds to an increase of 3.8 dB in sensitivity, some 3 dB less than predicted. This was attributed to the increased relative importance of the noise generated by Fresnel reflections at the connectors under large gain (> 20 dB) conditions. The authors state that the use of a two-stage preamplifier plus a booster at the transmitter suggested the possibility of an unrepeatered transmission span exceeding 200 km at 10 Gb/s.

The use of semiconductor optical amplifiers (SOAs) as preamplifiers was investigated throughly in the 1980s, especially before the development of

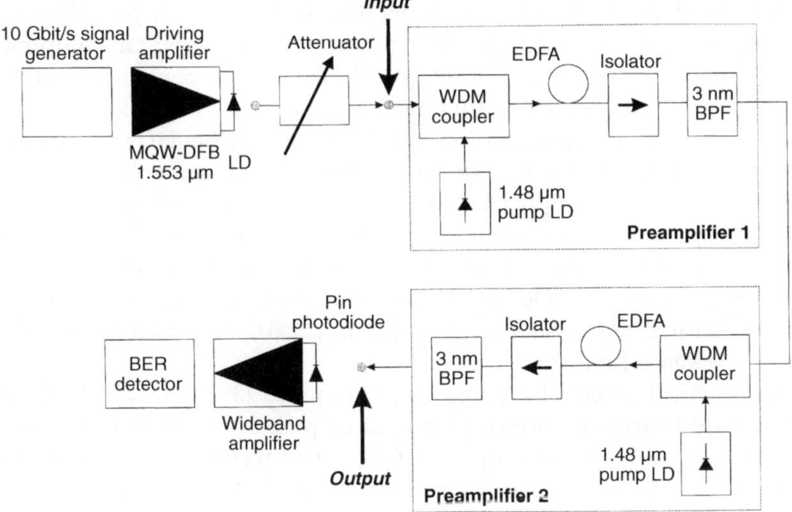

Fig. 8.16 A two-stage EDFA configuration. (After [27] with permission of IEE)

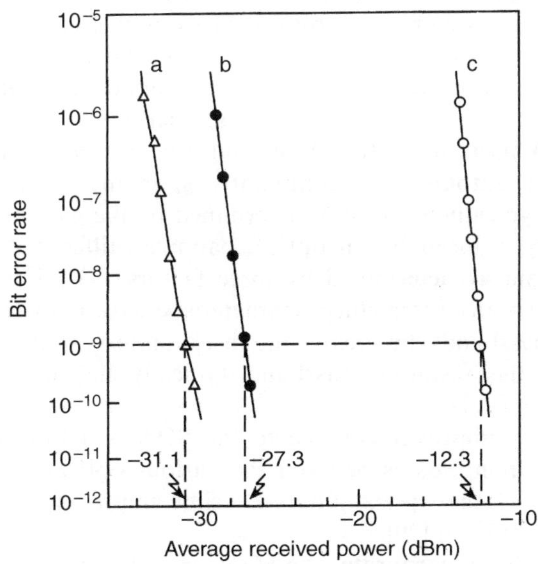

Fig. 8.17 BER performance at 10 Gb/s and 1553 nm: (a) with two-stage preamplifier; (b) with single-stage preamplifier; (c) without preamplifier. (After [27] with permission of IEE)

EDFA technology. In practical systems, the SOA gain is expected to vary by about ± 2 dB with changes in temperature, devices and component aging, source wavelength and input polarisation. Gain control is therefore required and a number of automatic gain control (AGC) systems have been described. Another limitation which applies particularly in intensity-modulated systems is due to saturation-induced crosstalk. Advantages of the SOA are inherent simplicity, low component count and the key feature of link transparency. Their small size makes them well suited for photonic integrated circuits, whereas fiber amplifiers are not. A practical example was reported where simple, easily manufactured, packaged, single-stage amplifiers with fiber-to-fiber peak gains of 21 dB and gain ripple of 1–2 dB were realised [28].

Some practical aspects have also been discussed [29]. The sensitivity and gain-bandwidth performance of preamplified p-i-n receivers were compared with APD receivers at rates up to 8 Gb/s. The technical interest in p-i-n receivers arises from the need for gain to offset the insertion loss of future optical networking elements such as switches, splitters and multiplexers being developed for optical fiber systems. The losses in these elements must be accommodated within the system power budget. This requirement, together with the need for ever higher transmission capacities, calls for receivers with improved sensitivities and bandwidths. APD receivers are more sensitive than p-i-n types at higher bit rates up to several gigabits per second, but the gain-bandwidth product of the APD sets an upper limit on sensitivity. Receivers employing p-i-n devices offer greater potential bandwidth but poorer sensitivity; however, the sensitivity can be enhanced, without affecting bandwidth, by employing an optical preamplifier.

An SOA-pin receiver can consist of a fiber tail coupled to the SOA input, the SOA itself, an optical filter and a p-i-n receiver in tandem. The amplifier noise output is predominantly spontaneous emission with a bandwidth of typically 30–40 nm, determined by the gain profile. This can be substantially reduced by an optical bandpass filter, with an optimum bandwidth primarily determined by three factors: (1) p-i-n receiver noise, (2) bit rate, (3) transmitter chirp characteristics. As is usual in filters, the narrower the passband, the greater the insertion loss, so a compromise has to be found; the resulting passband typically lies in the range 0.5–3.0 nm [29].

The sensitivity improvement due to the SOA is a function of internal cavity gain, coupling losses between the various optical components, and the bandwidth of the optical filter. Feasible improvements of 10–15 dB were typically obtained [29].

The receiver that incorporates the SOA, optical bandpass filter and front end is clearly wavelength selective, and may thus be employed as a wavelength demultiplexer.

The performance of receivers with more than a single-stage SOA has been analysed theoretically [30] using a lightwave simulation package which models laser chirp, fiber dispersion, SOA and receiver characteristics. The results showed that synchronously tuned stages gave better performance than stagger-tuned stages, despite having a larger gain ripple.

8.5 CIRCUITRY FOR CLOCK EXTRACTION

Clock recovery will be one of the key technologies in lightwave networks running at several gigabits per second. The transmission systems operating today use only electronic clock recovery. Two of the most common approaches are a phase locked loop (PLL) or a narrowband high-Q filter (e.g. dielectric resonators or SAW filters) [31]. In addition, when NRZ format is used, one needs a squaring circuit to extract from the received signal the spectral component at the bit rate. In the gigahertz range this electronic circuitry becomes difficult to be implemented. In order to avoid the speed limitations of the electronics, optical techniques for timing extraction are being extensively studied. In addition to the speed advantage, some of the optical techniques can be insensitive to the bit rate. This is an attractive feature in a future transparent optical network.

8.5.1 Electronic clock recovery

In general, if the timing is to be generated without any knowledge of the information content, it requiresan action that generates the bit frequency component independently of the bit pattern. This action is accomplished by an electronic unit, usually named the 'preprocessing unit'. It is essential that the received signal contains sufficient timing information and that the phase of the resulting spectral line remains constant. The bit frequency component is not directly available in the case of highly used communications systems, and must be generated by non-linear means. If the signal contains a spectral bit frequency line of sufficient amplitude, no preprocessing is required at all. Signals obtained by preprocessing still contain some unwanted frequency components, and these must be suppressed.

The preprocessing can be accomplished in two different ways. One is a passive method, consisting of an oscillating circuit to reduce noise effects followed by a hard limiter to eliminate amplitude variations. The other is an active method using PLL, which contains a local oscillator.

On the theoretical side, there is no important difference between a passive solution and a first-order PLL. A reduction of the timing information combined with insufficient stability of the components leads to increased phase fluctuations [30]. In a second-order PLL, this interrelationship can be reduced by proper design of the loop filter to any desired amount. On the other hand, this adversely affects the pull-in behavior. It is therefore necessary to make provision for an acquisition aid. A PLL equipped in this way remains unaffected by fluctuations of the clock content.

The PLL approach would is generally preferred since it is a closed-loop approach; that means it can adjust to small variations in the timing of the incoming signal without suffering from detuning or sacrificing filter bandwidth. However, the practical realisation of PLLs at gigabits data rates has proved difficult and the most common approach to date is to use an open-loop solution [31]. The key device is then a narrow high-Q filter. A standard device of these frequencies is the dielectric resonator. This device has been successfully been used for clock extraction in a 10 Gb/s experiment [32].

Design aspects for electronic clock recovery are discussed in literature [33]; a phase and frequency detector integrated circuit, operating at high bit rate (8 Gb/s) has also been reported [34].

8.5.2 Technology for electronic clock recovery

Silicon bipolar technology is a strong candidate for electronic circuits in optical receivers, such as the clock recovery circuits. The prime reason for choosing silicon is that process costs are lower than for GaAs circuits, based on MESFET (metal-semiconductor field-effect transistor) and on HBT (heterojunction bipolar transistor). Another reason is the future possibility of integrating faster and slower functions (memories, for instance) on the same chip (BiCMOS technology), which will provide wider scope for enhanced circuit functionality.

GaAs HEMT (high electron mobility transistor), SiGe HBT, InP HEMT, InP HBT and, possibly, CMOS are other potential technologies for the development of this type of circuit. But all these technologies have one drawback in common: they are less developed than the processes for bipolar silicon, hence they are more expensive. Some of them, e.g. InP HEMT, only allow production of circuits with a few transistors. CMOS is probably too slow for gigabits applications [30].

The special suitability of submicrometer self-aligning silicon bipolar technologies for optical communications ICs is demonstrated with data rates as high as 32 Gb/s [35]. It is also foreseen that the operating speed of digital functions can be further improved by a factor of about 3.

8.5.3 Optical clock recovery

Several promising clock recovery schemes have been proposed and demonstrated where most of electronic elements are replaced with optical ones.

For instance, a 6.3 GHz optical clock signal, corresponding to one of 16 channels in an optical time domain multiplexing (OTDM) scheme, is directly extracted from a 100 Gb/s optical pseudorandom bit stream (PRBS) datastream after 50 km of transmission using an optical PLL scheme [36]. Both the optical demultiplexer and the decision circuit are driven by the extracted clock. A traveling wave laser diode amplifier was used as a phase detector by optical gain modulation. Another interesting approach is to use a self-pulsating laser diode to extract the timing information [37]. An optical signal was extracted by synchronising a self-pulsating laser diode to a 5 Gb/s RZ data signal. The 5 Gb/s data stream was demultiplexed from a 20 Gb/s OTDM transmission system using the extracted clock signal. No penalty was observed compared to the measurements using the transmitter clock. An important advantage of this approach is that the resonance of the self-pulsating laser diode can be tuned continuously from 3 to more than 5 GHz by varying the bias current. In this way, the clock extraction circuit should be able to adapt automatically to the incoming bit rate.

REFERENCES

1. A. Yariv, *Optical Electronics*, Saunders, Philadelphia (1985).
2. S.M. Sze, *Physics of Semiconductor Devices,* John Wiley, New York (1981).
3. R.S. Muller and T.I. Kamins, *Device Electronics for Integrated Circuits*. John Wiley, New York (1977).
4. A.G. Milnes and D.L. Feucht, *Heterojunction and Metal-Semiconductor Junctions* Academic Press, New York (1972).
5. R. Sabella, and S. Merli, Analysis of InGaAs p-i-n photodiode frequency response, *J. Quantum Electron.* **29**, 906–916 (1993).
6. T.P. Pearsall *et al.*, Electron and hole mobilities in InGaAs, in *GaAs and related compounds*, Institute of Phisics, H.W. Thim, London (1980).
7. T.H. Windhorn *et al.*, Temperature dependent electron velocity-field characteristics for InGaAs at high electric fields, *J. Electron. Mater.* **11**, 1065–1082 (1982).
8. P. Hill *et al.*, Measurement of hole velocity in n-type InGaAs, *Appl. Phys. Lett.* **50**, 1260–1262 (1987).
9. D. Kuhl *et al.*, Influence of space charges on the impulse response of InGaAs metal-semiconductor-metal photo detetectors, *IEEE/OSA J. Lightwave Technol.* 753–759 (1992).
10. B.J. van Zeghroeck *et al.*, 105-GHz bandwidth metal-semiconductor-metal photodiode, *IEEE Electron. Dev. Lett.* 527–529.
11. D.L. Rogers, Integrated optical receivers using MSM detectors, *IEEE/OSA J. Lightwave Technol.* 1635–1638 (1991).
12. J.B.D. Soole and H. Schumacher, InGaAs Metal-semiconductor-metal photo detectors for long wavelength optical communications, *IEEE J. Quantum Electron.* **27**,. 737–752 (1991).

13. K. Kajiama, Y. Mizushima, and S. Sakata, Schottky barrier height of n-InGaAs diodes, *Appl. Phys. Lett.* **23**, 458–460 (1973).

14. C.L. Lin *et al.*, Compositional dependence of Au/InxAl1-xAs Schottky barrier heights, *Appl. Phys. Lett.* **49**, 1593–1595 (1986).

15. D. Rogers *et al.*, High-speed 1.3 μm GaInAs detectors fabricated on GaAs substrates, *IEEE Electron Dev. Lett.* **9**, 515–517 (1988).

16. Y.C. Lim and R.A. Moore, Properties of alternately charged coplanar parallel strips by conformal mappings, *IEEE Trans. Electron. Dev.* **ED15**, 173–180 (1968).

17. C.-K. Chang *et al.*, High-performance monolithic dual-MSM photodetector for long-wavelength coherent receivers, *Electron. Lett.* **25**, 1021–1023 (1989).

18. S.M. Sze, D.J. Coleman Jr., and A. Loya, Current transport in metal-semiconductor-metal (MSM) structures, *SolidState Electron.* **14**, 1209–1218 (1971).

19. B.L. Kasper, Receiver Design, in *Optical Fiber Telecommunications II*, Ed. by S.E. Miller and I.P. Kaminow, Academic Press, San Diego CA (1988).

20. C. Baack, *Optical Wideband Transmission Systems*, CRC Press, Boca Raton FL (1986).

21. E.M. Kimber, B.L. Patel, and A.P. Hadjifotiou, 12 GHz pin-HEMT optical receiver front-end, IEE Colloquium Digest, no. 173 (1993).

22. M.A.R. Violas *et al.*, The design and performance of a 10 GHz bandwidth low noise optical receiver using discrete commercial devices, *in Procedings of the 2nd Bangor Symposium on Communications*, University of Wales, Bangor, pp. 77–81 (1990).

23. M.A.R. Violas and D.J.T. Heatley, Microstrip and coplanar design considerations for microwave bandwidth pin-hemt optical receivers, paper presented at the IEE Colloquium on microwaves optoelectronics (1990).

24. W.S. Lee *et al.*, Hybrid optical receivers with integrated electronics, *IEE Colloquium Digest*, no. 173 (1993).

25. D. Simeonidou *et al.*, Erbium-doped fibre pre-amplifier with an integrated band-pass filter', *IEE Colloquium Digest*, no. 173 (1993).

26. W.S. Lee *et al.*, Practical implementation of high performance optical transmitter/receiver subsystems for 20 Gbit/s tdm operation, *IEE Colloquium Digest*, no. 120 (1994).

27. L.T. Blair, H. Nakano, High sensitivity 10 Gbit/s optical i.m.d.d. receiver using two cascaded edfa preamplifiers, *Electron. Lett.* **27**, 835–836 (1991).

28. R. Boudreau *et al.*, High gain (21dB) packaged semiconductor optical amplifiers, *Electron. Lett.* **27**, 1845–1846 (1991).

29. C.A. Hunter *et al.*, The design and performance of semiconductor laser pre-amplified optical receivers', IEE *IEE Colloquium Digest*, no. 119 (1989.

30. A. Djupsjöbacka, I. Andersson, and B. Rudberg, A 10 Gbit/s demonstrator, *Ericsson Rev.* **2**, 70–83 (1994).

31. O. Tveito *et al.*, Survey of optical receiver, paper presented at COST 239 meeting in Leidschendam (1993).

32. P. Monteiro *et al.*, 10 Gbit/s timing recovery circuit using dielectric resonator and active bandpass filters, *Electron. Lett.* **28**, (1992).

33. A. Buchwald and K. Martin, *Integrated Fiber-Optic Receivers*, Chs 4 and 10, Kluwer Academic, Norwell MA (1995).

34. A. Pottbaecker, U. Langmann, and H-U. Schreiber, A Si bipolar and frequency detector IC for clock extraction up to 8 Gb/s, *IEEE J. Solid State Circuits* **27**, 1747–1751 (1992).
35. L. Treitinger *et al.*, Silicon bipolar technology and circuits for optical communications at data rates above 10 Gbit/s, *Proc. OFC/IOOC '93*, paper FF3 (1993).
36. S. Kawanishi *et al.*, 100 Gbit/s, 50 km optical transmission experiment employing all-optical multi/demultiplexing and PLL timing extraction, *Proc. OFC '93*, San Jose CA, pp. PD2–1 to PD2–4 (1993).
37. P. Barnsley *et al.*, A 4 × 5 Gb/s transmission system with all-optical clock recovery, *IEEE Photon. Technol. Lett.* **4**, (1992)

9
Fundamentals of Fiber-Optic Transmissions

9.1 INTRODUCTION

The extraordinary advances in fiber and semiconductor technology have resulted in a very rapid growth in the performance of lightwave communications systems. The evolution of optical fiber transmission systems is illustrated in Fig. 9.1 [1], where the transmission capacity is shown for both commercial and experimental systems.

Distance and capacity (bit rate when considering digital signals) are the primary factors that influence optical system designs and the associated economic viability for their construction and operation. Historically, the evolution of optical fiber transmission systems over the past decades can be classified, according to the operating wavelength, the type of fiber and the transmission format:

1. 0.8 μm, multimode fiber. These types of system were limited in repeater spacing and bit rate by high fiber loss and excessive chromatic dispersion in the fiber caused by the use of light-emitting diodes (LEDs) as sources.
2. 1.3 μm, multimode fiber. The receiver spacing and the bit rate were better than in the previous generation. However, the use of multimode fibers limited the performance because there was modal dispersion.
3. 1.3 μm, single-mode fiber. The employment of single-mode fibers allowed the repeater spacing and the bit rate performance to be improved.
4. 1.5 μm, single-mode fiber. This generation of systems benefited from shifting the operating wavelength to 1.5 μm, which offers the minimum achievable attenuation for silica fibers. Furthermore, the introduction of single-mode lasers allowed a further improvement in the bit rate-span performance, due to the reduction of chromatic dispersion.
5. Coherent systems. So far, all the systems have used intensity modulation of the optical transmitter. This generation uses phase (or frequency) modulation of the optical transmitter and a coherent detection scheme which improves the receiver sensitivity, thereby increasing the repeater spacing.

However, the advent of erbium-doped fiber ampliers (EDFAs), which can amplify the radiation in the 1.5 μm region, has dramatically influenced the

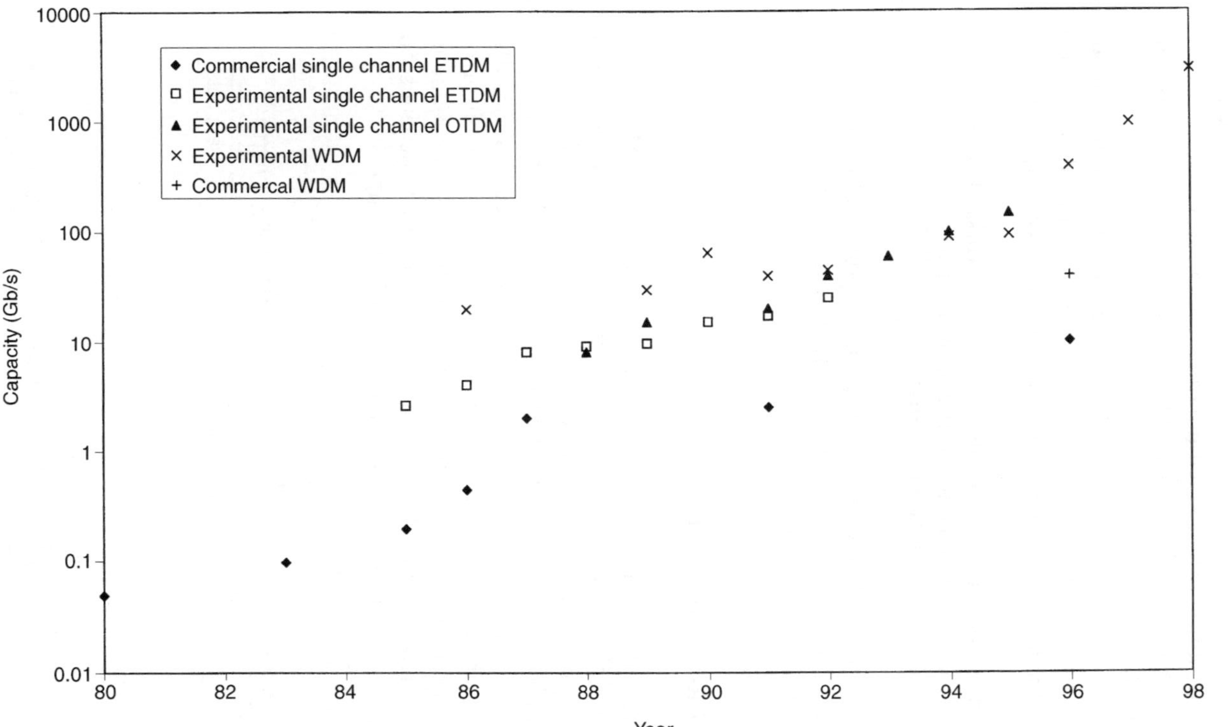

Fig. 9.1 Evolution of optical fiber systems in terms of transmission capacity. The capacity of optical fiber systems, whether commercial or experimental, has grown exponentially over the past 15 years. Commercial systems top out at about 40 Gb/s, whereas experimental sistems have exceeded 2 Tb/s. Experimental systems fall generally into three categories: TDM sistems, for which the time division multiplexing of the channels occurs in either the electronic domain (ETDM) or the optical domain (OTDM); and WDM systems, which use wavelength division multiplexing. Commercial systems use either ETDM or WDM. (Courtesy Tingye Li, AT&T Research Laboratories and Alan E. Willner)

development of telecommunication systems, favoring the realisation of intensity-modulated direct detection (IM-DD) systems (there is no point in developing complex systems if the higher receiver sensitivity can be recovered by an optical amplifier) and eliminating the presence of repeaters along a link. EDFA systems have virtually replaced coherent systems s in the sixth generation.

This progress has allowed optical fiber technology to be introduced in all the areas of telecommunication systems. In particular, the three broad categories, which relate to short, medium and long distance, are known as

(a)   distribution networks (local loop) and local area networks (LANs),
(b)   transport networks (interexchange traffic on wide areas),
(c)   long-haul (intercity or international traffic).

In this chapter the main characteristics of point-to-point links, belonging to the fourth generation, are discussed, together with some remarks on coherent systems (fifth generation).

9.2 BASIC FIBER-OPTIC LINK

The simplest fiber-optic system, for pulse code modulated (PCM) signals [2, 3], is shown in fig. 9.2a. It consists of a transmitter, a fiber transmission medium and a receiver. The transmitter converts incoming binary data to ON-OFF light pulses, which are launched into the fiber. At the receiver, the optical stream is detected and converted back into electrical signals. The primary measure of performance of a digital system is the bit error rate (BER), which represents the probability that an error will be made in the detection of a received bit. The maximum tolerable BER depends on the application (usually 10^{-9} or 10^{-10} for data communications).

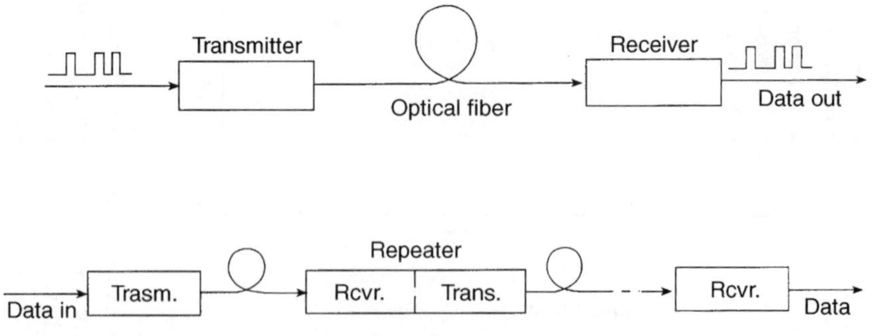

Fig. 9.2 Basic fiber-optic link: (a) simple digital optical link; (b) lightwave repeatered line composed of a cascade of simple links.

A useful measure of capability for a given transmission system, especially in trunk appplications, is given by the *BL* product, where *B* is the bit rate and *L* is the length of the fiber in a single repeater section (span). If we want to transmit a total bit rate of $Bt \gg B$ over a distance $L_t \gg L$, where *B* and *L* refer to each individual repeater section; then the total number of repeaters required for the system is

$$N = \frac{B_t \cdot L_t}{B \cdot L} \cdot$$ (9.1)

A high *BL* product implies small number of repeaters, and therefore low costs (repeaters are among the most expensive components).

The performance of a basic transmission system is mainly affected by two types of impairments: signal distortion and noise.

Signal distortion derives from the intermingling of laser source chirping (Chapter 3) and fiber chromatic dispersion (4). The transmitted pulses spread and overlap as they propagate, so generating the inter-symbol interference (ISI) which causes an increase of the BER. Note noted that only a limited number of neighboring pulses, within the dispersion or 'memory' of the communication channel[1], can contribute interference at a given sample time. The effect can be seen on an oscilloscope by means of an eye diagram, which is obtained by triggering the oscilloscope in synchronism with the bits of a random incoming data stream. The resulting display is a superposition of waveforms corresponding to a large number of different data sequences. Fig. 9.3 shows the received signal and the eye diagram in the ideal case (no dispersion, i.e. no ISI) and in the real case. The degree of eye closure is a convenient measure of system degradation. In particular, it is possible to define a power penalty due to the vertical eye degradation as the ratio, in decibels, between the ideal eye opening and the real eye opening. This power penalty can be measured or estimated, and it can be subtracted from the value of the *extinction ratio* (the ratio, in decibels, between the power levels corresponding to the ON and OFF states, respectively). In this way the performance of the system can be evaluated, considering the system as a noise-limited transmission link.

Note that not only vertical eye degradation affects the performance, but even horizontal degradation gives rise to the waveform jitter. Waveform jitter influences the receiver timing recovery, increases the variance of receiver sampling instants, and introduces timing error to the receiver. Its effects can also be translated into system penalty. However, the exact calculation of the system penalty due to timing jitter [4] depends on the

---

[1] The term channel, in this book as in the common literature, is used either to identify the communication system as a whole (the line), or to recognise the signal consisting of a time division multiplexed set of signals.

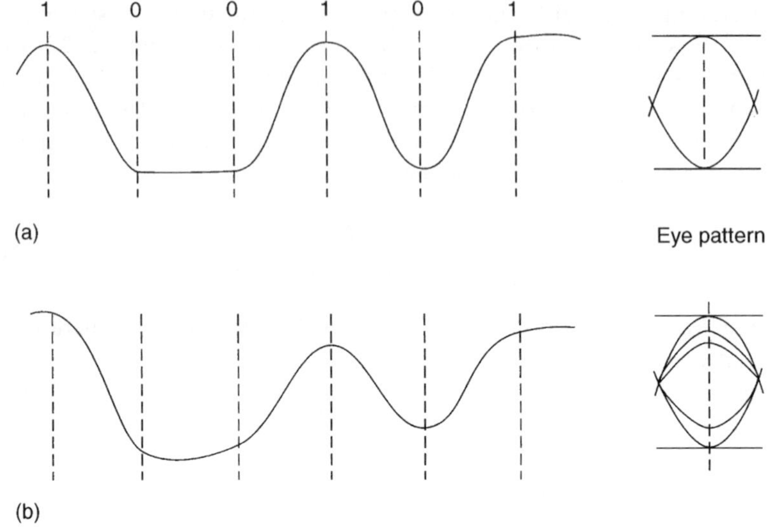

Fig. 9.3 Receiver output waveforms and corresponding eye diagrams: (a) *Ideal* case with no waveform distortions; (b) distorted waveform and resulting eye closure.

details of the timing recovery circuitry used in the receiver and is not pursued here.

A commonly used parameter for the receiver is its *sensitivity*. It is defined as the minimum optical power which is required in order to obtain a fixed performance in terms of BER. If a received optical power is greater than the sensitivity, the receiver will operate better; if it is smaller than the sensitivity, the BER will be worse.

The next section looks at the theory of IM-DD systems affected by noise.

9.3 ERROR PROBABILITY IN BINARY SYSTEMS AFFECTED BY NOISE

A PCM transmission consists of a sequence of binary pulses (logical 0 and 1, where for simplicity the zero is associated with 'dark', i.e. the absence of the pulse). The optical sequence is detected by the photodiode, thus converted into a sequence of electrical pulses, and sent to the decision circuit. Each received pulse is sampled and compared with a predetermined threshold. If the sampled pulse is greater than the threshold, the circuit 'decides' for a mark (logical 1), otherwise it decides for a 'space' (logical 0).

In the absence of distortions, the communication channel is perfect (the transfer function of the channel, as a whole, consists of a net gain and a

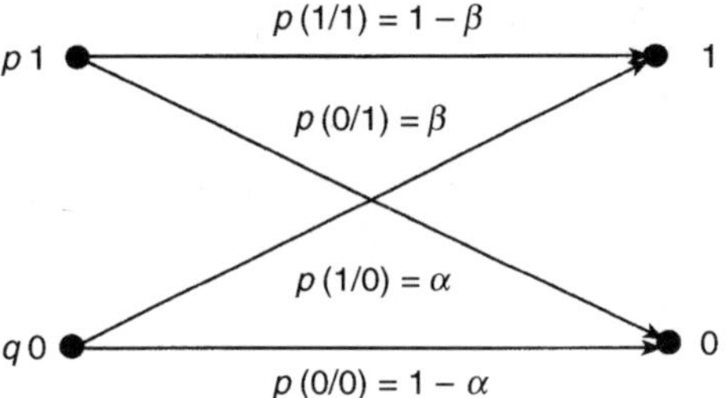

Fig. 9.4 Transmission in a binary channel.

phase delay). To evaluate the error probability in case of a perfect channel, we can refer to Fig. 9.4.

Figure 9.4 is the general transmission diagram of a binary channel. The source is completely evaluated by assigning the probabilities p and q ($= 1 - p$), relating to the transmission of a mark and a space, respectively. The channel is characterised by the probabilities α and β, respectively describing, the possibility that a space is transmitted but a mark is revealed, and vice versa. It is worth observing that α and β depend on the signal extinction ratio, on the threshold of the decision circuit of the receiver and on the noise characteristics of the transmission system.

The total error probability is given by the sum of the probability of revealing a wrong mark (that is a mark not present in the launched signal) and the probability of revealing a wrong space:

$$P_e = p \cdot \beta + q \cdot \alpha \tag{9.2}$$

The practical evaluation of the error probability requires a knowledge of the probability density function (p.d.f.) of the received signal y.

Usually it can be assumed that the system noise is additive, i.e.

$$y(t) = s(t) + n(t) \tag{9.3}$$

where $s(t)$ represents the transmitted signal, and $n(t)$ represents the random variable associated to the noise. Consequently with the p.d.f. of the received signal y can be obtained by convolution of the p.d.f.'s associated with s and n, as

$$p_y(x) = p_s(x) * p_n(x) \tag{9.4}$$

The p.d.f. p_s can be expressed as two Dirac functions centered at the levels corresponding to space and mark, and with weights q and p.

We can consider the system noise to be a zero-mean Gaussian function.

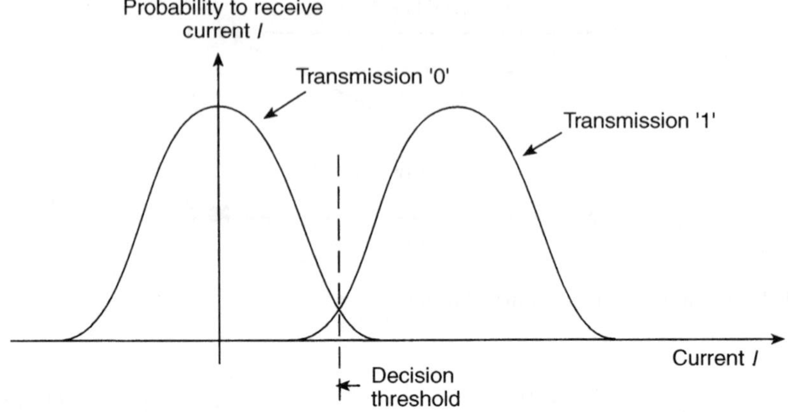

Fig. 9.5 Probability density functions for a typical IM-DD system.

And the error probability, from equation (9.2), is obtained by appling the following expressions (Fig. 9.5):

$$p(1/0) \equiv \alpha = \int_{\eta}^{+\infty} \frac{1}{\sigma\sqrt{2\pi}} e^{-\frac{x^2}{2\sigma^2}} dx \qquad (9.5)$$

$$p(0/1) \equiv \beta = \int_{-\infty}^{\eta} \frac{1}{\sigma\sqrt{2\pi}} e^{-\frac{(x-x_m)^2}{2\sigma^2}} dx \qquad (9.6)$$

where η represents the threshold of the decision circuit, x_m the level of mark and σ the standard deviation of the Gaussian noise.

9.4 THE PHOTODETECTION MECHANISMS

In an IM-DD scheme the photodiode acts simply as a 'photon counter'. The photogenerated current is proportional to the incident optical power, through a constant called responsivity. The main limitations related to this photodetection mechanism are described below.

9.4.1 The Poisson process and the quantum limit

The photodetection process can be described as a Poisson process. In particular, the probability of generation of n photocarriers, relating to the absorption of photons with energy E, is given by

$$p(n) = \left(\frac{\eta E}{h\nu}\right)^n \frac{\exp[-(\eta E/h\nu)]}{n!}, \qquad (9.7)$$

The mean value of photogenerated carriers is given by

$$<n> = \frac{\eta E}{h\nu}. \qquad (9.8)$$

If we write $p(n, <n>)$ the probability of detecting n photocarriers when the mean value is $<n>$, and if we assume that a transmitted space really correspond to $E = 0$, i.e. $<n> = 0$ the following probabilities are obtained when a space is transmitted:

$$p(0,0) = 1 \quad \text{and} \quad p(n > 0, 0) = 0 \qquad (9.9)$$

This suggests to place threshold 1 photon. Under these assumptions, the residual error probability is given by the possibility of observing 0 electrons as a response an incident light of energy different from zero. Assuming the same probability for mark and space emission, this leads to the following expression:

$$P_e \equiv p\beta = \frac{1}{2}p(0, <n>> 0) = \frac{1}{2}\exp\left(-\frac{\eta E}{h\nu}\right); \qquad (9.10)$$

The minimum energy E_{min} that is needed to obtain a determined error probability is called the quantum limit. If an error probability of 10^{-9} is requested, the quantum limit of $E_{min} \cong 21h\nu\,\eta$, i.e. at least 21 photons are required for each pulse!

The quantum limit represents an ultimate bound for an optical receiver, and it can be used as a reference value to measure the characteristics of any type of receiver.

9.4.2 The shot noise limit

By *shot noise* we mean a process

$$i_T(t) = \sum_{i=1}^{N_T} h(t - t_i) \qquad (9.11)$$

where $h(t)$ is a given time function, and t_i are random points in time, each is independent of the others and is uniformly distributed in the interval $(0, T)$, and N_T is the total number of events (photon absorptions in our case) occurring in time T. This process can be considered as the output of a linear time-invariant system with impulse response $h(t)$ driven by a Poisson sequence of impulses. In order to derive the power spectral density of this type of noise, one considers a carrier generated by photon absorption producing current i_e in the external circuit:

$$i_e(t) = \frac{e}{d}\nu_e(t) \tag{9.12}$$

where e is the electron charge, d the diode active region thickness, and $\nu_e(t)$ the instantaneous velocity. The Fourier transform of a single current pulse is

$$H(\omega) = \frac{e}{2\pi d}\int_0^{ta} \nu_e(t)e^{-i\omega t}dt \tag{9.13}$$

where t_a is the arrival time of an electron emitted at $t = 0$. If the transit time of an electron is sufficiently small that is $\omega t_a \ll 1$ (ω is the frequency of interest), we can replace exp $(-i\omega t)$ by unity and obtain

$$H(\omega) = \frac{e}{2\pi d}\int_0^{ta} \frac{dz}{dt}dt = \frac{e}{2\pi}, \tag{9.14}$$

since $x(t_a)$ is, by definition, equal to d. The Fourier transform of a train of randomly occurring events depicted in (9.11) is given as

$$I_T(\omega) = \sum_{i=1}^{N_T} H_i(\omega) = \sum_{i=1}^{N_T} \frac{1}{2\pi}\int_{-\infty}^{\infty} h(t-t_i)e^{-i\omega t}dt = $$
$$\sum_{i-1}^{N_T} \frac{e^{-i\omega t_i}}{2\pi}\int_{-\infty}^{\infty} h(t)e^{-i\omega t}dt = \sum_{i=1}^{N_T} e^{-j\omega t}H(\omega) \tag{9.15}$$

thus the spectral density function can be expressed as

$$|I_T(\omega)|^2 = \frac{|H(\omega)|^2}{4\pi^2}\sum_{i=1}^{N_T}\sum_{j=1}^{N_T} e^{-i\omega(t_i-t_j)} = \frac{|H(\omega)|^2}{4\pi^2}\left(N_T + \sum_{i\neq j}^{N_T}\sum_{j=1}^{N_T} e^{-i\omega(t_j-t_i)}\right). \tag{9.16}$$

By taking an ensemble average of (9.16), the second term on the right-hand side can be neglected in comparison with N_T. Hence

$$\overline{|I_T(\omega)|^2} = N_T\frac{|H(\omega)|^2}{4\pi^2} \equiv \overline{N}T\frac{|H(\omega)|^2}{4\pi^2}, \tag{9.17}$$

where the horizontal bar denotes ensemble averaging, and where $\overline{N}$ is the averaging rate at which the events occur, so $N_T = NT$. The spectral density function $\theta(\omega)$ of $i_T(t)$ is [5, 6]:

$$\Theta(\omega) = 4\pi\overline{N}|H(\omega)|^2 \, or \, \Theta(f) = 8\pi^2\overline{N}|H(2\pi f)|^2. \tag{9.18}$$

As a result, the shot noise spectral density can be expressed using (9.14) in (9.18):

$$\Theta(f) = 8\pi^2\overline{N}\left(\frac{e}{2\pi}\right)^2 = 2e\overline{I} \tag{9.19}$$

The output signal current of an ideal photodetector is given by $I_0 = eP_s/hf$, and the corresponding signal power can be written as

$$S = R_L(eP_s/hf)^2, \tag{9.20}$$

R_L being the load resistance. On the other side, the shot noise power can be expressed as follows [5, 6]:

$$N = R_L\Theta B = R_L 2eI_0 B = R_L(2e^2 P_s/hf)B. \tag{9.21}$$

As a result, the ideal signal-to-noise ratio for the detection mechanism can be derived as

$$\left(\frac{S}{N}\right)_{ideal} = \frac{P}{2hfB}. \tag{9.22}$$

In real cases the receiver noise is much greater than the shot noise from the detection mechanism.

9.5 ERROR PROBABILITY IN FIBER-OPTIC IM-DD RECEIVERS

A PCM receiver is outline in Fig. 9.6. After the photodetection (which we assume is performed by a p-i-n diode), the signal is amplified and filtered. The resulting detector output is compared to a threshold level once per time slot in order to determine whether or not a pulse of light was present at the detector in that time slot.

9.5.1 Performance calculation

The error probability is the probability that the filter output voltage at the sampling time exceeds the threshold level when a pulse of light is not

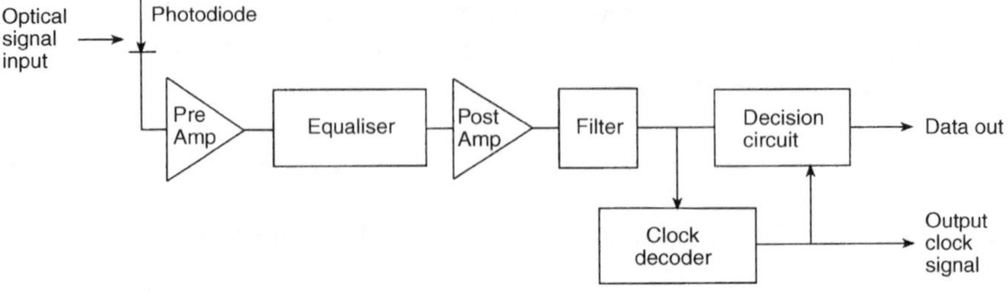

Fig. 9.6 Basic optical receiver for PCM signals.

present, plus the probability that it is below threshold when a pulse is present, all divided by 2 (for equally probable pulse, i.e. a mark has the same likelihood as a space). In order to calculate the error probabilities, the signal probability distribution should be known. The voltage signal can be expressed as

$$v_{out}(t) = \sum_{-\infty}^{\infty} em_k h_{lc}(t - t_k) + n(t) \qquad (9.23)$$

where $h_{lc}(t)$ is the linear channel impulse response, e is the electron charge, m_k is the number of secondaries produced by primary k (generated at time t_k), and $n(t)$ is the noise. The average value of $v_{out}(t)$, given the sequence of data digits, is

$$< v_{out}(t) >= \sum_{l} q_l h_{out}(t - lT), \qquad (9.23)$$

where q_l are the data digits and

$$h_{out}(t) = h_p(t)R * h_{lc}(t), \qquad (9.24)$$

$h_p(t)$ being the shape of an isolated optical power pulse and R the detector responsivity.

Since output pulses of the isolated filter may overlap, due to ISI, the error probabilities generally depend upon the values of adjacent data digits. However, we can assume that the filter is chosen to shape output pulses to have zero ISI at the sampling times (e.g. adopting a zero-forcing equalisation). In this way the error probabilities are determined by the statistical variations in the detection process and by the amplifier noise.

Since the exact calculation described above is quite tedious, various approximate techniques have been formulated. A good approximation, which in most real cases yields results very close to those obtained from exact theory, is the Gaussian approximation. The signal is assumed to be a Gaussian random variable, whose variance, corresponding to the noise power, can be easily calculated by summing the noise power contributions from the different noisy stages of the receiver. In this case the error probability can be evaluated as in the case of ordinary binary transmission systems. In ordinary cases the Gaussian approximation is a valid approach.

9.5.2 Noise and signal in an IM-DD receiver

In ordinary cases the Gaussian approximation is a good approach for the evaluation of the BER of an IM-DD system. This section derives the expressions for noise and signal.

Two basic types of noise are generated in an optical receiver: thermal noise from electronic circuitry and shot noise associated with different currents flowing into the photodiodes.

The most important noise term is represented by thermal noise (or circuit noise). It can be represented by the equivalent input current of thermal noise plus amplifier (front end) noise, as

$$\overline{i_{th}^2} = \frac{4kTF}{R_L}B, \tag{9.25}$$

where R_L is the load resistance, F is the equivalent noise figure of the amplifier, B is the receiver bandwidth, assumed equal to the signal bandwidth for out IM-DD receiver, in which a baseband amplifier is used.

The shot noise contributions related to signal light power P_s and to the background light power P_b, can be expressed as

$$\overline{i_{s,b}^2} = 2e\frac{\eta e}{hf}P_{s,b}B, \tag{9.26}$$

and the shot noise term associated with the dark current is

$$\overline{i_d^2} = 2eI_dB. \tag{9.27}$$

In the case of an avalanche photodiode (APD), the shot noise contributions calculated above have to be multiplied by M^{2+x}; where M is the avalanche multiplication factor and x is the excess noise factor of the APD.

Under the assumptions of the previous section, the total noise power, i.e. the total variance of the Gaussian process that represents the receiver noise as a whole, is given by

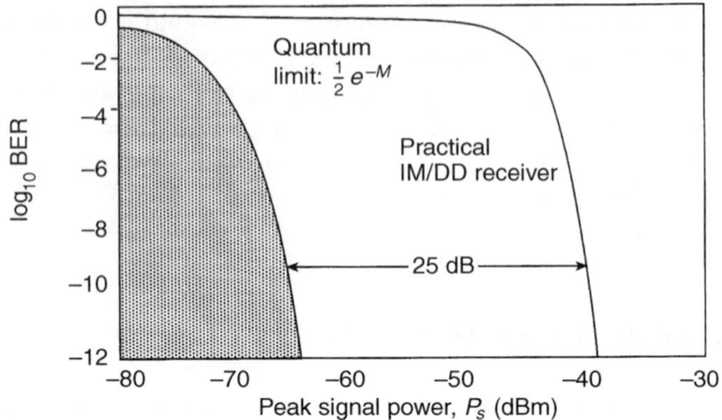

Fig. 9.7 BER curves for a practical IM-DD receiver at a bit rate of 100 Mb/s with $\lambda = 1500$ nm. They are compared with the curve for an ideal receiver.

$$\sigma_1^2 = \overline{i_s^2} + \overline{i_v^2} + \overline{i_d^2} + \overline{i_{th}^2} \text{ (for a transmitted mark)}, \qquad (9.28)$$

$$\sigma_0^2 = \overline{i_b^2} + \overline{i_d^2} + \overline{i_{th}^2} \text{ (for a transmitted space).} \qquad (9.29)$$

The p.d.f.'s for these two cases will be asymmetric, that is the Gaussian p.d.f. centered at marks will be broader than the p.d.f. centered at spaces. In most practical cases, the shot noise terms are negligible, unless a receiver is located very close to the transmitter.

Since the equivalent input signal current is given by

$$S = \frac{\eta e}{hf} P_s, \qquad (9.30)$$

the S/N ratio can be expressed as:

$$\frac{S}{N} = \frac{R_L S^2}{R_L \sigma^2}. \qquad (9.31)$$

As an example, Fig. 9.7 shows the BER curves for a practical IM-DD receiver at a bit rate of 100 Mb/s, at a wavelength of 1500 nm, compared with the curve for an ideal receiver [7]. Notice that only dark current and thermal noise cause a 25 dB penalty in sensitivity with respect to the quantum limit.

9.6 REMARKS ON COHERENT SYSTEMS

In a direct detection receiver the optical signal is directly converted to the baseband signal. In a heterodyne receiver, whose basic scheme is sketched

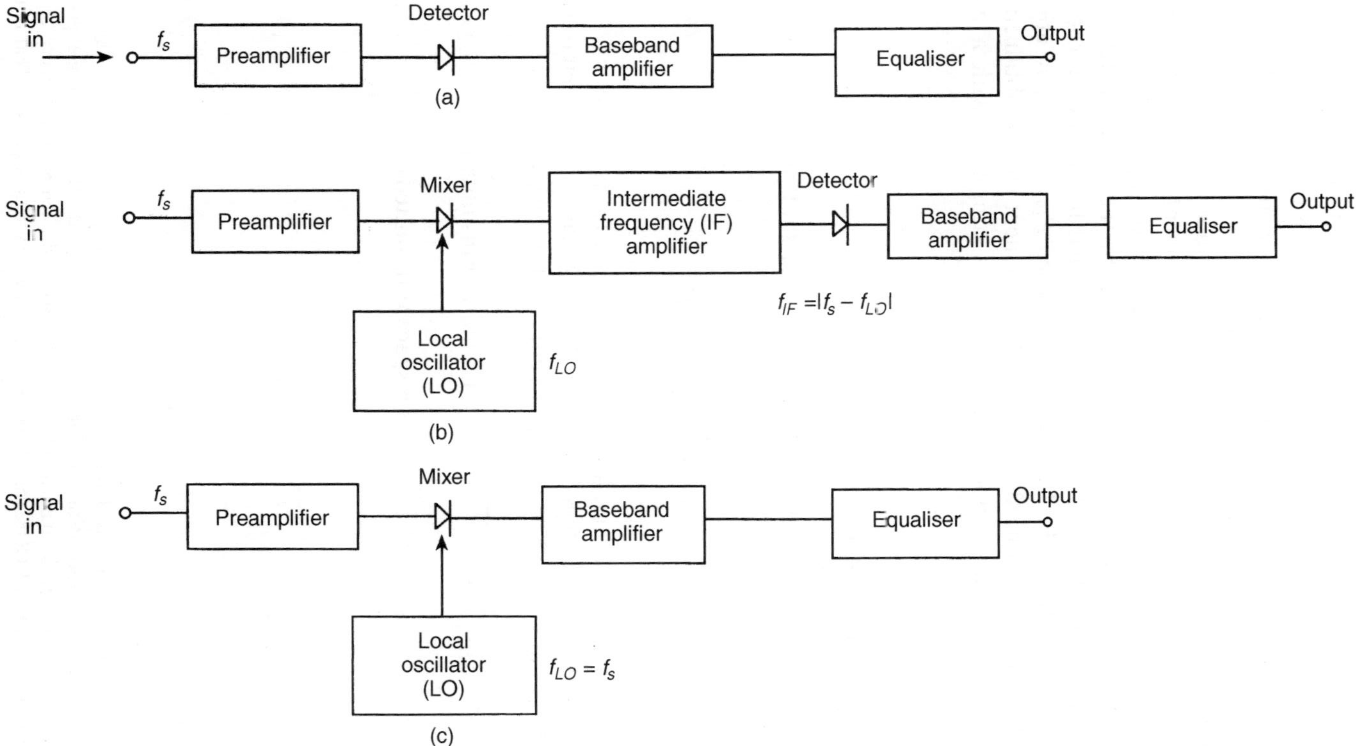

Fig. 9.8 Basic heterodyne/homodyne receiver.

in Fig. 9.8, the received signal is mixed with a local oscillator (LO). The intermediate frequency signal is then amplified and detected. A particular case is represented by the homodyne receiver in which the LO frequency as well as its phase are controlled so that they are always equal to the frequency and phase of the received signal carrier.

The greatest advantage of a heterodyne (homodyne) system is the improvement of receiver sensitivity. This improvement is basically due to the increase of S/N at the output end of the receiver preamplifier (for a given Ps). In fact, when the LO light is added to the received light, the time-varying part of the total power is proportional to the square root of the received power times the LO power P_{LO}. When P_{LO} is large, an amplification-like effect is produced. A further improvement can be obtained by a coherent modulation /demodulation scheme as compared with a non-coherent scheme.[2]

9.6.1 Noise and signal in heterodyne receiver

In heterodyne detection there exists an additional shot noise due to local oscillator light:

$$\overline{i_L^2} = 2e\left(\frac{\eta e}{hf}\right)P_{LO}(2B).$$

(9.32)

The bandwidth of the IF amplifier is now assumed to be 2B, to make the system evaluation consistent with the IM-DD case. In a heterodyne receiver, the total variance of the equivalent input noise currents are

$$\sigma_{H1}^2 = \overline{i_s^2} + \overline{i_c^2} + \overline{i_L^2}, \text{(for a mark transmitted)}$$

(9.33)

$$\sigma_{H0}^2 = \overline{i_c^2} + \overline{i_L^2}, \text{(for a space transmitted)}$$

(9.34)

where

$$\overline{i_s^2} = 2e\frac{\eta e}{hf}P_s(2B)$$

(9.35)

$$\overline{i_c^2} = \frac{4kTF}{R_L}(2B)$$

(9.36)

---

[2]The term *coherent* is generally used for the systems using heterodyne/homodyne detection schemes, whatever modulation/demodulation scheme is adopted. However, some people believe that a scheme has to be named coherent if it exploits the temporal coherence of the carrier. For instance, a PCM-OOK (on-off keying) heterodyne system is non-coherent in this sense. Here say that a system is coherent when a heterodyne/homodyne scheme is adopted

In the heterodyne detection scheme, the p.d.f.'s for the mark and space states have almost identical shapes, since the contribution of $\overline{i_s^2}$ in (9.33) and (9.34) is negligible compared with $\overline{i_L^2}$.

On the other hand, the input IF current (amplitude value) in a heterodyne detection system can be derived as follows. The instantaneous current waveform can be expressed as

$$i(t) = R_0 \left[\sqrt{2P_s} \cos(\omega, t) + \sqrt{2P_{LO}} \cos(\omega_{LO}{}^t) \right]^{ms}, \tag{9.37}$$

where $[\ \ ]^{ms}$ denotes a mean (time-averaged) square value, and R_0 is the responsivity of the photodiode. Eq. (9.37) can be rewritten as

$$i(t) = R_0 \left[\Re\left\{ \sqrt{2P_s} \exp(i\omega_s t) + \sqrt{2P_{LO}} \exp(i\omega_{LO}t) \right\} \right]^{ms}$$

$$= R_0 \left[\Re\left\{ \exp(i\omega_{LO}t) \left[\sqrt{2P_{LO}} + \sqrt{2P_s} \exp(i(\omega_s - \omega_{LO})t) \right] \right\} \right]^{ms}$$

$$= \frac{R_0}{2} \left| \sqrt{2P_{LO}} + \sqrt{2P_s} \exp(i\omega_{IF}t) \right|^2 .$$

$$= \frac{R_0}{2} \left[\left(\sqrt{2P_{LO}} + \sqrt{2P_s} \cos(i\omega_{IF}t) \right)^2 + \left(\sqrt{2P_s} \sin(o\omega_{IF}t) \right)^2 \right]$$

$$= R_0 \left(P_{LO} + P_s + 2\sqrt{2P_s P_{LO}} \cos(i\omega_{IF}t) \right) \tag{9.38}$$

hence the input IF current (amplitude value) is a heterodyne detection scheme and is given as

$$S_H = 2\left(\frac{\eta e}{hf} \right) \sqrt{P_{LO}P_s}. \tag{9.39}$$

The S/N ratio for this case (at the IF stage) can be expressed as

$$\frac{S}{N} = \frac{1/2R_L S_H^2}{R_L \sigma_H^2}. \tag{9.40}$$

9.6.2 BER performance in heterodyne/homodyne receivers

In heterodyne receivers, the analyses of signal and noise phenomena are more complicated than in an IM-DD, because the output of the photodetector appears as an IF signal, not a baseband signal. Here we consider

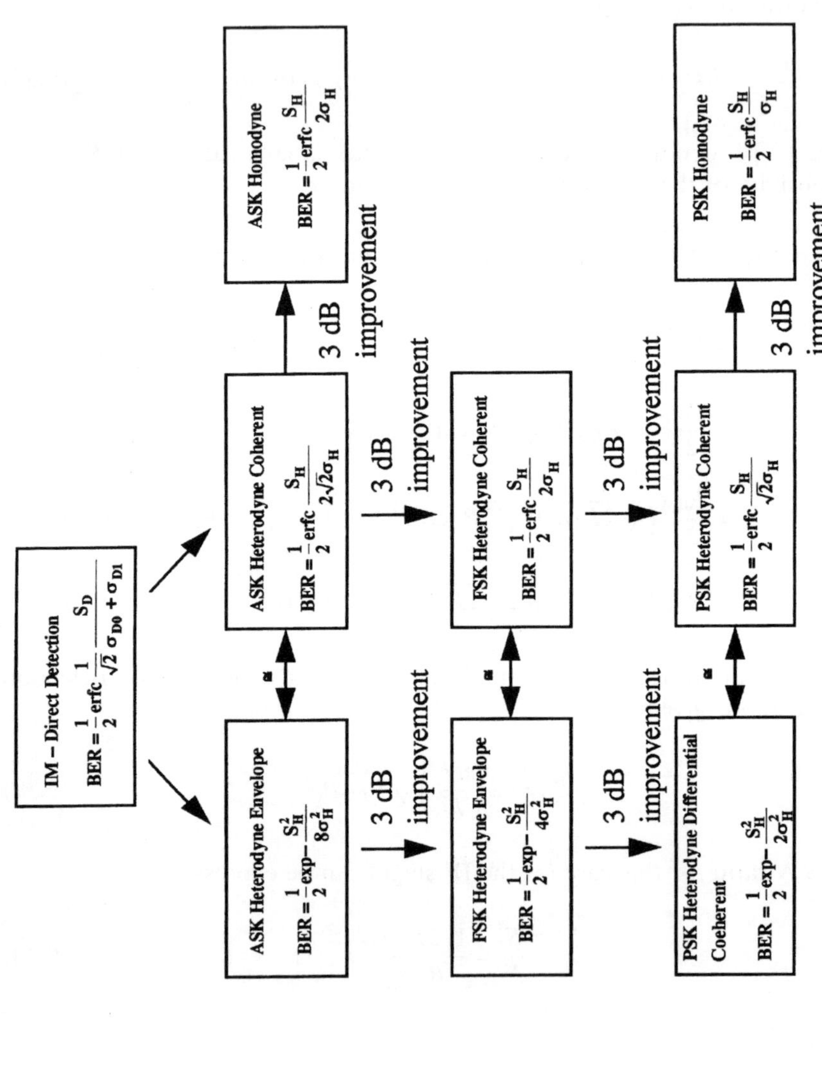

Fig. 9.9 Receiver sensitivities for optical communication systems employing different modulation/demodulation schemes.

three different coherent schemes: amplitude shift keying (ASK), frequency shift keying (FSK), and phase shift keying (PSK).

The analytical determination of the receiver sensitivies is reported in the literature [8]. Here we illustrate the comparison of the receiver sensitivities of optical communications systems employing various modulation/demodulation schemes. In particular, Fig. 9.9 illustrates such a comparison in terms of BER expressions for the different cases.

The improvement from IM-DD to ASK system depends strongly upon the noise parameters of the photodetector and the front-end amplifier. In general it can be estimated as between 10 and 25 dB [8]. On the other hand, the difference between ASK heterodyne-coherent and ASK hetero-dyne-envelope detection schemes is small. The FSK and PSK heterodyne systems show respective sensitivity improvements of more than 3 dB and 6 dB over an ASK heterodyne system. In any case, the use of a homodyne scheme allows a further improvement of 3 dB to be obtained with respect to the heterodyne scheme.

In the heterodyne schemes, the coherent and envelope (or differential) detection schemes make little practical difference at low BER.

Consider how the shot-noise limited S/N values are represented in terms of the number of photons per signal bit. Assuming an LO power high enough to make the amplifier noise negligible, the following shot noise limited expression for S/N can be derived for an ideal heterodyne receiver:

$$\left(\frac{S}{N}\right)_{shot-noise\ limit} = \frac{\eta P_s}{2hfB}. \qquad (9.41)$$

For an ideal homodyne receiver this ratio assumes half the value of the heterodyne case, since the bandwidth can be reduced from $2B$ to B.

Equation (9.41) can be rewritten in terms of the number N_{ph} of received photons per bit. In fact, assuming the signal bandwidth B is a half the signal bit rate (Shannon's essential limit), we obtain

$$N_{ph} = \frac{P_s}{2Bhf} \qquad (9.42)$$

Combining the BER formulas given in Fig. 9.9, and computing the required number of photons for the various schemes to achieve BER = 10^{-9}, assuming $\eta = 1$, we obtain the results summarised in Table 9.1. The required value N_{ph} ranges between 10 and 80 for various coherent schemes.

However, the receiver sensitivity improvements described in this section are achievable only when the phase fluctuations of both the transmitter and the local oscillator lasers are neglected. In real cases the spectral width of semiconductor lasers is usually rather wide (e.g. 1–10 MHz or more), and so the receiver sensitivity is degraded.

Table 9.1 The required number of photons for the various schemes to achieve BER $= 10\text{--}9$ ($\eta = 1$)

Coherent schemes		Number of photons required
	ASK	80
Heterodyne	FSK	40
	PSK/DPSK	20
Homodyne	ASK	40
	PSK	10

APPENDIX 9A

The matched-filter

A network whose frequency response function $H(f)$ maximises the ratio of output peak signal to mean noise (power) ratio is called a matched filter. If the bandwidth of the receiver is wide compared with that occupied by the signal energy, extraneous noise is introduced by the excess bandwidth, which lowers the input signal-to-noise ratio (SNR). On the other hand, if the receiver bandwidth is narrower than the bandwidth occupied by the signal, the noise energy is reduced along with a considerable part of the signal energy. The net result is a lower SNR. Thus there is an optimum bandwidth at which the SNR is a maximum.

Derivation of the matched filter characteristic

Let $s(t)$ be the received waveform, the ratio we wish maximise is

$$R_f = \frac{|s_0(t)|^2_{\max}}{N} \tag{A.1}$$

where $|s_0(t)|_{\max}$ is the maximum value of output signal voltage and N is the mean noise power at the receiver output. The output voltage of a filter with frequency response function $H(f)$ is

$$|s_0(t)| = \left| \int_{-\infty}^{\infty} S(f)H(f)\exp(j2\pi ft)df \right| \tag{A.2}$$

where $S(f)$ is the Fourier transform of the input (received) signal. The mean output noise power is

$$N = \frac{N_0}{2} \int_{-\infty}^{\infty} |H(f)|^2 df \tag{A.3}$$

where N_0 is the input noise power per unit bandwidth. The factor $1/2$ appears because N_0 is defined as the noise power per cycle of bandwidth over positive values only. Substituting (9A.2) and (9A.3) into (9A.1), and assuming that the maximum value of $|s_0(t)|^2$ occurs at time $t = t_1$, the ratio R_f becomes

$$R_f = \frac{|\int_{-\infty}^{\infty} S(f)H(f)\exp(j2\text{o}ft_1)df|^2}{\frac{N_0}{2} \int_{-\infty}^{\infty} |H(f)|^2 df}. \tag{A.4}$$

Schwartz's inequality states that if P and Q are two complex functions, then

$$\int P^* P \, dx \int Q^* Q \, dx \geq \left| \int P^* Q \, dx \right|^2 ; \tag{A.5}$$

where the equality sign applies when $P = kQ$, where k is a constant. Letting

$$P^* = S(f)\exp(j2\pi f t_1) \text{ and } Q = H(f) \tag{A.6}$$

and applying the Schwartz inequality to the numerator of (9A.4), we obtain the expression

$$R_f \leq \frac{\int_{-\infty}^{\infty} |H(f)|^2 df \int_{-\infty}^{\infty} |S(f)|^2 df}{\frac{N_0}{2} \int_{-\infty}^{\infty} |H(f)|^2 df} = \frac{\int_{-\infty}^{\infty} |S(f)|^2 df}{\frac{N_0}{2}} \tag{A.7}$$

From Parseval's theorem,

$$\int_{\infty}^{\infty} |S(f)|^2 df = \int_{\infty}^{\infty} s^2(t) dt = E \tag{A.8}$$

where E is the signal energy. Therefore, we have

$$R_f \leq \frac{2E}{N_0}. \tag{A.9}$$

The frequency response function which maximises the ratio R_f may be obtained when $P = kQ$, that is

$$H(f) = G_a S(f)^* \exp(-j2\pi f t_1) \tag{A.10}$$

where the constant k has been set equal to $1/G_a$.

The interesting property of the matched filter is that no matter what the shape of the input signal waveform, the maximum ratio of the peak signal power to the mean noise power is simply twice the energy E contained in the signal divided by the noise power per unit bandwidth N_0.

The output of the matched filter is proportinal to the input signal cross-correlated with a replica of the transmitted signal, except for the time delay t_1. In other words, the matched filter forms the cross-correlation between the received signal corrupted by noise and a replica of the transmitted signal. The replica of the transmitted signal is 'built in' to the matched filter via the frequency response function.

It can easily be seen that the matched filter minimises the probability of making an erroneous decision [9, 10].

Matched filter with non-white noise

In the derivation of the matched filter characteristic, the spectrum of the noise accompanying the signal was assumed to be white, i.e. independent of the frequency. If this assumption were not true, the filter which maximises the output SNR would not be the same as the matched filter in (9A.10). It has been shown [9, 10] that if the input power spectrum of the interfering noise is given by $[N_i(f)]^2$, the frequency response function of the filter which maximises the output SNR is

$$H(f) = \frac{G_a S(f)^* \exp(-j2\pi f t_1)}{[N_i(f)]^2} \tag{A.11}$$

When the noise is non-white, the filter maximising the output SNR is called the NWN (non-white noise) matched filter. It is interesting to rewrite (9A.11) as follows:

$$H(f) = \frac{1}{N_i(f)} \cdot G_a \left(\frac{S(f)}{N_i(f)} \right)^* \exp(-j2\pi f t_1) \tag{A.12}$$

This signifies that the NWN matched filter can be considered as the cascade of two filters. The first, whose frequency response function is $1/N_i(f)$, acts to make the noise spectrum uniform, or white (usually named *whitening filter*). The second is the matched filter described before.

APPENDIX 9B

The optimum linear receiver (equaliser)

A data communication system of finite complexity but not constrained to

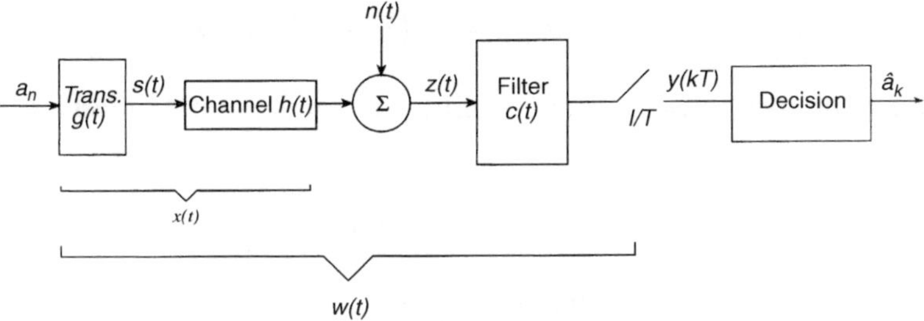

Fig. 9B.1

linear processing can (at least theoretically) realise a maximum likelihood estimate of the transmitted data sequence, i.e. it can produce the sequence most likely to have been transmitted on the basis of an observation of the noisy received signal. Since this is the best we could hope to do, any other structure will be suboptimal.

However, the best reason for using relatively simple linear receiver structures is that they usually offer entirely adequate performance, and can be used at high data rates.

A baseband data transmission system with linear receiver is depicted in Fig. 9B. The transmitted waveform is

$$s(t) = \sum_n a_n g(t - nT) \tag{B.1}$$

where the $\{a_n\}$ are independent data values. The receied waveform, after the signal has passed through a channel $h(t)$, is

$$z(t) = \sum_m a_m x(t - mT) + n(t) \tag{B.2}$$

where $n(t)$ is white Gaussian noise with two-sided power density $N_0/2$. The pulse presented to the receiver is

$$x(t) = g(t) * h(t) \tag{B.3}$$

We wish to determine the receiver filter $c(t)$, such that the sampled receiver output sequence $\{y(kT)\}$, differs as little as possible from the data sequence $\{a_n\}$, and where

$$y(t) = z(t) * c(t), \tag{B.4}$$

Therefore we wish to minimise the men-square error (MSE):

$$MSE = \langle [y(kT) - a_k]^2 \rangle = \langle y^2(kT) - 2a_k y(kT) + a_k^2 \rangle. \tag{B.5}$$

Defining the overall pulse shape from transmitter input to receiver output as

$$w(t) = x(t) * c(t) \tag{B.6}$$

we see that the sampled receiver output is

$$y(kT) = \sum_m a_m w(kT) - mT) + \nu(kT) \tag{B.7}$$

where $v(kT) = n(t) * c(t)$ is the noise component of the receiver output. With $P = \langle a_n^2 \rangle$, the MSE is given by

$$MSE = \left\langle \left[\sum_m a_m w_{k-m} + \nu_k \right]^2 - 2a_k \sum_m a_m w_{k-m} + a_m^2 \right\rangle. \tag{B.8}$$

So the following expression can be obtained:

$$MSE/P = \int\int [A(t,\tau) + \sigma^2 \delta(t - \tau)] c(t) c(\tau) dt d\tau - 2 \int x(-t) c(t) dt + 1 \tag{B.9}$$

where

$$\langle n(t)n(\tau) \rangle = \frac{N_0}{2} \delta(t - \tau)$$

$$\sigma^2 = N_0/P \tag{B.10}$$

$$A(t,\tau) = \sum_m x(mT - t)x(mT - \tau)$$

The minimum MSE with respect to the function $c(t)$ is found by equating the first variation of MSE (with respect to c) to zero. Doing that we obtain

$$\int [A(t - \tau) + \sigma^2 \delta(t - \tau)] c(\tau) d\tau = x(-t) \tag{B.11}$$

From (9B.6), the sample of the equalised pulse at time nT is

$$w_n = \int x(nT - t)c(t)dt \tag{B.12}$$

so that the left-hand side of (9B.11) can be rewritten as

$$\int \left[\sum_m x(nT - t)x(nT - \tau) + \sigma^2\delta(t - \tau) \right] c(\tau)d\tau = \sum_m w_n x(nT - t) + \sigma^2 c(t)$$

(B.13)

Hence (9B.11) becomes

$$\sum_m w_n x(nT - t) + \sigma^2 c(t) = x(-t)$$

(B.14)

Solving for $c(t)$,

$$c(t) = \frac{x(-t)}{\sigma^2} - \sum_n \frac{w_n}{\sigma^2} x(nT - t) = \sum_n c_n x(nT - t)$$

(B.15)

where the coefficient c_n are defined as

$$c_0 = (1 - w_0)/\sigma^2$$
$$c_n = -w_n/\sigma^2, n \neq 0$$

(B.16)

Recall that a matched filter, having impulse response $x(t)$, is the front end of a receiver designed for optimum detection of the signal $\sum_n a_n x(t - nT)$ in additive Gaussian noise, and taking the Fourier transform of (9B.15), we obtain

$$C(f) = \sum_n c_n e^{-j2\pi fnT} X^*(f).$$

(B.17)

Note that (9B.17) represents a cascade of the matched filter, $x(-t)$, and a tapped delay line. In other words, the optimum filter $c(t)$, of (9B.15) is the weighted sum of outputs of the matched filter delayed by different amounts nT. The structure that produces a weighted sum of time-delayed versions of a signal is a transversal filter, and is pictured along with the matched filter that precedes it. Figure 9B.2 shows the structure of the optimum linear receiver. Note that, since the equaliser tap spacing is the same as the sampling rate, the sampler could be moved to the input of the equaliser (which is shown as an analog filter), and the resulting arrangement, would have an analog-to-digital converter (A/D) followed by a digitally implemented equaliser. The transversal filter alone is often called a synchronous equaliser.

For the sake of brevity, we do not continue to explain the optimum linear receiver further details can be found in communications books [9,

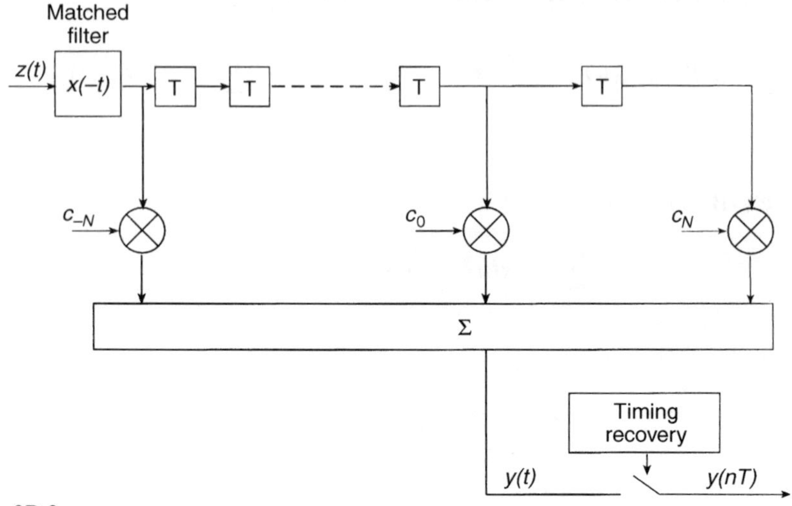

Fig. 9B.2

10]. Nevertheless, we report the expression which can be used to evaluate the optimum tap weights, i.e. the coefficient c_n of (9B.15), without reporting the intermediate mathematical steps. In fact, the Fourier transform of the set of tap weights is

$$C_T(f) = \frac{1}{1/T \sum_n |X(f + n/T)|^2 + \sigma^2} \qquad (B.18)$$

The tap weights can be obtained numerically by inverse Fourier transforms of (9B.18). Note how this equation is derived on the assumption that the transversal filter of Fig. 9B.2 is preceded by a filter exactly matched to the preceding channel $x(t)$.

The overall optimum receiver transfer function (matched filter plus transversal filter), as specified by (9B.17), can be expressed as

$$C(f) = \frac{X^*(f)}{1/T \sum_n |X(f + n/T)|^2 + \sigma^2} . \qquad (B.19)$$

Notice that the receiver strikes a balance between inverting the folded magnitude-squared channel response, $\sum_n |X(f + n/T)|^2$, over those frequency regions where the noise is small, i.e.

$$\sigma^2 << 1/T \sum_n |X(f + n/T)|^2$$

and matching the channel over the frequency regions where the noise is large

$$\sigma^2 >> 1/T \sum_n |X(f + n/T)|^2.$$

If the receiver simply inverted the channel, the receiver would not properly enhance the noise over those frequency regions where the noise power greatly exceeded the signal power; for a flat noise noise spectrum, this noise enhancement would occur over frequency regions where the channel had a relative null. Thus, the optimum linear receiver strakes a balance between noise enhancement and channel inversion.

A very important property of the optimum linear receiver is the lack of dependence of the receiver on the phase characteristics of the channel, including the timing phase; not surprisingly, the overall mean square error will also be independent of these characteristics. Combining the above equations, we find that the overall channel, from transmitter input to receiver output, has the transfer function

$$W(f) = X(f)C(f) = \frac{|X(f)|^2}{1/T \sum_n |X(f + n/T)|^2 + \sigma^2} \qquad (B.20)$$

Note that the overall transfer function, $W(f)$, is independent of the phase characteristics of the channel. There are two points to be highlighted:

- Since the sampling phase of the receiver using a fractionally spaced equaliser may be absorbed as part of the channel, the performance of a system that implements the optimum linear receiver will be independent of the sampling epoch (timing phase).
- In general, the performance of a system with an optimum linear receiver will be independent of the channel phase characteristics. Incorporation of the conjugate channel phase characteristics in the optimum linear receiver ensures this independence, and since the noise power cannot be affected by pure phase compensation, compensation for the channel phase characteristics is achieved without enhancing the noise.

REFERENCES

1. A. Willner, Mining the optical bandwidth for a terabit per second, *IEEE Spectrum* **34**, (1997).
2. P.S. Henry, R.A. Linke, and A.H. Gnauck, Introduction to Lightwave Systems, in *Optical Fiber Telecommunications II*, Eds. S.E. Miller and I.P. Kaminow, Academic Press, San Diego CA (1988).

3. Technical staff of CSELT, *Optical Fibre Communication*, CSELT Turin (1980).
4. P.R. Trischitta and E.L. Virima, *The Effect of Jitter on Transmission Quality*, Ch. 4, Artech House, Norwood MA (1988).
5. A. Papoulis, *Probability, Random Variables, and Stochastic Processes*, McGraw-Hill, New York (1965).
6. A. Yariv, *Optical Electronics*, Saunders, Philadelphia PA (1985).
7. J.R. Barry and E.A. Lee, Performance of coherent optical receivers, *Proc. IEEE* **78**, 1369–1393 (1990).
8. T. Okoshi, and K.Kikuchi, *Coherent Optical Fiber Communications*, Kluwer Academic, Tokyo (1988).
9. J.G. Proakis, *Digital Communications*, 3rd edn, McGraw Hill., New York (1995).
10. M. Schwartz, W. R. Bennet, and S. Stein, *Communication Systems and Techniques*, IEEE Press, New York (1996).

10

High-Speed Fiber-Optic Transmissions

10.1 INTRODUCTION

Due to the strong increase in telecommunication traffic, 2.5 Gb/s systems (STM-16, OC-48)[1] are available today as interoffice and long-haul links by network operators and interexchange carriers. At the next level in the synchronous digital hierarchy, 10 Gb/s, field trials were begun a few years ago and are still continuing. Hence 10 Gb/s (STM-64 or OC-192) systems are going to be available as commercial products. Since it has been demonstrated that digital electronic signal processing, such as multiplexing and demultiplexing up to 40 Gb/s, can be accomplished with silicon bipolar technology [1], we will consider here electrical timedivision multiplexing (TDM) up to 10 Gb/s. Any further increases in transmission capacity increasing can be achieved through the use of optical time division multiplexing (OTDM), or by wavelength division multiplexing (WDM), as described in the next chapter. The evolution of fiber-optic transmission systems with time is shown in Fig. 10.1.

For the link span, the advent of optical amplifiers has allowed an increase in repeaters spacing, or even their elimination, besides the relaxation of optical power budget restrictions, as illustrated in Fig. 10.1. However, both fiber dispersion and non-linearities limit the transmission span in high bit rate transmission systems. Techniques for overcoming such impairments have to be taken into account. The ultimate vision is the realisation of an optical network, as described in Chapter 12.

10.2 LIMITATIONS FOR MULTIGIGABIT TRANSMISSION

The main physical limitations for high-speed transmission systems result from the transport properties of optical fibers. Conventional single-mode fibers (SMFs) exibits zero dispersion around 1310 nm, i.e. in the center of the second optical window. In the past, nearly all long-haul high-speed lightwave systems were supplied with SMFs. However, due to the rapid progress of EDFAs which operate only in the 1530–1560 nm wavelength band, the third optical transmission window has gained enourmus importance. This technology, which is now commercially available, yields a prac-

---

[1] STM refers to the European SDH (synchronous digital hierarchy) and OC refers to the American SONET hierarchy.

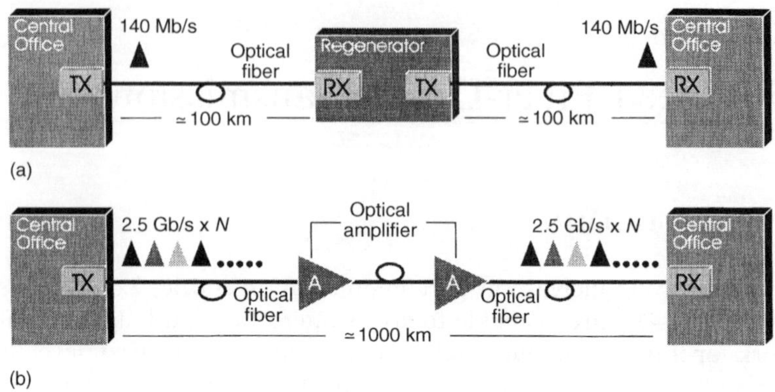

Fig. 10.1 Evolution of optical fiber transmission systems: (a) repeater transmission system; (b) optically amplified transmission systems with no repeaters.

tically unlimited bit rate-span length product, as shown in many transmission experiments.

Doped-fiber amplifiers for the 1310 nm window amplifiers (e.g. PDFA: praseodimium-doped fiber amplifiers) are still under research, and may not achieve the high performance characteristics of EDFAs. Semiconductor optical amplifiers, as mentioned in Chapter 5, are potentially available for both optical windows, but they are not considered here as options for high-speed transmission applications in the near term, since they exhibit unwanted properties such as polarisation sensitivity, gain ripples and relatively high fiber-coupling loss. Non-linear optical amplifiers [2], such as the fiber-Raman amplifier or the fiber-Brillouin amplifier are principally applicable, but they require high pump power and/or pump wavelength control [1].

As soon as attentuation limitations are eliminated through optical amplification, other fiber properties become important obstacles. This is shown in Table 10.1, where the dispersion and attentuation limits for high bit rate systems (10 ÷ 40 Gb/s) operating in the second and third windows are compared. The chromatic dispersion limits have been calculated for IM-DD systems [3] with a bandwidth-limited, chirp-free optical signal source. This type of transmitter may consist of a narrow CW source then an external optical modulator. For the attenuation and dispersion values, typical tolerances have been used.

The performance comparison shown in Table 10.1 demonstrates that the main link length limitation for 10 Gb/s transmission results from the fiber link losses (including splice/connector losses and system margins), even if the high dispersion of SMFs is taken into account. Therefore, only transmission in the 1550 nm window, where EDFAs are avaliable, is a feasible

Table 10.1   Link length limitations of high-speed systems.

	Attenuation limits (km)			Dispersion limits (km)	
Bit rate (Gb/s)	1550 nm (SMF)	1550 nm (DSF)	1300 nm (SMF)	1550nm	1300nm
10	58	283	400	47	31
20	14.5	70	100	37	34
40	3.6	18	25	27	18

Dispersion limitations are evaluated for 1 dB eye closure penalty. SMF = single-mode fiber (the dispersion at 1550 nm is 17 psec/km/nm, and at 1300 nm it is 3.5 psec/km/nm). Attenuation limitations are calculated with the following assumption: optical transmitter power – 3 dBm; receiver sensitivities –20 dBm at 10 Gb/s, –17 dBm at 20 Gb/s, –14 dBm at 40 Gb/s; fiber attenuation (1550 nm) 0.3 dB; and system margin 3 dB.

way for approaching the dispersion limit. However, especially in SMF-based systems, when operating in the third window at bit rates beyond 10 Gb/s, fiber dispersion will soon become the dominant limitation factor, and methods for overcoming the dispersion limitations will be of key importance for reaching the highest system performance, including the dispersion-shifted fibers (DSFs).

Polarisation mode dispersion (PMD) [4] is not considered to be a major concern for the high-speed system applications discussed here, assuming an extremely low PMD existing in conventional single-mode fibers ($0.1\mathrm{ps}/\sqrt{km}$) and a maximum targeted regenerator spacing of 160 km.

Furthermore, the availability of optical booster amplifiers, allows for a large increase of signal power transmitted over long SMFs. These power levels could give rise to a number of detrimental non-linear effects.

10.3 IMPACT OF SOURCE CHIRPING AND FIBER DISPERSION

The direct current modulation of single-longitudinal mode semiconductor lasers causes a dynamic variation of the peak emission wavelength, especially at high bit rates. The intermingling of such linewidth broadening (chirping) with the chromatic dispersion characteristics of single-mode optical fibers causes signal distortion and eventually generates intersymbol interference with consequent transmission penalties. As a result, the bit rate span product is reduced. This effect is described in Chapter 1. It is clear that, in order to achieve multiple gigabit transmission over large distances, the use of low-chirp light sources is necessary. An external intensity modulation of the laser source is the right choice in many cases, because it allows the laser to operate in CW, while the signal is applied as

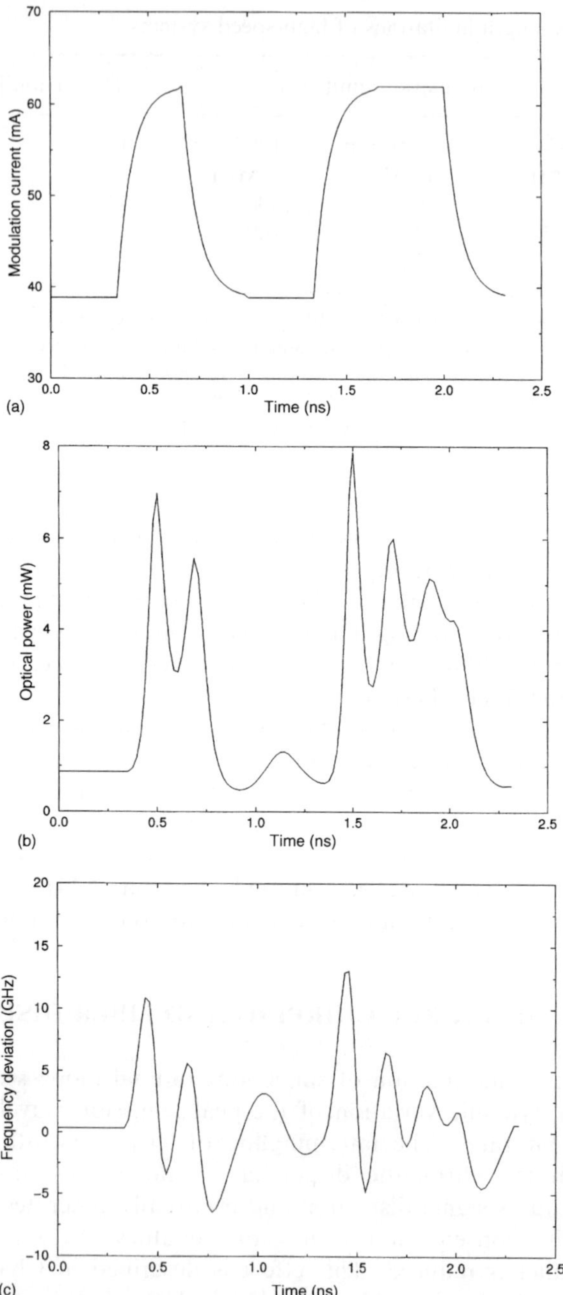

Fig. 10.2 Dynamic behaviour of a directly modulated DFB laser source: (a) electrical signal, (b) optical signal at the transmitter output, (c) instantaneous frequency shift.

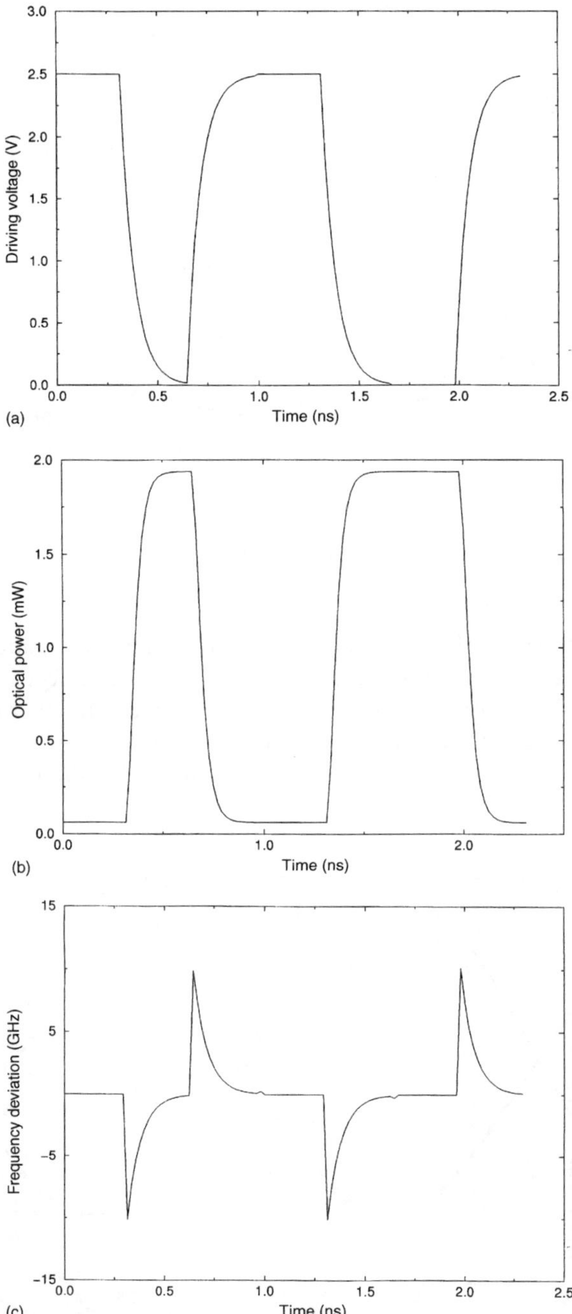

Fig. 10.3 Dynamic behaviour of an externally modulated laser source: (a) electrical signal, (b) optical signal at the transmitter output, (c) instantaneous frequency shift.

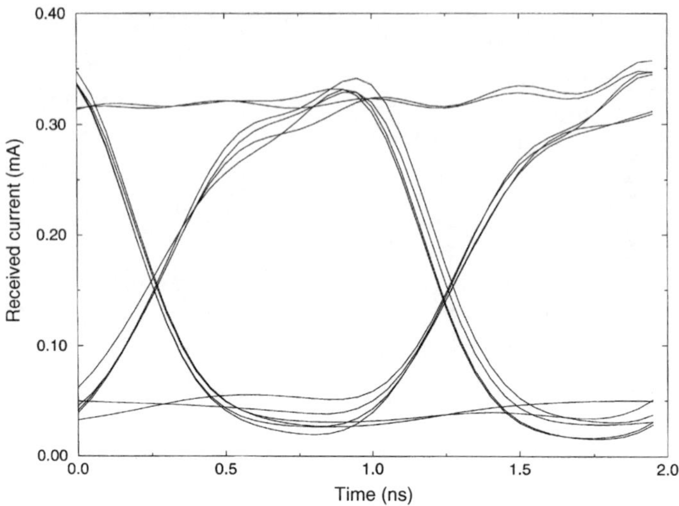

Fig. 10.4 Simulated eye diagrams, after 150 km of standard fiber propagation and at 2.5 Gb/s, related to a direct modulation of the laser source.

a driving voltage of the modulator. Unfortunately, a residual chirping effect is still present, but this is much smaller than any effect related to the intensity modulation of the laser. As a matter of fact, the bit-rate and span product is increased when using external modulators.

These effects are shown in Figs. 10.2 and 10.3, where the electrical

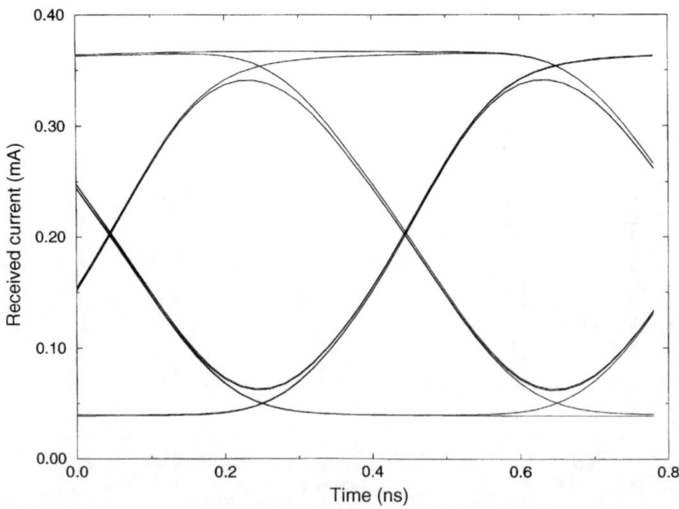

Fig. 10.5 Simulated eye diagrams, after 150 km of standard fiber propagation and at 2.5 Gb/s, related to an external modulation of the laser source.

signal, the optical signal at the transmitting source output, and the instantaneous frequency shift are reported for a directly modulated DFB laser source (Fig. 10.2) and an externally modulated laser source (Fig. 10.3), at a bit rate of 2.5 Gb/s, for devices whose characteristics are reported in the literature [5, 6].

Note that, when the laser is externally modulated, the chirp is perceivable only during transitions between mark and space, and vice versa. The impact of chirping and dispersion can be directly noticed through the eye diagrams. Figures 10.4 and 10.5 shows the eye diagrams, after 150 km of standard fiber propagation and at 2.5 Gb/s, respectively related to direct and external modulation of the laser source. Although the eye closure is quite evident when the laser is directly modulated, when using external modulation the eye still remains sufficiently open.

A quantitative assessment is the eye penalty, defined in Chapter 9 as the vertical opening of the eye divided by the opening obtained from the eye diagram relating to the transmitting signal, in decibel. Tthe cyc penalty, relating to both direct and external modulation, is shown in Fig. 10.6 as a function of the fiber length (i.e. the dispersion).

As a result, external modulation allows the transmission performance to be improved or, from another point of view, the bit rate-span product to be increase. However, when a transmission over longer distances or higher bit rates is required, dispersion compensation is needed; this is treated in section 10.7.

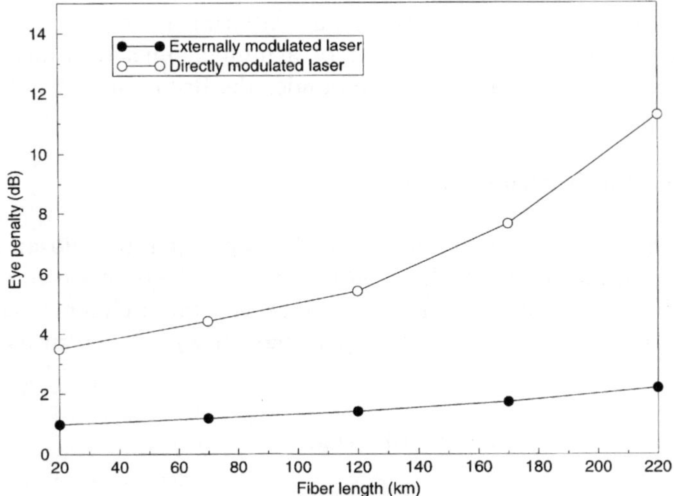

Fig. 10.6 Eye penalty versus the fiber length for both direct and external modulation.

10.4 IMPACT OF FIBER PROPAGATION NON-LINEARITIES

Besides the chirping/dispersion limitations, other factors impair the transmission performance, due to non-linear behavior of fiber propagation in particular conditions.

Stimulated Brilluoin scattering (SBS) relates to an interaction between light and sound waves in the fiber. It causes frequency conversion and reversal of the propagation direction of light.

Stimulated Raman scattering (SRC) relates to an interaction between light and vibrations of silica molecules. It causes frequency conversion of light and results in excess attenuation of short-wavelength channels in wavelength-multiplexed systems.

The Kerr effect relates to the dependence of the fiber refractive index on the propagating electric field intensity.

The impacts of these effects on transmission are briefly discussed below.[2]

10.4.1 Impact of Brillouin scattering

This effect can be suppressed in most transmission systems, since its threshold power depends on the source bandwidth [7]. In fact, broadening the laser linewidth, the threshold increases by a factor $\Delta\nu_s/\Delta\nu_B$, where $\Delta\nu$ and $\Delta\nu_B$ are the laser linewidth and the Brillouin bandwidth (typically $\Delta\nu_B = 20\ MHz$). Since the spectral bandwidth increases with the transmission speed, the increase in broadens the the bit rate induces a broadening of the curve of the Brillouin gain, which reduces the peak.

It is worth noting that the Brillouin scattering is independent of the number of channels and, in the case of a multichannel transmission, the power of each channel has to be taken under the Brillouin threshold.

10.4.2 Impact of Raman scattering

The Raman effect is characterised by a sharp power threshold and a very large gain bandwidth ($> 5\ THz$) with low gain. This characteristic can severely affect the transmission performance of a multi-channel system. In particular, Raman threshold power per channel P_{Ram} is given as follows [8]:

$$P_{Ram} = 9 \cdot 10^{15} \alpha L_{amp}/(N_c B_{opt} L_t) \tag{10.1}$$

---

[2] These effect have already been introduced in Chapter 4.

where α is the fiber attenuation (m^{-1}), N_c is the number of channels, L_t is the link length, L_{amp} is the amplifiers spacing, and B_{opt} is the total optical bandwidth. If the peak channel power is about 3 dB below the Raman threshold, this effect can be neglected; if it is above threshold, correct transmission is practically impossible. The Raman effect can therefore be avoided by limiting the peak optical power along the link 3 dB below the threshold.

10.4.3 Impact of the Kerr effect

The origin of the Kerr effect is the dependence of the fiber refractive index on the optical intensity:

$$n(E) = n_0 + n_2|E|^2, \tag{10.2}$$

where n_0 is the linear index and $n_2 = 3\chi^{(3)}/(8n_0)$ represents the non-linear index. From equation (10.2) if a monochromatic signal, at wavelength λ_0 propagates under the influence of the Kerr effect, its phase ϕ varies along the propagation direction z, according to the following relation:

$$\phi(z) = \frac{2\pi n_0 z}{\lambda_0} + \frac{2\pi n_2|E|^2 z}{\lambda_0}. \tag{10.3}$$

This effect can manifest itself in different ways.

It is very important when the intensity-modulated signal power is such that it produces a phase modulation (chirp) on itself. This effect is named *self phase modulation* (SPM). In the absence of chromatic dispersion, SPM induces a broadening of a signal pulse, in the spectral domain and in the time domain. This effect, in the presence of chromatic dispersion, can be enhanced or diminished, depending on the sign of the dispersion coefficient. In the normal dispersion zone, SPM and dispersion reinforce each other, causing a broadening in the time domain; in the anomalous region, the two effects tend to oppose each other. As a matter of fact, it is possible to find conditions in which the two effects cancel out, allowing the pulse to propagate without distortion (Chapter 13).

In multichannel transmission at different frequencies, the power variation of a signal can induces a phase modulation on the othr siganls. This effect, named *cross-phase modulation* (CPM), can be understood by observing that during propagation the refractive index 'seen' by a signal also depends on the power of the other channels.

Another important effect relating to multichannel transmission *four-wave mixing* (FWM) arises from the that beating of different channels at different frequencies. This effect produces crosstalk that is particularly importan for some types of wavelength-multiplexed systems. The evaluation of thiese impairments is treated in the next chapter.

10.5 EDFA TECHNOLOGY FOR OPTICAL TRANSMISSION

The considerable progress of EDFA technology has led to innovative developments and has opened up a new era in the design of multigigabit optical transmission systems, as depicted in Fig 10.1.

Important features of EDFAs which make them such a powerful tool for the optimisation and cost reduction of long-haul links are

- optical wideband gain (typically 1530–1560 nm),
- high small-signal gain,
- high output power,
- low noise Figure (3–6 dB) allowing cascadability and preamplifier applications,
- polarisation-independent gain.

EDFAs can be placed within fiber-optic systems, in three possible configuration: booster and preamplifier.

Inline amplifier

In this configuration (Fig. 10.1) the EDFA is used for compensating all types of link losses (due to fiber, splice/connectors, etc.) in order to increase the system margins or overall transmission lengths which have to be below the dispersion limit. The target is to realise an almost bit rate transparent light channel between regenerator sites. This technique allows the minimisation, or even the elimination of electrical regenerators, which can only deal with one fixed bit rate. However, if narrowband filters are inserted in the amplified link to prevent noise accumulation, the transparency of the light channel may no longer be ensured. Therefore, optimisation of the combination of simply amplifying and fully-regenerating sites is of crucial importance today.

Booster amplifier

In this configurationto increase the EDFA is used the power level of the optical transmitter. Power levels of more than 22 dBm [9] can be realised in order to enlarge the power budget or span length of unrepeatered systems. The main limitations for this postamplifier option arise from the critical power levels of non-linear optical effects in single-mode fibers.

Preamplifier

In this arrangement the EDFA is used to increase the received signal power before detection, in order to improve the sensitivity of direct detection receivers. It has been shown theoretically [10] that an optical preamplifier with an appropriate postfilter for amplified spontaneous emission noise (ASE) suppression can enhance the sensitivity of a standard direct detection receiver front end (PIN) approaching the quantum noise limit (e.g. 40 photons per bit average power) achievable with an ideal coherent DPSK receiver. As an example, a receiver sensitivity of –27.4 dBm was achieved at 10 Gb/s using an APD/GaAs IC receiver [11]. In another experiment at 10 Gb/s, using an optical preamplifier, the sensitivity of a p-i-n diode receiver with preamplifier was increased from –12.5 dBm to –38.4 dBm, corresponding to 112 photons per bit [12]. In this EDFA, a Fabry-Perot filter with a 20 GHz passband was included in order to reduce the noise. This nearly quantum noise limited detection can only be achieved with a filter bandwidth equal to the modulation bandwidth (i.e. around 0.1 nm optical filter bandwidth for a 10 Gb/s system), which requires narrow optical filters (like Fabry-Perot filters) with integrated wavelength control and stabilisation.

The application of bit rate transparent optical amplifiers drastically reduces the requirements on the electronic circuitry for the communication equipment. As an example, in 10–40 Gb/s receiver electronics, the demand on an analog broadband electronic amplifier and signal processing can be relaxed considerably if an optical preamplifier is used [13]. This could allow for the realisation of moderate-cost multigigabit subsytems. Furthermore, optical amplifiers avoid the introduction of OTDM techniques [14]. And all-optical demultiplexing of 64 Gb/s signals using a non-linear loop mirror (NOLM) has been reported [15].

10.5.1 Impact of EDFAs on transmission performance

The characteristics and modeling of EDFAs are discussed in chapter 5. However, in many practical cases it can be assumed that the EDFA has unity coupling efficiency, uniform gain G over an optical bandwidth B_o, and that only the noise, it introduces affects the transmission performance. Ithis section looks at the impact of EDFA noise on IM-DD system performance.

Let's assume that the input power P_{in}, at the optical frequency ω_0 centered on the optical passband B_o, enter the amplifier. Since the spontaneous emission power density can be expressed as in (5.47), the electric field E_{sp}, representing the spontaneous emission, can be expressed as the sum of 2M cosine terms:

$$E_{sp}(t) = \sum_{k=(-B_0/2\delta\nu)}^{B_0/2\delta\nu} \sqrt{2N_{sp}(G-1)hf\delta\nu}\,\cos[(\omega_0 + 2\pi k\delta\nu)t + \Phi_k] \qquad (10.4)$$

where ϕ_k is a random phase for each component of spontaneous emission, and $M = B_o/2\delta\nu$. Assuming that

$$N_0 = N_{sp}(G-1)hf, \qquad (10.5)$$

the total electric field at the output of the amplifier is

$$E(t) = \sqrt{2GP_{in}}\,\cos(\omega_0 t) + \sum_{k=-M}^{M} \sqrt{2N_0\delta\nu}\,\cos[(\omega_0 + 2\pi k\delta\nu)t + \Phi_k] \quad (10.6)$$

The photodiode generates a photocurrent which is proportional to the number of photons; that is, regarding the light as an electromagnetic radiation, proportional to the square of the electric field:

$$i(t) = \overline{E^2(t)}\,\frac{e}{hf} \qquad (10.7)$$

where the bar indicates time averaging over optical frequencies. Therefore

$$\begin{aligned} i(t) = {}& GP_{in}\frac{e}{hf} + \frac{4e}{hf}\sum_{k=-M}^{M} \sqrt{GP_{in}N_0\delta\nu}\,\cos(\omega_0 t)\cos[(\omega_0 + 2\pi k\delta\nu)t + \Phi_k] \\ & + \frac{2eN_0\delta\nu}{hf}\left[\sum_{k=-M}^{M}\cos((\omega_0 + 2\pi k\delta\nu)t + \Phi_k)\right]^2 \end{aligned}$$

$$(10.8)$$

The three terms in (10.8) represent signal, signal-spontaneous beat noise, and spontaneous-spontaneous beat noise, respectively.

Signal-spontaneous beat noise

The first term, representing the beating between the signal and the spontaneous noise contributions, can be rewritten as

$$i_{s-sp}(t) = \frac{4e}{hf}\sqrt{GP_{in}N_0\delta\nu}\sum_{k=-M}^{M}\cos((\omega_0 + 2\pi k\delta\nu)t + \Phi_k) \qquad (10.9)$$

where terms with which average to zero have been neglected. For each frequency $2\pi k\delta\nu$ in (10.8) the sum has two components but with a random phase. Therefore, the power spectrum of $i_{s\text{-}sp}(t)$ is uniform in the frequency interval from 0 to $B_o/2$ and has a density of

$$N_{s-sp} = \frac{4e^2}{(hf)^2}GP_{in}N_0 \cdot \frac{1}{2} \cdot 2 = \frac{4e^2}{hf}P_{in}N_{sp}(G-1)G \qquad (10.10)$$

Spontaneous-spontaneous beat noise

This noise contribution, arising from the beating of spontaneous noise components with themselves, is represented by the third term of (10.8), which can be rewritten as

$$i_{sp-sp}(t) = \frac{2eN_0\delta\nu}{hf}\left[\sum_{k=-M}^{M}\cos(\beta_k)\sum_{j=-M}^{M}\cos(\beta_j)\right], \qquad (10.11)$$

where

$$\beta_k = (\omega_0 + 2\pi k\delta\nu)t + \Phi_k \quad \text{and} \quad \beta_j = (\omega_0 + 2\pi j\delta\nu)t + \Phi_j. \qquad (10.12)$$

(10.11) can be written as

$$i_{sp--sp}(t) = \frac{2eN_0\delta\nu}{hf}\left[\sum_{k=-M}^{M}\sum_{j=-M}^{M}\frac{1}{2}\cos(\beta_k - \beta_j) + \frac{1}{c}\cos(\beta_k + \beta_j)\right], \quad (10.13)$$

However, the terms with $\cos(\beta_k + \beta_j)$ have frequencies $2\omega_0$ and averaged to zero. Therefore the next expression becomes

$$i_{sp-sp}(t) = \frac{2eN_0\delta\nu}{hf}\left[\sum_{k=0}^{2M}\sum_{j=0}^{2M}\cos((k-j)2\pi\delta\nu t + \Phi_k + \Phi_j)\right]. \qquad (10.14)$$

The DC term is obtained for $k = j$, and there are $2M$ of these terms:

$$I_{sp}^{dc} = \frac{eN_0\delta\nu 2M}{hf} = N_{sp}(G-1)eB_o \qquad (10.15)$$

By arranging the terms in (10.14) according to their frequencies, and considering that terms with the same absolute frequency but of opposite sign add in phase, we obtain that the power spectrum extends from 0 to B_o with a triangular shape and a power density near DC of

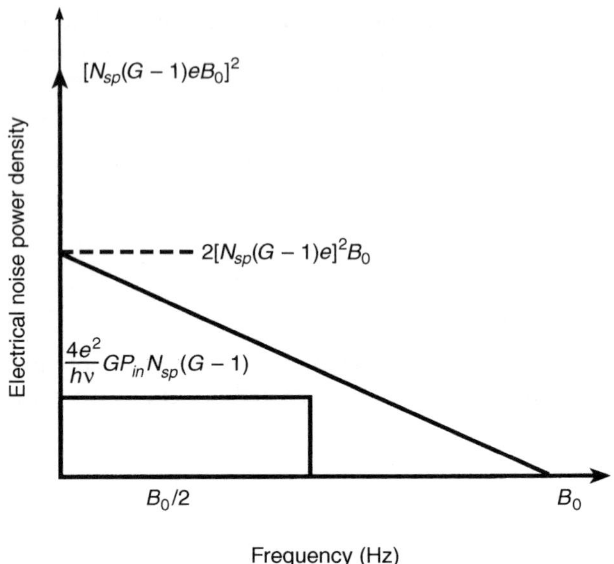

Fig. 10.7 Electrical noise power spectra of the different beating terms: spontaneous-spontaneous (triangular) beat noise, signal-spontaneous (rectangular) beat noise, and DC spontaneous emission power.

$$N_{sp-sp} = \frac{4e^2 N_0^2 \delta\nu 2M}{hf^2}\left(\frac{B_o}{\delta\nu} - 1\right)\cdot\frac{1}{2} = 2N_{sp}^2(G-1)^2 e^2 B_o. \tag{10.16}$$

To show the noise characteristics of the received signal (i.e. after photodetection), Fig. 10.7 illustrates the electrical noise power spectrum of the different beating terms.

10.5.2 Performance analysis of systems employing optical amplifiers

It is common to evaluate the BER transmission performance of a system employing optical amplifiers simply by adopting the Gaussian approximation for the noise added to the received signal. However, this is not correct, because the ASE noise can be treated as Gaussian additive noise during the fiber propagation but not after the photodetection process. Consequently, to evaluate the BER of a system affected by ASE noise, it is necessary to alter the analysis. This analysis [16] is briefly reported below for a direct detection receiver[3], and makes use of the characteristic function method [17].

---

[3] The case of a coherent receiver is treated in the literature [18].

The signal transmitted by each local transmitter can be formulated as

$$\mathbf{E}_I(t) = A_t m(t) e^{i(\omega_t - \phi)} \boldsymbol{\xi}. \tag{10.17}$$

The amplitude, phase and polarisation vector of the signal are indicated by A_t, ϕ and $\boldsymbol{\xi}$, respectively, and the transmitted message is indicated by $m(t)$. It is assumed that, neglecting the finite rise-time of the modulator and the non-ideal shape of the electrical modulating function, $m(t)$ can be written as

$$m(t) = \sum_{j=-\infty}^{\infty} b(j) rect_T(t - jT), \tag{10.18}$$

where T is the bit interval, $b(j) = 0,1$ is the bit transmitted in the jth bit interval and $rect_T(t)$ is equal to one for $0 = t = T$ and zero elsewhere.

The ASE noise introduced by EDFAs (and by other amplifiers if present in the link) is represented by the variable $\boldsymbol{\eta}(t)$. The ASE noise term can be decomposed in its quadratures as

$$\boldsymbol{\eta}(t) = [\eta_{Ix}(t) + i\eta_{Qx}(t)]\mathbf{x} + [\eta_{Iy}(t) + i\eta_{Qy}(t)]\mathbf{y}. \tag{10.19}$$

The four random processes $\eta_{Ix}(t)$, $\eta_{Qx}(t)$, $\eta_{Iy}(t)$, $\eta_{Qy}(t)$ can be modeled as white Gaussian noises. The power spectral density of each quadrature of $\rho_\eta(t)$ is given by ahalf the power spectral density of the ASE noise reaching the receiver, because the ASE power is equally divided between the two polarisation modes.

After optical filtering and detection by a p-i-n photodiode, the photocurrent $I_{pc}(t)$ can be written as follows:

$$I_{pc}(t) = R_p P_0 m(t) + N_{ng}(t) = x_1^2(t) + x_2^2(t) + x_3^2(t) + x_4^2(t) \tag{10.20}$$

where R_p is the photodiode responsivity, P_0 is the received optical power, and $N_{ng}(t)$ represents a *non-Gaussian* noise term. From the last equation, the photocurrent can be also written as the sum of the squares of four random processes $x_j^2(t)$ ($j = 1, 2, 3, 4$), whose expressions are

$$\begin{aligned}
x_1^2(t) &= \sqrt{R_p P_0 m(t)}\,\cos(\theta) + \sqrt{R_p}\eta_{Ix}(t) \\
x_2^2(t) &= \sqrt{R_p P_0 m(t)}\,\cos(\theta) + \sqrt{R_p}\eta_{Qx}(t) \\
x_3^2(t) &= \sqrt{R_p P_0 m(t)}\,\sin(\theta) + \sqrt{R_p}\eta_{Iy}(t) \\
x_4^2(t) &= \sqrt{R_p P_0 m(t)}\,\sin(\theta) + \sqrt{R_p}\eta_{Qy}(t)
\end{aligned} \tag{10.21}$$

If after detection the photocurrent is processed by an integrate and dump circuit to obtain the decision variable I_d, can be expressed as

$$I_d = \frac{1}{T}\int_0^T [x_1^2(t) + x_2^2(t) + x_3^2(t) + x_4^2(t)]dt \qquad (10.22)$$

where the time origin is chosen such that $t = 0$ is the beginning of the considered bit interval, and T is the bit interval.

The photocurrent $I_{pc}(t)$ can be considered a band-limited process with a bandwidth equal to the optical channel bandwidth. Under this hypothesis, $I_{pc}(t)$ can be expressed as a function of its samples at the sampling instants t_j. The sampling period can coincide, in this case, with the integration time T_0 of the optical filter in front of the receiver, which can be expressed, in general, as $T_0 = 2/B_o$, where B_o is the optical bandwidth. If the expression of $I_{pc}(t)$ is substituted in (10.19), the following expression for the decision variable can be obtained:

$$I_d = \frac{1}{M}\sum_{j=1}^{M}[x_1^2(t) + x_2^2(t) + x_3^2(t) + x_4^2(t)] = \sum_{j=1}^{M}\Xi_j, \qquad (10.23)$$

where M is the number of samples in the bit interval.

The variables $x_k^2(t_j)$ (with $k = 1, 2, 3, 4$) have Gaussian probability density, so Ξ_j is a Hermitian quadratic form of independent Gaussian variables. The characteristic function (bilateral Laplace transform of the probability density function) forHermitian quadratic form of independent Gaussian variables can be evaluated in closed form [19]. It can be expressed as

$$G_j(\xi, b) = \frac{\exp\left(-(S/N)_0\dfrac{b\xi}{1+\xi}\right)}{(1+\xi)^2}, \qquad (10.24)$$

where $(S/N)_o$ is the optical signal-to-noise ratio taking into account optical noise in both polarisation modes, and it is given by

$$(S/N)_o = \frac{P_o}{\dfrac{\pi}{2}B_o 2 S_\eta(b)}, \qquad (10.25)$$

bering $\pi B_o/2$ is the overall noise bandwidth of the optical filter in front of the receiver, and $S_\eta(b)$ is the power spectral density of each quadrature.

Since the photocurrent is a band-limited signal, the variables ξ_j corresponding to different values of j are statistically independent. Henceforth, the characteristic function of I_d can be simply evaluated by rising $G_j(\xi,b)$ to the Mth power. That is

$$G_d(\xi,b) = \frac{\exp\left(-(S/N)_o\dfrac{Mb\xi}{1+\xi}\right)}{(1+\xi)^{2M}}. \tag{10.26}$$

From the characteristic function of the decision variable, the error probability conditioned to the transmitted bit can be evaluated by means of the Cauchy formula. In fact, assuming equiprobable emission of mark and space ($b = 0$ and $b = 1$, respectively), the error probability can be written as

$$\begin{aligned}
P_e &= \frac{1}{2}P_e(0) + \frac{1}{2}P_e(1) = \frac{1}{2}\int_{c-j\infty}^{c+j\infty} \frac{G_d(\xi/b=0)}{\xi} \cdot e^{-\xi C_{t0}}\,d\xi \\
&\quad + \frac{1}{2}\int_{c-i\infty}^{c+i\infty} \frac{G_d(\xi/b=1)}{\xi} \cdot e^{\xi C_{t1}}\,d\xi
\end{aligned} \tag{10.27}$$

where $P(0)$ and $P(1)$ are the error probabilities conditioned to the transmission of a space and a mark, respectively c is a positive real number within the extremes of the convergence strip of $G_d(\xi,b)$ and C_{t0}, C_{t1} are the normalised thresholds. They are defined as functions of the threshold current I_{th} of the threshold device at the receiver as

$$C_{t0} = \frac{I_{th}/R_p}{2\pi RS_\eta(0)}; \quad C_{t1} = \frac{I_{th}/R_p}{2\pi RS_\eta(1)} \tag{10.28}$$

where R is the bit rate and $S_\eta(0)$, $S_\eta(1)$ are the power spectral densities related to the space and mark, respectively.

The integral involved in the evaluation of the error probability conditioned to the transmission of a space $P_e(0)$ can be evaluated in closed form by using the residues theorem [19]; this gives

$$P_e(0) = e^{-(C_{t0}/2)} \sum_{j=0}^{M=1} \frac{C_{t0}^j}{2^j j!}. \tag{10.29}$$

On the other hand, the evaluation of $P_e(1)$ is much less straightforward; the Cauchy integral involving $G_d(\xi,b = 1)$ cannot be solved exactly. However, since the values of the optical signal-to-noise ratio $(S/N)_o$ that are of interest in optical communications must be sufficiently high to assure an error probability of the order of 10^{-9} to 10^{-10}, an asymptotic approximation of the Cauchy integral can be adopted. In fact, thesaddle point approximation [20] can be used to evaluate the Cauchy integral appearing in the expression of $P_e(1)$ and the results have been compared with those obtained by numerical evaluation of the integral itself. The error in the value of $(S/N)_o$ corresponding to $P_e(1) = 10^{-10}$ is contained within 0.5 dB in all the analysed cases [21]. Using this approximation, the following expression is obtained:

$$\frac{1}{2}\int_{c-i\infty}^{c+i\infty} \frac{G_d(\xi/b=1)}{\xi} \cdot e^{\xi C_{t1}} d\xi$$

$$= \frac{\left[\sqrt{\dfrac{C_{t1}}{(S/N)_o M}}\right]^M}{\left[\sqrt{\dfrac{(S/N)_o M}{C_{t1}}} - 1\right]\sqrt{4\pi C_{t1}\sqrt{\dfrac{C_{t1}}{(S/N)_o M}}}} \cdot \exp\left\{-\frac{C_{t1}}{2}\left[1 - \sqrt{\frac{(S/N)_o M}{C_{t1}}}\right]^2\right\}$$

$$(10.30)$$

10.6 FIBER TRANSMISSION SYSTEMS AFFECTED BY DISTORTIONS

A variety of effectsintroduce distortions on a transmitting signal that can seriously affect the transmission performance. In particular, the intermingling of source chirping and fiber dispersion, fiber propagation under non-linear effects, and the non-ideal behavior of optical devices encountered by the signal have to be taken into account when analysing the performance. Apart from some particular instances relating to idealised and oversimplified cases, it is not possible to use analytical (formula-based) modeling of communication systems when distortion affects the transmission. In these cases it is necessary to use simulation approaches.

Different approaches have been reported in the literature. Here we treat a semi-analytical technique which is based on the simulation of the signal transmission from the optical transmitter up to the decision circuit placed in the receiver after the sampling circuit. Figure 10.8 sketches this type of modeling. As a matter of fact, the performance evaluation algorithm can be summarised as follows:

(1) The evolution of the optical signal along the transmission path is numerically simulated, taking into account the modeling of the relevant optical devices. Also the phase noise accumulation along the link, due to optical sources and some kinds of devices (e.g. wave-

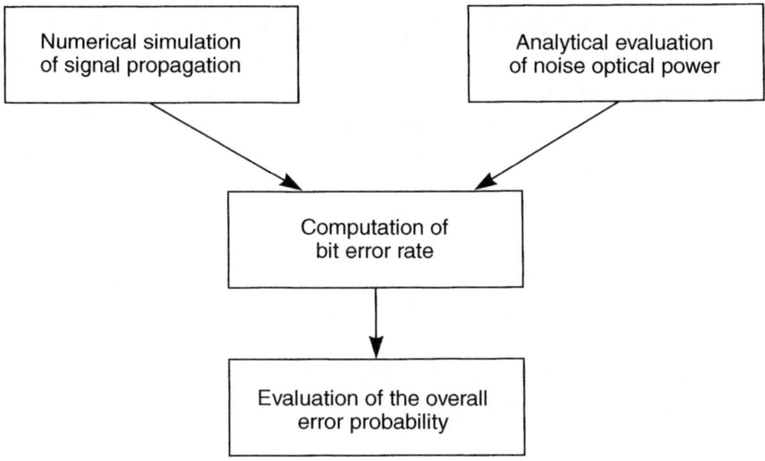

Fig. 10.8 A semi-analytical scheme for evaluating transmission performance.

length converters based on FWM in semiconductor amplifiers, as reported in Chapter 7), can be taken into account, by including random phase generator whose statistics is governed properly.[4]

(2) The optical noise power is analytically evaluated starting from the noise characteristics of the active devices that are present along the optical path. The thermal noise of the receiver is also evaluated.

(3) The error probability is evaluated, bit by bit, as described in the previous section, since the signal power corresponding to that bit, which is sampled before the decision circuit, has been evaluated by simulation.

(4) The overall error probability is evaluated by averaging the probability calculated for a random signal pattern of a number of bits *Nbit*. The value of N_{bit} is chosen in order to consider the pattern dependency of the frequency chirp and dispersion. In this way N_{bit} has to be formed by all of the possible *M*-bit sequences, with *M* larger than the memory of the transmission channel, to account for all possible intersymbol interferences.

The first point is carried out by using the physical modeling reported in the first part of the book. In particular, the signal exiting from the transmitter can be evaluated by solving the laser rate equations in Chapter 2

---

[4] In the case of wavelength translators based on FWM in semiconductor amplifiers, the phase is characterised by a Wiener process, whose variance is known.

where the dynamic behavior of the laser is treated, in the case of a direct modulation of a laser source, or by using the proper modulator model, in the case of external modulation. Once the signal is evaluated, its spectrum can be obtained by fast Fourier transform (FFT). Then its evolution along the fiber can be evaluated straighforwardly by using the fiber transfer function (Chapter 4), if we can assume a linear behavior for the fiber in that condition, or by solving the fiber propagation according to the split-step method of Chapter 4. The spectrum of the signal passing through a filter or through whatever device, can be evaluated using the device transfer function, if it exists, or by using the physical modeling of the device itself.

It is worth noting that the present method is open for different modeling approaches. Examples on this type of performance evaluation are reported in the literature [22-25].

10.7 METHODS FOR OVERCOMING DISPERSION LIMITATIONS

Since optical amplifiers may go well beyond the limitation of a conventional systems power budget, fiber dispersion remains as the main obstacle for multigigabit transmissions. Expressly in 1550 nm transmission systems, operating at bit rates beyond 10 Gb/s over SMFs, dispersion is more restrictive than attenuation.

Besides the adoption of external modulation, which allows sensitive reduction of the chirping/dispersion penalties, there are different methods for overcoming the fiber dispersion limitations: using dispersion-compensating fibers [26, 27], dispersion-supported transmission [28], non-linear optical transmission [29], pre-chirping techniques [30], midway optical phase conjugation [31–33]. These methods are now briefly reviewed.

10.7.1 Dispersion-compensating fibers

The overall dispersion of a fiber link can be eliminated at a certain wavelength if dispersion-compensating fibers (DCF) are added to the conventional SMFs. As an example, a special DCF fiber with –65 ps/nm/km dispersion has been realised [26], and an error-free transmission of 10 Gb/s signals, from a direclty modulated laser source, has been achieved. In this experiment, the DCF has a loss of 0.48 dB/km that has to be overcome by additional optical amplification. In another experiment, carried out at 2.5 Gb/s, a DCF with a dispersion of –45 ps/nm/km and an attenuation of 0.74 dB/km at 1550 nm was used [27].

10.7.2 Dispersion-supported transmission

Dispersion-supported transmission (DST), reported in [28], utilises FSK transmission. In particular, two cases have been reported [34]: pure FSK and FSK with residual amplitude modulation.

Pure FSK transmission

The principle of this method is sketched in Fig. 10.9. The optical transmitter generates an FSK signal. The dispersive fiber link is used to convert the frequency modulation into an amplitude modulation. Such an amplitude modulation is detected at the optical receiver. The pulse shapes of the transmitter driving signal I, the optical frequency ν, and the optical power $Popt$ of the transmitter are shown in Fig. 10.9a. Assuming that the optical transmitter is not modulated in amplitude, the optical frequency ν is switched by the incoming non-return-to-zero (NRZ) binary data between two values with the frequency shift $\Delta\nu$, corresponding to the wavelength shift $\Delta\lambda = -\Delta\nu\,\lambda^2/c$. Due to fiber dispersion, the different signal components with different wavelengths arrive at different times at the output of the fiber of length L, as shown in Fig. 10.9b. The time difference is given

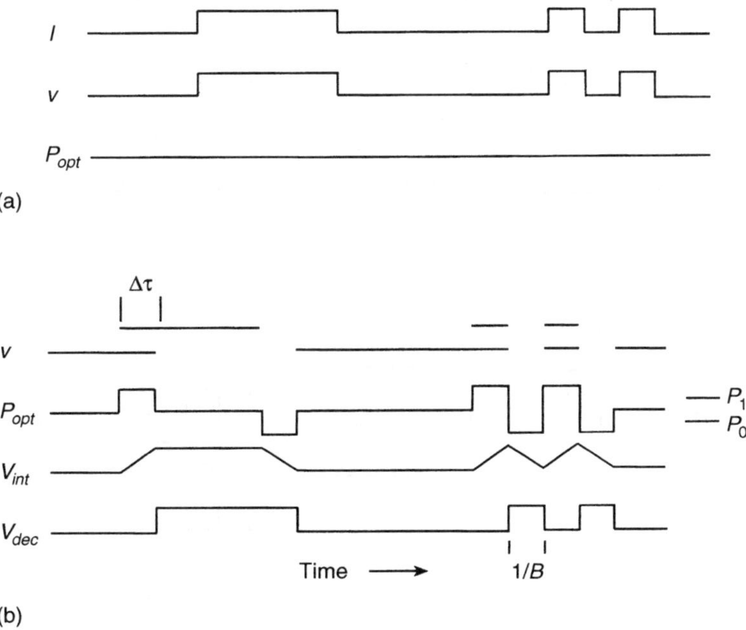

Fig. 10.9 Dispersion-supported transmission pure FSK: (a) transmitter signals, (b) receiver signals after dispersive fiber length L.

by $\Delta\tau = -\Delta\lambda\, DL$, where D is the fiber dispersion. Hence the resulting amplitude modulation of the optical signal P_{opt}, shown in the same figure, is not longer constant. In the region where the two signals components with different optical frequencies overlap, interference effects are expected. For constructive interference, a positive peak of $P_{opt}(t)$ is assumed here. In the region where no signal component is present, $P_{opt}(t)$ will show a negative peak. If the optical transmitter is modulated with a data signal of bit rate B and $\Delta\tau$ is adjusted to $|\Delta\tau| = 1/B$, the three-level signal $P_{opt}(t)$, sketched in the figure, is expected at the receiver input. After direct detection with a p-i-n diode, in principle, the original NRZ data signal can be derived from this signal in the optical receiver by using an electrical integrator in conjunction with a decision circuit. The figure also displays the corresponding signal V_{int}, after the integration, and the output voltage V_{dec} after the decision circuit.

FSK transmission with residual amplitude modulation

The principle of this method is illustrated in Fig. 10.10. In real systems, if a directly modulated laser diode is used, the optical signal at the transmitter output is modulated both in frequency and amplitude, as shown in the figure. Let's assume that the optical frequency increases when the optical power increases, and that the signal component with the higher

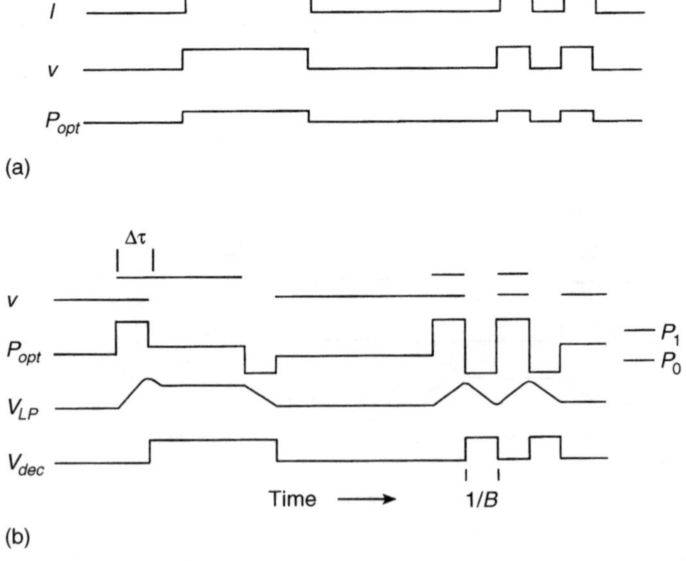

Fig. 10.10 Dispersion-supported transmission FSK with residual amplitude modulation: (a) transmitter signals, (b) receiver signals after dispersive fiber length L.

optical frequency propagates faster on the dispersive fiber, as in practical cases. On the receiver side, a four-level optical signal is generated. The original data signal can be recovered in different ways by using (1) a low-pass filtering of the signal in conjunction with a decision circuit, or (2) a dual threshold detection.

As a result, it has been shown [34] that unregenerated optical transmission at 10 Gb/s via a standard single-mode fiber can be achieved in the whole range of fiber lengths from 0 km up to the record span of 253 km.

10.7.3 Non-linear optical transmission

Non-linear effects related to fiber propagation can be exploited for overcoming the dispersion problem. At high optical power levels, a noticeable self-phase modulation occurs due to Kerr effect, as mentioned in section 10.4. This effect can compensate the dispersive pulse broadening, and stable optical pulses, i.e. solitons, can propagate along the fiber. Since this technique is of considerable importance for long-haul optical communications, it will be detailed in Chapter 13.

10.7.4 Prechirping techniques

This method has been investigated both theoretically and experimentally [30]. Its principle can be summarised as follows. Fiber dispersion distorts the waveform after transmission, as sketched in Fig. 10.11a, where an ideal intensity-modulated light waveform is depicted. When this intensity-modulated light pulse is transmitted through an anomalous dispersion optical fiber, the pulse envelope width becomes broadened and high-frequency power components in the pulse are pushed toward the leading edge of the pulse, as shown in Fig. 10.11b.

The broadened transmitted pulse through a dispersive optical fiber introduces an intersymbol interference (ISI) and causes power penalty. The prechirp method is based on a predistortion technique [30]. A light waveform is set in such a way to have the low-frequency power components in the leading edge of the pulse, and the high-frequency components in the trailing edge. When the ideal modulated waveform is transmitted through the normal dispersion optical fiber, the frequency components in the pulse distribute themselves as shown in Fig. 10.11c. If this waveform is transmitted through an anomalous dispersion optical fiber, it becomes narrower (Fig. 10.11d), during the first half of the transmission distance and subsequently becomes broadened after further transmission (Fig. 10.11e). Further transmission introduces an ISI with consequent power penalty (Fig. 10.11f). As a result, the allowable transmission length for a

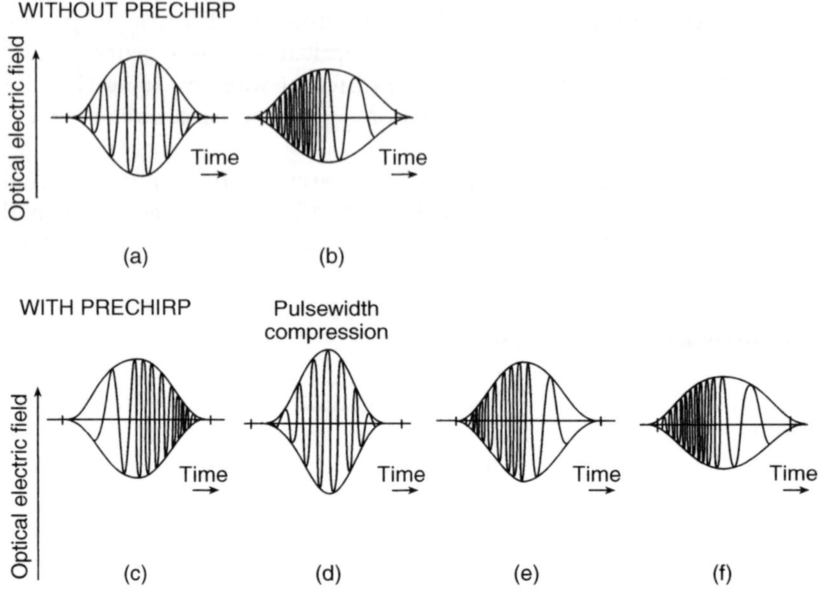

Fig. 10.11 The prechirping technique: (a) ideal intensity-modulated waveform (b) received optical waveform (c) prechirped optical waveform (d, e) middle of a transmission fiber, (f) received optical waveform.

system with prechirp will be greater than that for a system using an ideal external modulator.

10.7.5 Midway optical phase conjugation

Fiber dispersion can be completely compensated if a phase conjugation is performed in the middle of an optical transmission link [31]. As a matter of fact, the
dispersion in the first half of the link can be compensated by the second half, if the signal, in the middle of the link, is spectrally inverted. Such an optical phase conjugation, or spectral inversion, can be accomplished in different ways. Among them, the spectral inversion obtained by the four-wave mixing effect (FWM), either in optical fibers or in semiconductor optical amplifiers, represents a good solution, and it has been reported in different papers. The FWM effect is analysed in Chapter 7, when describing the operation of wavelength translators exploiting FWM in SOAs. In practice these types of device can be used as wavelength converters and as phase conjugators. The performance of these types of device as optical phase conjugators has been reported in the literature [32, 33].

REFERENCES

1. R. Heidemann, B. Wedding, and G. Veith, 10 Gb/s transmission and beyond, *Proc. IEEE* **81**, 1558–1567 (1993).
2. D. Cotter, Fiber nonlinearities in optical communications, *Opt. Quantum Electron.* **19**, 1–17 (1987).
3. A.F. Elrefaie *et al.*, Chromatic dispersion limitations in coherent lightwave transmission systems, *IEEE J. Lightwave Technol.* **6**, 704–709 (1988).
4. Y. Namihira and J. Maeda, Comparison of various polarisation mode dispersion measurement methods in optical fibers, *Electron. Lett.* **28**, 2265–2266 (1992).
5. R. Sabella, P. Lugli, D. Meglio, and G. Alcini, Numerical evaluation of bulk electroaabsorption modulator transmission performances, in *Proc. EFOC/N '94*, pp. 179–183, Heidelberg, Germany (1994).
6. D. Meglio, P. Lugli, R. Sabella, and O. Sahlén, Analysis and optimization of bulk electro-absorption modulators, *IEEE J. of Quantum Electron.* **31**, 261–268 (1995).
7. A.R. Chraplyvy, Limitations on lightwave communications imposed by optical-fiber nonlinearities, *IEEE J. Lightwave Technol.* **8**, 1548–1557 (1990).
8. E. Iannone, F. Matera, A. Mecozzi, and M. Settembre, *Nonlinear Optical Communication Systems*, John Wiley, New York (1998).
9. H. Takenaka *et al.*, Compact size and high output power Er-doped fiber amplifier modules pumped with 1.48 mm MQW LDs, *Technical Digest on Optical Amplifiers and their Applications,* Snowmass Village CO, paper FD2 (1991).
10. O.K. Tonguz, Impact of spontaneous emission noise on the sensitivity of direct-detection lightwave receivers using optical amplifiers, *Electron. Lett.* **26**, 1343–1344 (1990).
11. Y. Miyamoto *et al.*, 10 Gb/s 1310-nm optimized fiber link over 80 km using laser-diode transmitters and avalanche-photodiode receivers with high manufacturability, *Technical Digest OFC/IOOC '93*, San Jose CA, paper TuD2 (1993).
12. R.I. Laming *et al.*, High sensitivity optical pre-amplifier at 10 Gb/s employing a low noise composite EDFA with 46 dB gain, *TechnIcal Digest on Optical Amplifiers and their Applications*, Santa Fe NM, paper PD13 (1992).
13. K. Hagimoto *et al.*, 'Twenty-Gb/s signal transmission using simple high-sensitivity optical receiver, in *Proc. OFC '92*, San Jose CA, paper Tul3 (1992).
14. B. Wedding and T. Pfeiffer, 20 Gb/s optical pattern generation, amplification and 115 km fiber propagation using optical time division multiplexing, in *Proc. ECOC '90*, Amsterdam Netherlands, pp. 453–456 (1990).
15. P.A. Andrekson *et al.*, Ultra-high speed demultiplexing with the nonlinear optical loop mirror', in *Proc. OFC '92*, San Jose, CA, paper PD8 (1992).
16. E. Iannone, R. Sabella, Performance evaluation of an optical multi-carrier network using wavelength converters based on FWM in semiconductor optical amplifiers, *IEEE J. Lightwave Technol.* **13**, 312–324 (1995).
17. M. Schwartz, W.R. Bennett, and S. Stein, *Communication Systems and Techniques*, Mc Graw-Hill, New York (1966).
18. E. Iannone and R. Sabella, Analysis of wavelength switched high-dense WDM

networks employng wavelength converters based on FWM in semiconductor amplifiers, *IEEE J. Lightwave Technol.* **13**, 1579–1592 (1995).

19. S. Betti, G. De Marchis, and E. Iannone, *Coherent Optical Communications Systems*, John Wiley, New York (1995).

20. J.E. Mazo and J. Salz, Probability of error for quadratic detectors, *Bell Syst. Tech. J.* **44**, 2165–2187 (1965).

21. K. Shumacher and J.J. O'Reilly, Relationship between the saddle-point approximation and the modified Chernov bound, *IEEE Trans. on Commun.* **38**, 270–272 (1990).

22. R. Sabella and E. Iannone, Performance of high speed-long distance optical links using wavelength converters based on FWM in semiconductor optical amplifiers, *Fiber and Integrated Optics* **15**, 7–14 (1996).

23. E. Iannone, R. Sabella, L. de Stefano, and F. Valeri, All-optical wavelength conversion in multi-carrier networks, IEEE Trans. on Commun. **44**, 716–724 (1996).

24. R. Sabella, E. Iannone, and E. Pagano, Optical transport networks employing all-optical wavelength conversion: limits and features, IEEE J. Select. Areas in Commun. **14**, (1996).

25. R. Sabella and E. Iannone, Wavelength conversion in optical transport networks, *Fiber and Integrated Optics.* **15**, 167–192 (1996).

26. J.M. Dugan *et al.*, All-optical, fiber-based 1550 nm dispersion compensation in a 10 Gb, 150 km transmission experiment over 1310 nm optimized fiber, in *Proc. OFC '92,* San Jose CA, paper PD14 (1992).

27. H. Izadpanah *et al.*, Dispersion compensation for upgrading interoffice networks built with 1310 nm optimized SMFs using an equalizer fiber and 1310/1550 nm WDM, in *Proc. OFC '92*, San Jose CA, paper PD15 (1992).

28. B. Wedding, New method for optical transmission beyond dispersion limit, *Electron. Lett.* **28**, 1298–1300 (1992).

29. M. Nakazawa *et al.*, Distortion-free single-pass soliton communication over 250 km using multiple Er3+ doped optical repeaters, in *Proc. OFC '90*, San Francisco CA, paper PD5 (1990).

30. N. Henmi, T. Saito, and T. Ishida, Prechirp technique as a linear dispersion compensation for ultrahigh-speed long-span intensity modulation directed detection optical communication systems, *IEEE J. Lightwave Technol.* **12**, 1706–1719 (1994).

31. M.C. Tatham, G. Sherlock, and L.D. Westbrook, Compensation of fibre chromatic dispersion by mid-way spectral inversion in a semiconductor laser amplifier, in *Proc. ECOC '93*, Montreaux, paper Th12.3 (1993).

32. R. Sabella, E. Iannone, and E. Pagano, Impact of wavelength conversion by FWM in semiconductor amplifiers in long distance transmissions, in *Proc. CLEO '95*, 306–307, Chiba, Japan (1995).

33. R. Sabella and E. Iannone, Long distance optical communication systems adopting wavelength conversion by FWM in semiconductor amplifiers, *Opt. Commun.* **18**, 109–113 (1997).

34. B. Wedding, B. Franz, and B. Junginger, 10-Gb/s optical transmission up to 253 km via standard single-mode fiber using the method of dispersion-supported transmission, **12**, 1710–1727 (1994)

11
Multichannel Optical Systems

11.1 INTRODUCTION

The advances in single-frequency lasers, tunable lasers, narrow-band optical filters (tunable or not) have increased the interest in wavelength division multiplexing (WDM) techniques. The strong interest in these techniques can be divided into two basic areas: very high capacity transmissions and optical networks [1]. As a matter of fact, WDMs allow the very large fiber bandwidth to be efficiently exploited by adopting mostly existing technologies and provides the possibility of transmitting many channels simultaneously. Thus, the transmission capacity may be massively increased, with respect to single time division multiplexed (TDM) channels.

Furthermore, by means of high-speed tunable transmitters and receivers, WDM can be used for circuit and packet switching, as well as for wavelength routing. In fact, WDM is expected to be a key issue for the realisation of the optical layer of the transport network, thanks to photonic switching techniques that take advantage of this multiplexing technique. Interest in WDM has also been expanded to other new areas, such as optical interconnect in computer and neural networks.

The progress of WDM networks is greatly facilitated by significant advances in optical amplifiers, which provide the extra budget needed to compensate for the insertion loss of multiplexing components and also for power splitting or tapping.

This chapter looks at the WDM technique in high-capacity transmissions and some applications of WDM in optical networks.

11.2 WDM FOR OPTICAL COMMUNICATION SYSTEMS

A fundamental property of single-mode fibers is their enormous low-loss bandwidth of many terahertz. Single-channel transmission is limited in speed to much less than the fiber capacity due to limitations in optoelectronic component speed and dispersive effects. However, a widespread and straightforward approach, for more effectively utilising the fiber bandwidth, is to transmit different channels simultaneously on a single fiber, with each channel placed at a different wavelength. This technique not only enables significant capacity enhancements (Fig 9.1), but will also enable new networks in which the routing path is wavelength dependent.

A crucial question for WDM systems is related to the maximum number of wavelengths which can be accommodated on a single fiber. This issue is directly related to wavelength separation. In principle, the available optical bandwidth in the third region (where optical amplifiers are available) is approximately 100 nm. However, on the assumption that most advanced systems will make use of EDFAs, the maximum bandwidth available is about 30 nm, which is the bandwidth of a single EDFA. The equivalent frequency range is found by using the relationship $c = f\lambda$, which on differentiation leads to

$$\Delta f = -\frac{c}{a^2}\Delta\lambda \qquad (11.1)$$

where λ is the operating wavelength and c the speed of light. Therefore at 1550 nm, the available bandwidth corresponding to the EDFA region of $\Delta\lambda = 30\,\text{nm}$ is $\Delta f = 3.7 \cdot 10^{12}\ Hz$.

Chapter 6 explained that the maximum number of channels which can be accommodated in a fiber is theoretically 740. But this is unrealistic since the avoidance of crosstalk between channels with this separation would demand exceptional laser source wavelength stability, and extreme selectivity at the receiver to separate the individual channels. Any laser source chirping would also spread spectral components into adjacent channels.

Besides crosstalk, the arrangement of many channels onto the same fiber leads to very high power in the fiber. To ensure an adequate BER, a certain power per channel is required, so a high total power has to be to be conveyed on the same fiber. Besides the danger from an operating viewpoint, such high powers expose the fundamental non-linear nature of optical fibers, which gives rise to distortions and crosstalk (Chapter 10).

Channel separation, which has to be accomplished before the single-channel detection, is a crucial issue. Selectivity is important in determining the channel wavelength separation and is strictly related to the available technology. Thus it is necessary to use different names for WDM systems, according to the wavelength separation technology. Even if there is no standardised version, a common form is as follows [2]:

- WDM systems: the spacing is greater than 1 nm (e.g. 4 nm) using direct detection techniques. Optical filtering can be used before detection to select wavelength. These systems typically use a relatively small number of channels (28).
- HD-WDM systems: high-density WDM means, in general, spacing less than or equal on 1 nm. They generally use direct detection together with predetection narrowband filtering for wavelength selection.
- CMC systems: coherent multichannel systems use coherent heterodyne detection techniques to provide very high selectivity together with good detection sensitivity, thus channel separation can be very small, e.g. 0.1 nm. This technique, which would allow more than 100 channels, is

the best one for utilising the fiber bandwidth, but also the most complex and costly.

11.3 WDM-BASED HIGH-SPEED TRANSMISSION SYSTEMS

For PCM signals, the transmission capacity can be easily expressed as the product of the number of channels carried onto the same fiber by the bit rate of the single channel. Thus, two possibilities exist to increase the transmission capacity: increase either the bit rate or the number of WDM channels. Naturally the system goal is to realise the longest distance possible.

Unfortunately, there are several factors inherent to optical fiber transmission which can seriously degrade WDM system performance. These factors can be summarised as follows:

(1) Fiber dispersion limits the bit rate-span product.
(2) Linear crosstalk is due to non-ideal filtering associated with the demultiplexing process.
(3) Non-uniform EDFA gain which causes a signal-to-noise ratio differential among many channels and limits the available wavelength range [2].
(4) Several non-linear effects, as considered in Chapter 10.

11.3.1 Dispersion limitations

These limitations are discussed in detail in Chapter 10. WDM can overcome dispersion. In fact, although a 10 Gb/s system would be limited to around 4050 km, the use of 2.5 Gb/s on four channels would increase the maximum distance by a factor of 16, i.e. < 600 km [2].

11.3.2 Linear crosstalk limitations

Linear crosstalk principally two effects: (1) optical filters do not have ideal stopbands, and adjacent channel interference will always be present; (2) the spectral tails of adjacent channels enter the bandwidth of a selected channel. Hence, in the analysis or design of a WDM system, it is necessary to evaluate the amount of crosstalk that can be tolerated to achieve a given BER, and investigate the appropriate filter type.

11.3.3 Limitations due to nonuniform gain of EDFAs

The passband of an EDFA is characterised by a gain peak at a wavelength of approximately 1531 nm. The deviation in gain across the band

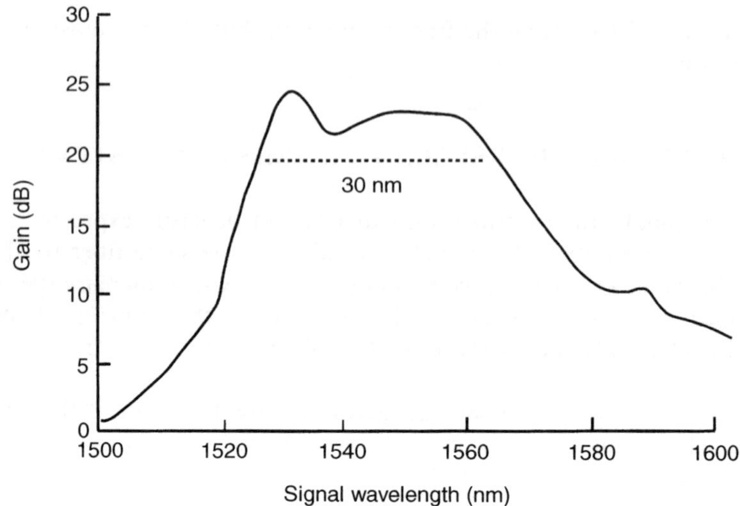

Fig. 11.1 Transmission characteristic of an EDFA.

15301560 nm is roughly 3 dB. For a WDM system this gain variation is of much greater significance than in a single-wavelength system. By way of illustration, with a cascade of 10 amplifiers, each with a transmission characteristic as shown in Fig.11.1, it is possible that the intensity of a signal at one wavelength at the end of the cascade may be 30 dB lower relative to another at a different wavelength this would be deleterious. As a consequence, in WDM systems it could be necessary to use gain-flattened amplifiers [3]. However, in practical cases, after a certain number of cascaded amplifiers, the bandwidth reduces considerably. For example, after 50 gain-flattened amplifiers [2, 3], the bandwidth has reduced from about 30 nm to less than 10 nm. Therefore in the design of an optically amplified WDM system the end-to-end bandwidth has to be considered when deciding on the number of wavelengths and the wavelength separations.

11.3.4 Non-linear system limitations

The impact of non-linear effects on system performance was mentioned in the previous chapter. In the WDM systems the effects of SPM, CPM and FWMF (FWM in fiber), related to Kerr or Raman effects, are of particular concern, since they arise from the interaction betwen two or more channels.[1] One method to significantly reduce the effect of dispersion is to use a dispersion-shifted fiber (DSF), which presents a dispersion parameter

---

[1] Stimulated Brillouin scattering is practically negligible in the considered case.

four times lower than a standard fiber. Unlike conventional SMFs, which have a dispersion zero point located near 1300 nm, DSFs have a dispersion zero point located near 1550 nm, enabling transmission in the third window where EDFAs can be used. Even if dispersion is negligible, a single or multiple channel located near the dispersion zero point will still experience the non-linear effect of FWMF upon transmission. A good solution is to use different types of fiber for the transmission span, with one type of fiber having a positive dispersion and the other one having a negative dispersion for a particular wavelength. As a result, the total accumulated dispersion is zero after some distance, but the absolute dispersion per unit length is non-zero at all points along the fiber. The result is that the total effect of dispersion is negligible for that wavelength and the non-zero absolute dispersion causes a phase mismatch among channels, thereby destroying any efficient FWM production.

11.4 ANALYSIS OF WDM TRANSMISSIONS

To analyse the transmission of high-speed WDM channels through a cascade of EDFAs, when dispersion and non-linear effects are present, it could be insufficient to analyse the transmission of a single channel. One method is to simulate, simultaneously, the transmission of all channels. In particular, it is possible to simulate different high-speed channels (e.g. 2–8 channels at 10 Gb/s), through a cascade of EDFAs, experiencing dispersion, SRS, SPM and CPM. In the evaluation of system performance, the contribution of FWMF can be treated as a crosstalk contribution. The following two sections describe the simulation model [4], and the crosstalk analysis of both linear [5] and non-linear contributions [6, 7].

11.4.1 Modeling of WDM channel transmission

The following assumptions are made:

(1) The WDM channels are externally intensity modulated (e.g. at 10 Gb/s), with no initial chirp (e.g. assuming the use of a Mach-Zehnder modulator).
(2) The wavelength range has been chosen in order to operate in the most uniform portion of the EDFA gain for short lengths and near the gain peak generated in long-distance EDFA cascade links [8].
(3) The receiver has an electronic bandwidth of 7.5 GHz.
(4) Each wavelength transmits a different 64-bit pseudorandom NRZ data pattern.
(5) The temporal shape of the bit is super-Gaussian to assure a smooth

transition in the Fourier transform between the time domain and frequency domain.[2]

The propagation of the optical signals through non-linear fibers is accounted for by solving the non-linear Schrödinger equation described in Chapter 4. In fact, the modeling of transmission consists of a system of N equations , where N is the number of propagating signals. For $N = 4$ the following equations are obtained:

$$\frac{\partial A_1}{\partial z} + \frac{1}{\nu_1}\frac{\partial A_1}{\partial t} + \frac{i}{2}\beta(\lambda_1)\frac{\partial^2 A_1}{\partial t^2} + \frac{1}{2}\alpha A_1$$
$$= i\gamma_1\left[|A_1|^2 + 2|A_2|^2 + 2|A_3|^2 + 2|A_4|^2\right]A_1 \qquad (11.2)$$
$$+ \left[\frac{g_{12}}{2}|A_2|^2 + \frac{g_{13}}{2}|A_3|^2 + \frac{g_{14}}{2}|A_4|^2\right]A_1$$

$$\frac{\partial A_2}{\partial z} + \frac{1}{\nu_2}\frac{\partial A_2}{\partial t} + \frac{i}{2}\beta(\lambda_2)\frac{\partial^2 A_2}{\partial t^2} + \frac{1}{2}\alpha A_2$$
$$= i\gamma_2\left[2|A_1|^2 + |A_2|^2 + 2|A_3|^2 + 2|A_4|^2\right]A_2 \qquad (11.3)$$
$$+ \left[\frac{g_{21}}{2}|A_1|^2 + \frac{g_{23}}{2}|A_3|^2 + \frac{g_{24}}{2}|A_4|^2\right]A_2$$

$$\frac{\partial A_3}{\partial z} + \frac{1}{\nu_3}\frac{\partial A_3}{\partial t} + \frac{i}{2}\beta(\lambda_3)\frac{\partial^2 A_3}{\partial t^2} + \frac{1}{2}\alpha A_3$$
$$= i\gamma_3\left[2|A_1|^2 + 2|A_2|^2 + |A_3|^2 + 2|A_4|^2\right]A_3 \qquad (11.4)$$
$$+ \left[\frac{g_{31}}{2}|A_1|^2 + \frac{g_{32}}{2}|A_2|^2 + \frac{g_{34}}{2}|A_4|^2\right]A_3$$

$$\frac{\partial A_4}{\partial z} + \frac{1}{\nu_4}\frac{\partial A_4}{\partial t} + \frac{i}{2}\beta(\lambda_4)\frac{\partial^2 A_4}{\partial t^2} + \frac{1}{2}\alpha A_4$$
$$= i\gamma_4\left[2|A_1|^2 + 2|A_2|^2 + 2|A_3|^2 + |A_4|^2\right]A_4 \qquad (11.5)$$
$$+ \left[\frac{g_{41}}{2}|A_1|^2 + \frac{g_{42}}{2}|A_2|^2 + \frac{g_{43}}{2}|A_3|^2\right]A_4$$

---

[2] The super-Gaussian shape can be obtained by passing rectangular bits through a 20th-order Bessel filter of 20 GHz bandwidth.

where A_i is the propagating optical field amplitude at the wavelength λ_i, v_i the group velocity, $\beta(l_i)$ the linear group velocity dispersion, which is given by $\beta(\lambda_i) = \lambda_i^2 D / 2\pi c$, D being the dispersion parameter, α the fiber loss coefficient (assumed independent of the wavelength), γ_i the non-linear coefficient defined as $\gamma_i = 2\pi n_{nl} / (\lambda_i A_{eff})$, A_{eff} being the effective fiber core area and n_{nl} the non-linear index coefficient; and g_{ij} ($j = 1, 2, 3, 4$) the Raman gain coefficients between channel i and j.

In each equation there are six terms. Starting from the left-hand side, they represent respectively:

(1) the evolution of the optical field amplitude with distance of propagation in the fiber,
(2) the change of the field amplitude with time,
(3) the wavelength-dependent dispersion,
(4) the wavelenght-independent fiber loss,
(5) the self-and cross-phase modulation of the optical field,
(6) the Raman gain provided to a single channel.

The equations above can be solved using the split-step method, as described in Chapter 4.

Polarisation effects can be neglected since polarisation-dependent gain of an EDFA can be eliminated by using a polarisation scrambler. The signal and ASE power accumulation along the EDFA chain can be determined by using the model in Chapter 5. Once the signal and noise at the receiver have been obtained, the eye penalty and the error probability can be evaluated using the model in Chapter 10.

11.4.2 Crosstalk analysis

Two crosstalk mechanisms can be envisaged in WDM signal transmission: linear crosstalk from non-ideal selection processes operated by the selection filter before the receiver, and non-linear crosstalk due to FWMF. The two effects can be modeled separately.

Linear crosstalk

The selection process, accomplished through an optical filter, inevitably introduces a distortion effect on the selected channel and a disturbance due to the presence, in the filter bandwidth, of tails of channels different from the selected one. This is because the spectral tails of each channel theoretically extend over all the frequency axis. As an example, Fig. 11.2 sketches the optical spectrum of a WDM set of signals, where an ideal filter transfer function is also indicated. Moreover, the non-ideal optical

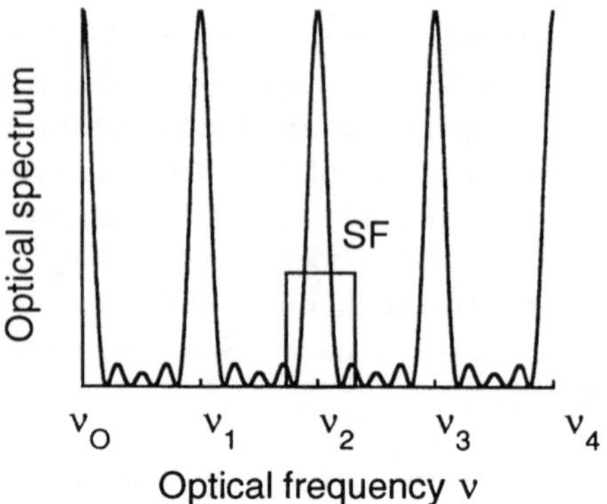

Fig. 11.2 Optical spectrum of a WDM set of signals; an ideal filter transfer function is also indicated. The spectral tails of any signal fall into the bandwidth of adjacent signals.

filtering catches contributions from other channels. All these mechanisms can be schematised as crosstalk. Eventually crosstalk causes a sensitivity penalty that increases with decreasing channel spacing so as to fix the minimum channel spacing conforming with the system performance requirements. The crosstalk-induced penalty can be modeled as follows. The optical signal at the receiver end, before the filter, can be written as

$$\mathbf{E}(t) = \sum_{k=-N/2}^{N/2-1} A_k \sqrt{m_k(t)}\, e^{i(k\Delta\omega t + \Psi_k)} \boldsymbol{\xi}_k e^{i\omega, st}, \tag{11.6}$$

where A_k, ψ_k, $m_k(t)$, and $\boldsymbol{\xi}$ are respectively amplitude, phase modulating message and polarisation vector. N is the number of WDM signals, ω_s the middle angular frequency of the optical spectrum and $\Delta\omega$ the channel spacing.

After optical filtering, the optical signal assumes the following expression:

$$E(t) = \sum_{k=-N/2} A_k \left[\sqrt{m_k(t)}\, e^{i(k\Delta\omega t + \Psi_k)} \right] * h(t) \boldsymbol{\xi}_k\, e^{i\omega_s t}$$

$$= \sum_{k=-N/2}^{N/2-1} A_k s_k(t) \boldsymbol{\xi}_k e^{i\omega_s t} \tag{11.7}$$

where * indicates convolution, the channel is indicated by index 0, and $h(t)$ represents the impulse response of the optical filter. After the optical signal detection, through a photodiode, and electrical baseband filtering, the electrical signal can be denoted as

$$c(t) = c_{pc}(t) * h_R(t) + \eta(t), \tag{11.8}$$

where $c_{pc}(t)$, $h_R(t)$, and $\eta(t)$ represent the average value of the photocurrent, the electrical filter impulse response and the noise process, assumed to be a white Gaussian process. If, is usual, the only appreciable contributions are from the two contiguous channel, we have to consider only three bits, and the expression $c_{pc}(t)$ can be written as

$$\begin{aligned}
c_{pc}(t) = R_p &\left[A_0^2 |s_0(t)|^2 + A_{-1}^2 |s_{-1}(t)|^2 + A_1^2 |s_1(t)|^2 \right] \\
&+ 2R_p \Re \left\{ A_0 A_1 s_0(t) s_1^*(t) \vec{\xi}_0 \cdot \vec{\xi}_1^* \right. \\
&\left. + A_0 A_{-1} s_0(t) s_{-1}^*(t) \vec{\xi}_0 \cdot \vec{\xi}_{-1}^* + A_1 A_{-1} s_{-1}(t) s_1^*(t) \vec{\xi}_{-1} \cdot \vec{\xi}_1^* \right\}
\end{aligned} \tag{11.9}$$

where $\Re$ stands for the real part.

The error probability can be expressed as a function of the error probability conditioned to the plural bit transmitted on the other channels. If b_k ($k = -1, 0, 1$) denotes the bit transmitted on the kth channel in the considered bit interval, and assuming equiprobable bits, the error probability can be expressed as

$$P_e = \frac{1}{8} \sum_{b_0=0}^{1} \sum_{b_1=0}^{1} \sum_{b_{-1}=0}^{1} P_e(b_{-1}, b_0, b_1), \tag{11.10}$$

$P_e(b_{-1}, b_0, b_1)$ being the error probability conditioned to the bit transmitted on the three channels.

Write N_h for the power spectral density of the Gaussian noise $\eta(t)$, write C for the sample of $c_{pc}(t) * h_R(t)$ taken at the center of the bit interval, and write C_{th} for the decision threshold, than the following expression can be derived for $P_e(b_{-1}, b_0, b_1)$:

$$P_e(b_{-1}, b_0, b_1) = \frac{1}{2} erfc\left(\frac{|C - C_{th}|}{\sqrt{2} B_r N_v} \right), \tag{11.11}$$

where B_R is the noise bandwidth of the baseband electrical receiver.

As described in (11.9), the crosstalk is related to the relative polarisation of the transmitted channels and to bandwidth/shape of the optical and electrical filters. The instantaneous phases of the optical carriers Ψ_k ($k = -1, 0, 1$), and the relative phases of the modulating functions $m_{-1}(t)$, $m_0(t)$ and $m_1(t)$ are other parameters to be considered when evaluating the crosstalk-induced penalty. The dependence on the modulation functions can be

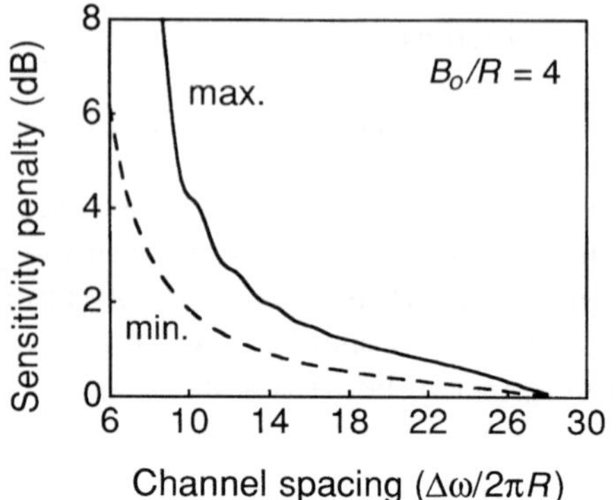

Fig. 11.3 Penalties for the best and worst cases versus the channel spacing, relative to an optical Fabry-Perot filter with a bandwidth equal to four times the bit rate R.

expressed considering the time delays between a given transition of the signal $m_0(t)$ and the immediately following transitions of $m_{-1}(t)$ and $m_1(t)$. An evaluation of the crosstalk-induced penalty for different values of the phase W_k shows that the dependence is negligible, unlike the dependence on such time delays.

Just to gain some feeling about the crosstalk-induced penalties in practical systems design, it is useful to evaluate the penalties relative to the best and the worst cases. In order to obtain the worst-case penalty, the scalar products of the polarisation vectors are assumed to equal unity, which means that all channels present the same polarisation. On the other hand, the best-case penalty can be obtained assuming these scalar products to be zero, which is equivalent to assuming the polarization state of the selected channel at the receiver to be orthogonal with respect to its nearest neighbors. The time delays can be assumed equal to zero, which means perfect synchronisation of the modulation functions.

The penalties for the best and worst cases, plotted versus the channel spacing, are shown in Fig. 11.3, relative to an optical Fabry-Perot filter with a bandwidth equal to four times the bit rate R. It can be seen that for having a 1 dB maximum sensitivity penalty, the channel spacing has to be wider than about $20R$.

For optically amplified WDM systems, the performance analysis should take into account the optical amplification of the multiplexed signals. In this case the error probability has to be evaluated as in section 10.5.2.

Non-linear crosstalk due to FWMF

Power exchange among copropagating channels, due to FWM which arises from the dependence of refractive index on the optical field intensity, produces a crosstalk contribution. In particular, from the interaction of three optical fields at frequencies $\omega 1$, ω_2, and ω_3, a fourth optical field is generated at frequency ω_4, if phase matching conditions are verified [9]. This fourth frequency is generated with an efficiency which is given as follows [10, 11]:

$$\nu = \frac{\alpha^2}{\alpha^2 + (\Delta\beta)^2} \left[1 + \frac{4e^{-\alpha z}\sin^2(\Delta\beta)}{(1 - e^{\alpha z})^2} \right], \qquad (11.12)$$

where the phase mismatch $\Delta\beta$ is defined as

$$\Delta\beta = [\beta(\omega_1) - \beta(\omega_2)] - [\beta(\omega_3) - \beta(\omega_4)]. \qquad (11.13)$$

Using a series expansion for $\Delta\beta$, truncated to second order, it is possible to obtain

$$\Delta\beta = 2\pi\beta_2(\omega_1 - \omega_2)(\omega_3 - \omega_4) \qquad (11.14)$$

In practical cases, the WDM channels are equally spaced, so the phase mismatch can be written as a function of the channel spacing $\Delta\omega$ as

$$\Delta\beta = 2\pi\beta_2(\Delta\omega)^2. \qquad (11.15)$$

A significant expression for the efficiency can be written by adopting two approximations which hold in practice:

(1) low attenuation, i.e. $\alpha \to 0$,
(2) not very long distances, i.e. $z\Delta\beta \ll 1$

Hence

$$\eta = \frac{4}{(\Delta\beta)^2 \cdot z^2} = \frac{4}{(2\pi\beta_2)(\Delta\omega)^4 \cdot z^2}. \qquad (11.16)$$

The last expression shows that efficiency decreases very quickly with channel spacing. Furthermore, $\beta_2 = -\lambda^2/2\pi c\, D$, with D the dispersion coefficient, the efficiency is inversely proportional to the square of the dispersion coefficient. Thus, if the fiber dispersion is drastically lowered, the FWMF effect becomes more important. For this reason, in actual dispersion management systems, normal fibers are adopted (e.g. DFS) and dispersion compensation techniques based on passive fibers or midway optical phase conjugation are used. This is how the dispersion is actually compensated, and the local dispersion is sufficiently high the FWMF effect is comparatively insignificant.

A meaningful parameter, in this case, is the FWMF bandwidth, which can be defined as the channel spacing for a 3 dB reduction in efficiency. Notice that the FWMF bandwdth can be expressed as

$$\Delta\omega_{-3dB} = \frac{\sqrt[4]{2}}{\sqrt{\pi\beta_2 z}} = \frac{0.45}{\sqrt{\beta_2 z}} \ . \tag{11.17}$$

And notice that if the dispersion increases, the FWMF bandwidth decreases. This means that the channel spacing of a WDM system can be reduced if the dispersion increases.

The analysis of FWMF-induced crosstalk can be achieved through a statistical approach [12]. This approach allows the BER to be calculated taking into account the FWMF.

If we consider N equally spaced channels, with equal power, the FWM interaction between them can be described by

$$\frac{dE_j}{dz} = \frac{\alpha}{2} E_j - i\beta \sum_{l,m} E_1 E_m E^*_{1+m-j} e^{i\Delta k_{lmj} z} \tag{11.18}$$

where E_j is the complex amplitude of the electric field at frequency f_j, α the linear attentuation coefficient, $\beta = 12\pi^2 / \lambda n \, \chi^{(3)}$, λ the wavelength, n the refractive index, $\chi^{(3)}$ the third-order non-linear susceptibility. The phase mismatch is $\Delta k_{lmj} = 2\pi \, \lambda^2 \, D(\Delta f)^2 / c$, D being the group velocity dispersion and Δf the channel spacing.

The ensemble of crosstalk contributions can be considered as a noise. If the number of independent contributions is large enough, the central limit theorem can be applied, and noise can be schematised as a Gaussian noise [12, 6].

In particular, it is possible to discern two types of FWM contribution: degenerate and non-degenerate. The non-degenerate contribution represents the general case in which three optical fields interact and generate a fourth field (four waves altogether); whereas the degenerate contribution relates to the possibility that two fields (one twice the energy of the other) interact to generate a third field. These two contributions can be taken into account independently. As a result, the variances of both the schematised Gaussian noises can be expressed as follows [6]:

$$\sigma^2_{dg} = \frac{\Psi_{FWM}}{2} \int_{-\infty}^{\infty} |H_s(f)|^2 \cdot \sum_{j\neq k} \theta(j,j,k)|L_{j,j,k}|^2 D_{dg}(f) df \tag{11.19}$$

$$\sigma^2_{nd} = \frac{\Psi_{FWM}}{2} \int_{-\infty}^{\infty} |H_s(f)|^2 \cdot \sum_{j\neq k} \sum_{h\neq j\neq k} 4\theta(j,h,k)|L_{j,h,k}|^2 D_{nd}(f) df, \tag{11.20}$$

where $H_s(f)$ is the transfer function of the selection filter before the receiver; W_{FWM} is given as follows [6]:

$$\Psi)FWM = \left(\frac{5}{6} \frac{96\pi^3}{\lambda nc} \chi^{(3)} \frac{A_t^3}{A_{eff}} \right) \tag{11.21}$$

A_{eff} being the fiber mode effective area and A_t the peak field amplitude. The function $\theta\_(j, h, k)$ is equal to one if the number N of WDM channels is $N \geq j + h - k \geq j$, and zero elsewhere. $L_{j,h,k}$ is the effective length of the fiber in the presence of dispersion. The functions $D_{dg}(f)$ and $D_{nd}(f)$ are given as follows [6]:

$$D_{dg}(f) = \frac{1}{4}\delta(f) + \frac{1}{2R} \frac{\sin^2\left(\frac{\pi f}{R}\right)}{\left(\frac{\pi f}{R}\right)^2} + \frac{R}{4\pi^2 f^2} \left[1 - \frac{\sin^2\left(\frac{2\pi f}{R}\right)}{\left(\frac{2\pi f}{R}\right)^2} \right] \tag{11.22}$$

$$D_{nd}(f) = \frac{1}{8}\delta(f) + \frac{3}{8R} \frac{\sin^2\left(\frac{\pi f}{R}\right)}{\left(\frac{\pi f}{R}\right)^2} + \frac{3R}{8\pi^2 f^2} \left[1 - \frac{\sin\left(\frac{2\pi f}{R}\right)}{\left(\frac{2\pi f}{R}\right)} \right]$$

$$+ \frac{3R}{16\pi^2 f^2} \left[\frac{1 - \sin^2\left(\frac{2\pi f}{R}\right)^2}{\left(\frac{2\pi f}{R}\right)^2} \right] \tag{11.23}$$

where $\delta(f)$ is the Dirac function.

Once the variances have been calculated, they can be added to those for the ASE noise and the BER can be evaluated as in section 10.5.2.

Application of the central limit theorem, which allowed the crosstalk noise to be schematised as a Gaussian noise, is ligitimate when the number of crosstalk contributions is large enough. However, in practical cases, the FWMF effect is not so strong as to generate such a large number. Consequently, the p.d.f. is not a Gaussian curve, and the noise power evaluation becomes much more complicated. Nevertheless, the Gaussian approximation, which yields simple calculations, represents a good conservative approximation and can be used to estimate the impact of FWMF in WDM transmission.

11.5 APPLICATIONS OF WDM IN OPTICAL NETWORKS

So far this chapter has considered WDM to increase the transmission capacity, trhough better utilisation of the fiber bandwidth. However, the WDM technique has a growing impact on the realisation of optical networks. Optical networks are those in which the path between any pairs of nodes, crossing several other nodes, remain entirely optical from end to end. These paths are called *lightpaths*. In fact, it is possible to take advantage of the huge fiber bandwidth by using wavelength to perform network and system-oriented functions such as routing, switching and service segregation. Different utilisations of WDM-based network architectures have been reported in the literature. The following section consider the basic principles and applications of WDM techniques and the next chapter considers relevant optical networks.

11.5.1 Principles and applications of WDM in optical networks

The two basic architectural forms that have been most commonly used in WDM networks are *broadcast-and-select networks* and *wavelength routing networks* [13], whose fundamental schemes are shown in Figs. 11.4 and 11.5, respectively.

The broadcast-and-select form of network operates by assigning a single optical frequency to the transmitting side of each port in the network, blending all the transmitted signals at the center of the network in an optical star coupler, and then broadcasting the complete frequency comb to the receiving sides of all ports. In principle, it is possible to place the wavelength tunability required for dynamic access in all receivers, all transmitters, or in both transmit and receive sides of each port. Figure 11.4

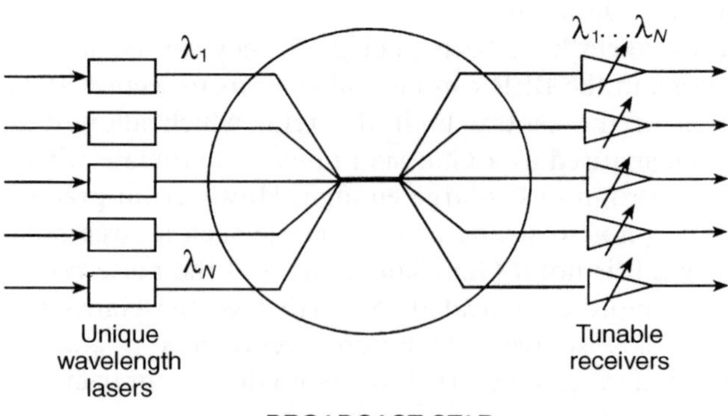

Fig. 11.4 The broadcast-and-select network.

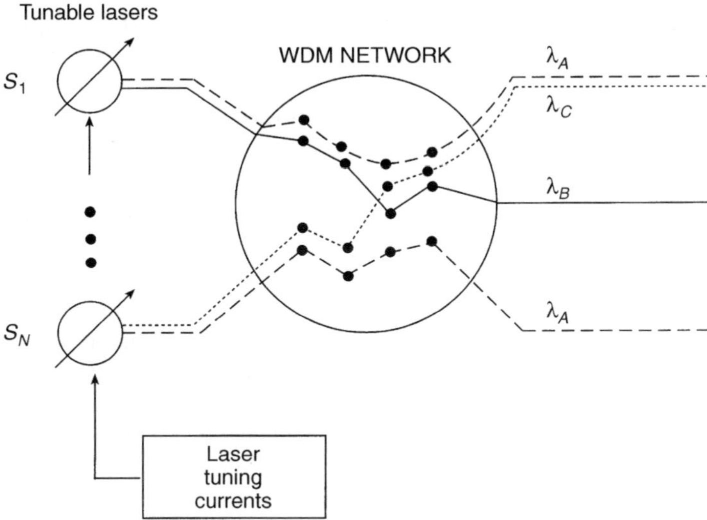

Fig. 11.5 The wavelength routing network.

shows the structure of a typical broadcast-and-select node that uses the fixed tuned sources and tunable receivers that represent commercially available technology. Using a suitable media access control (MAC) protocol [14], shown running as an application, when one node wants to communicate to another (either by setting up a fixed circuit or by exchanging packets), in the case that only the receivers are tunable, the destination receiver tunes for each interchange to the source transmitted wavelength and vice versa. The entire inner structure of the network (e.g. the star coupler, perhaps with further combining or splitting outside the star) is completely passive and unpowered, hence extremely reliable and easy to manage.

The design issues which have to be addressed with this type of networks imclude not only the access protocol, but the technology for tuning the receivers and making sure that the correspondence is maintained between the tuned receiver wavelength and that of the transmitter.

Aside from high cost, which is currently a problem with any optical network, there are two other problems: (1) the power from each transmitter is mostly wasted on receivers that do not use it, since it is broadcasted to all receivers; (2) the maximum number of nodes the network can manage is equal to the number Nrw of resolvable wavelengths. At present, the wavelength resolving technology allows the realisation of networks only up to 100 wavelengths [15]. And a network that allows only 100 nodes does not constitute a revolution. Technology must provide more wavelengths, and also some way must be found to assure additional scalability, that is the possibility of adding more nodes when required, by using each wavelength in many places in the network at the same time, so reusing wavelengths.

The wavelength routing network is composed of one or more wavelength-selective elements. There the path taken by the signal through the network is uniquely determined by the wavelength of the signal and the port through which it enters the network. In practice, at each intermediate node between the end nodes, light coming in one port at a given wavelength gets routed out of one and only one port by a *wavelength router* component that has purely optical paths. This a component could in principle be fixed or switchable; it could involve no change in the wavelengths of the optical signals passing through or it could enforce a pattern of *wavelength translators*.

Static routers provide flexibility for the inevitable change of traffic pattern, while wavelength-translating routers are costly and complex.

A wavelength routing network accomplishes the function of *wavelength-reuse*, which represents a problem for broadcast-and-select networks, and also avoids wasting the transmitted power, by channeling the energy trafrom each node along a restricted route to the receiver instead of letting it spread out over the entire network, as with the broadcast-and-select architecture.

To complete the lightpaths between end users, the settings of all the routers along the path need to be coordinated, either form a single central controller or by distributed and coordinated action of a set of controllers, one located at each router.

11.5.2 Broadcast-and-select networks

These types of network represent viable solutions for costructing local area netwroks (LANs) or even metropolitan area networks (MANs) with tens of high bit rate nodes utilising commercially available technology. The first system to be realised was the Lambdanet network, built by Bellcore in 1987 [16]. The Lambdanet design combines three basic elements: (1) it associates a unique optical wavelength with each transmitting node in a cluster of nodes; (2) the physical topology is that of a broadcast star; (3) each receiving node identifies transmitting nodes based on the transmission wavelength through wavelength demultiplexing. With 18 transmitters at 1.5 Gb/s per transmitter and 16 receiving nodes, the largest point-to-point bandwidth-distance product was 1.56 Tb/s · km, and the point-to-multi-point figure of merit was 21.5 Tb/s · km · node.

Other examples are the Rainbow-1 [17] and Rainbow-2 [18] networks, built by IBM and Los Alamos National Laboratory. The more recent Rainbow-2 is a MAN supporting 32 nodes each at 1 Gb/s over a distance of 10–20 km. Each node uses a separate fixed wavelength for transmitting data and a tunable receiver for receiving one of several data streams.

The all-optical network realised by the AON consortium of AT&T Bell Labs, MIT and DEC, is another relevant example. In this network, instead of the classical form of star coupler fashioned from many biconical tapered 3 dB couplers, the star coupler was of the more recent planar waveguide type. AON was able to cover 50 km.

11.5.3 Wavelength routing networks

Wavelength routing networks can be used for wide area networks (WANs) or backbone networks. The first demonstration of a wavelength routing network was achieved by British Telecom Laboratories [19]. Later, due to the scalability provided by wavelength reuse [20], wavelength routing became the main nucleus of ARPA's support of optical networking in the United States, and of many of the RACE programs in Europe, such as the network realised by the consortium MWTN (Multi-Wavelength Transport Network).

Besides WAN and backbone networks, it is worth mentioning the Passive photonic loop (PPL) [21], shown in Fig. 11.6, as an example of WDM-based access networks. It represents a costly solution for access networks, but could be a viable solution for future broadband access networks.

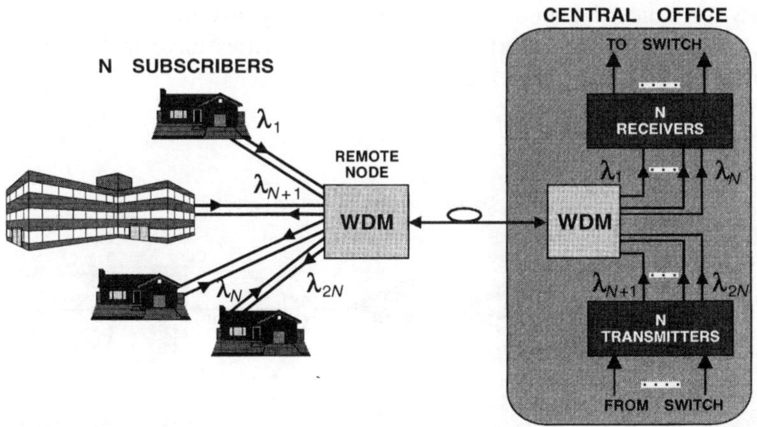

Fig. 11.6 An access network based on WDM: the passive photonic loop (PPL) [21].

11.6 SUB-CARRIER MULTIPLEXED (SCM) NETWORKS

So far, WDM has been considered in LANs, MANs, WANs and backbone networks. However, another multichannel approach has been very successful inaccess networks, the subcarrier multiplexed networks (SCM). This approach represent the most commonly used solution for realising cable television distribution networks [22, 23]. The SCM basic network is sketched in Fig. 11.7. A number of frequency-separated RF carriers (e.g. CATV channels), in either analog or numerical form, are frequency division multiplexed (FDM) in the RF domain. The resulting FDM signal modulates an optical laser source (by either direct or external modulation). At the remote node, e.g. placed in the curb (fiber-to-the-curb), the optical signal is detected, and the resulting RF signal is sent to all the receivers. Then each receiver can select any channel by an electric tunable filter.

Two main problems are to be taken into account when designing such networks: signal distortion due to the non-linearity of the laser source modulation, and splitting losses.

There are two modulation possibilities: direct or external modulation of the laser source. In the first case, the non-linear transcharacteristic of the laser (primarily due to the presence of the laser threshold, and secondarily due to gain saturation of the laser), introduces intermodulation distortions which affect the analog / numerical transmission. This effect, known as *clipping* [24], is the main impairment which has to be controlled in these applications. Besides clipping, the combined effect of laser chirping and fiber dispersion introduces further distortion [25]. When the laser source is externally modulated, perhaps by a Mach-Zehnder modulator, the intrinsic non-linear characteristic of the modulator still causes intermodulations distortions. In particular, third-order distortions affect the performance of the network, since the second-order contributions are minimised due to the characteristic of the modulator. Such non-linearities have to be minimised by introducing linearisation techniques.

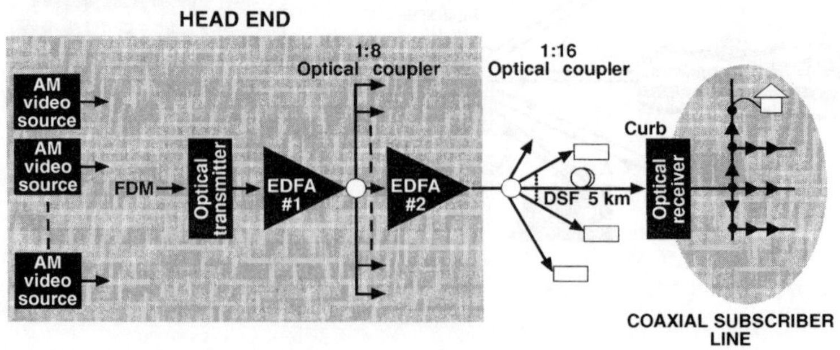

Fig. 11.7 A network based on SCM.

In any case, the high splitting losses often require, the adoption of EDFAs. A detailed analysis of this kind of network is beyond the scope of this book.

REFERENCES

1. Dense wavelength division multiplexing techniques for high capacity and multiple access communication systems, *IEEE J. Select. Areas in Commun.* **8**, (1990).
2. M.J. O'Mahony, Wavelength/optical frequency division multiplexing, in *High capacity optical transmissions explained,* Eds. D.M. Spirit and M.J. O'Mahony, John, New York (1995).
3. P.W. France, *Optical Fibre Lasers and Amplifiers*, Blackie, Glasgow (1991).
4. X.Y. Zou, M. I. Hayee, S.-M. Hwang, and A. Willner, Limitations in 10 Gb/s WDM optical-fiber transmission when using a variety of fiber types to manage dispersion and nonlinearities, *J. Lightwave Technol.* **14**, 1144–1152 (1996).
5. S. Betti, G. De Marchis, and E. Iannone, *Coherent Optical Communications Systems*, John Wiley, New York (1995).
6. E. Iannone and R. Sabella, Analysis of wavelength switched high-dense WDM networks employng wavelength converters based on FWM in semiconductor amplifiers, *IEEE J. Lightwave Technol.* **13**, 1579–1592 (1995).
7. R. Sabella, E. Iannone, and E. Pagano, Optical transport networks employing all-optical wavelength conversion: Limits and Features, IEEE J. Select. Areas in Commun. **14**, (1996).
8. A. Willner and S.-M. Hwang, Transmission of many WDM channels through a cascade of EDFA's in long-distance links and ring networks, *IEEE J. of Lightwave Technol.* **13**, (1995).
9. E. Iannone, F. Matera, A. Mecozzi, and M. Settembre, *Nonlinear Optical Communication Systems*, John Wiley, New York (1998).
10. N. Shibata, R.P. Braun, and R.G. Waarts, Phase-mismatch dependence of efficiency of wave generation through four-wave mixing in a single-mode optical fiber, *IEEE J. Quantum Electron.* QE23, 1205–1210 (1987).
11. R.W. Tkach *et al.*, Four-photon mixing and high-speed WDM systems, *IEEE J. Lightwave Technol.* **13**, 841–849 (1995).
12. E. Lichtman, Performance degradation due to four-wave mixing in multichannel coherent optical communications systems, *J. Opt. Commun.* **12**, 53–58 (1991).
13. C.A. Brackett, Dense wavelength division multiplexing networks: principles and applications, *IEEE J. Select. Areas in Commun.* **8**, 948–964 (1990).
14. R. Ramaswami, Multiwavelength lightwave networks for computer communication, *IEEE Comm. Mag.* 78–88 (1993).
15. H. Toba *et al.*, 100-channel optical FDM transmission/distribution at 622 Mb/s over 50 km using a waveguide frequency selection switch, *Electron. Lett.* **26**, 376–377 (1990).
16. M.S. Goodman *et al.*, The LAMBDANET multiwavelength network: architecture, applications, and demonstrations, *IEEE J. Select. Areas in Commun.* **8**, 995–1004 (1990).

17. F.J. Janniello, R. Ramaswami, and D.G. Steinberg, A prototype circuit-switched multi-wavelength optical metropolitan-area network, *IEEE J. Lightwave Technol.* **11**, 777–782 (1993).
18. W.E. Hall *et al.*, The Rainbow-II gigabit optical network **14**, 814–823 (1996).
19. H.J. Westlake *et al.*, Reconfigurable wavelength routed optical networks: a field demonstration, *ECOC '91*, pp. 753–756 (1991).
20. C.A. Brackett *et al.*, A scalable multiwavelength multi hop optical network: a proposal for research on all-optical networks, *IEEE J. Lightwave Technol.* **11**, 736–753 (1993).
21. S.S. Wagner *et al.*, A passive photonic loop architecture employing wavelength-division multiplexing, *Proc. GLOBECOM '88*, pp. 1569–1573 (1988).
22. T.E. Darcie and G.E. Bodeep, Lightwave subcarrier CATV transmission systems, *IEEE Trans. Microwave Technol.* **38**, 525–533 (1990).
23. T.E. Darcie, Subcarrier multiplexing for multiple access lightwave networks, *IEEE J. Lightwave Technol.* **LT5**, 1103–1110 (1987).
24. N.J. Frigo, M.R. Phillips, and G.E. Bodeep, Clipping distortion in lightwave CATV systems: models, simulations, and measurements, *IEEE J. Lightwave Technol.* **11,** 138–146 (1993).
25. R. Sabella, M. Paciotti, and A. Di Fonzo, Analysis of clipping and chirping effects on AM-VSB CATV subcarrier multiplexed optical systems, *J. Opt. Commun.* **18**, (1997)

12
All-Optical Networks

12.1 INTRODUCTION

The principal motivations for all-optical networking arise from the ability of optical fiber technology to fulfil the growing demand for bandwidth per user, protocol transparency, higher path reliability, and simplified operation and management. In all these areas, established approaches realised via electronic circuitry, based on time division multiplexing (TDM) are beginning to prove insufficient, as they cannot perform the required operations as cheaply as the all-optical techniques, assuming they can perform them at all.

The traditional TDM technique, which in its electronic form has served so well throughout the history of digital communications, is in fact becoming inadequate, due to increasing operational speed. In fact, TDM demands that each port handles not only its own bit streams but also those belonging to many or most of the other ports on the network, whatever the system topology. The bit rate of the front-end electronics then scales as the product of the number of ports and the per-port bit rate [1]. As networks evolve to higher values of this product, the values inevitably exceed that of the fastest available digital technology. The TDM *bottleneck* can be overcome by adopting wavelength division multiplexing (WDM) and space division, which demand that each electrical port handles only its own bit streams.

The development of very short pulse technology has allowed the possibility of realising all-optical TDM (OTDM) networks [2] for very localised networks in which dispersion is kept small, and for larger networks by use of soliton propagation, in which dispersion effects are neutralised by nonlinear effects [1]. Compared to WDM networks, OTDM networks are in their infancy, partly due to the very primitive and expensive nature of the required devices. Besides dispersion effects, the OTDM approach of transmitting a number of bits per frame equal to the number of bits per node times the number of nodes supported has two other disadvantages: it destroys the protocol transparency by dictating the framing format, and it exacerbates synchronisation problems, since timing must be more accurate by a factor equal to the number of nodes. Code division multiaccess (CDMA, or spread spectrum) intensifies the problems of dispersion and synchronisation even further [1].

As a result, WDM still represents a valid solution for optical networking, providing several advantages, including higher aggregate bandwidth per fiber, new flexibility for automated network management and control, noise immunity, transparency to different data formats and protocols, low bit error rates, and better network configurability and survivability, all leading to more cost-effective networks.

The main issues relating to WDM networking are discussed in section 12.2. The main objective of this chapter is to analyse large-scale networks, which cover large geographical areas, for high-speed telecommunications. At present, two basic approaches have been adopted by several large consortia, e.g. ARPA in the United States and RACE in Europe: multihop networks and photonic transport networks. Multihop networks are covered in section 12.3 and the rest of the chapter is devoted to photonic transport networks, which represent the core of network evolution towards a global telecommunication infrastructure and information superhighways.

12.2 BASIC ASPECTS OF WDM NETWORKING

Optical networking began with a few very simple and basic concepts, and is evolving dramatically toward solving the real-world issue of constructing large-scale networks that are robust, regardless of failure and traffic rising, and that will evolve smoothly with time and size. The advances in optical technology have been allowing the realisation of wavelength routing elements, such as highly flexible optical cross-connects (OXCs), exploiting the concept of wavelength routing, discussed in the previous chapter. There are several key topics, relating to optical networking, which have to be analysed when dealing with optical networks [3]:

(1) structure of the wavelength routing elements,
(2) scalability of the network,
(3) wavelength translation,
(4) transparency of the network,
(5) network layering,
(6) network operation and management (OAM) and control.

12.2.1 Wavelength routing elements

Wavelength routing is defined to be the selective routing of optical signals according to their wavelengths as they travel throughout the network elements between source and destination. Wavelength routing is accomplished by realising, in some way, wavelength-selective elements at the nodes. In general we can distinguish two possible approaches: fixed wavelength routing and flexible wavelength routing.

Fixed wavelength routing would most probably use WDM multiplexers in a back-to-back configuration, to allow interchange of wavelengths

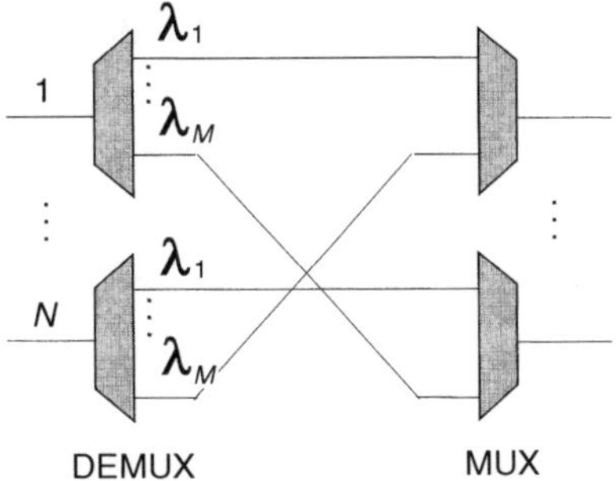

Fig. 12.1 Configuration of the WDM cross-connect.

between input and output ports (fibers) in a prearranged pattern. This configuration (Fig. 12.1) is usually called WDM cross-connect and allows no rearrangeability.

Rearrangeability can be introduced by inserting a space division switching function by using space switching matrices (Fig. 12.2). Here each wavelength, incoming from any input fiber, can be routed to any output fiber that is not already using that wavelength. This OXC presents a bandwidth proportional to NMB, where N is the number of input/output fiber ports, M is the number of wavelengths carried by each fiber, and B is the bit rate per wavelength.

The constraint that two channels, carried on two different fibers at the same wavelength, cannot be routed simultaneously onto a single outgoing link can be tolerated or not, depending on the network topology, dimensions, traffic, OAM functions, etc. However, this constraint can be eliminated by using wavelength translators in conjunction with a large switch inside the optical node, as shown in Fig. 12.3. On the other hand, this configuration adds significant complexity to the routing node structure, but permits better wavelength reuse. To avoid large space division

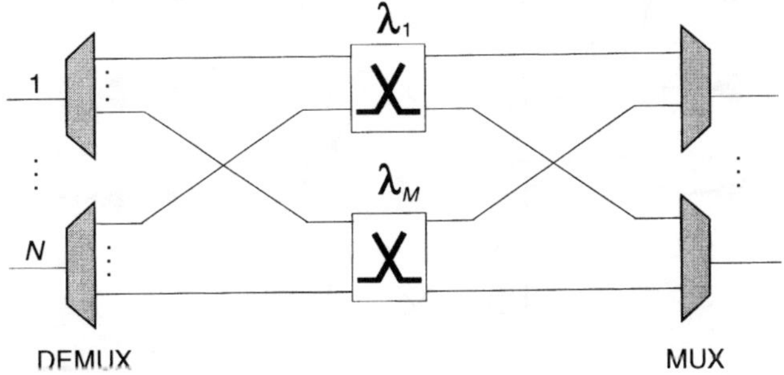

Fig. 12.2 Configuration of a rearrangeable cross-connect.

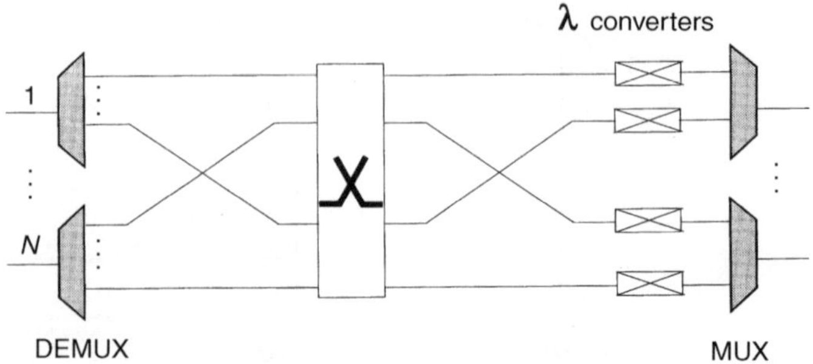

Fig. 12.3 A strictly non-blocking optical cross-connect using wavelength converters and a large switch matrix.

switches, which are impractical, especially for large dimensions of the cross-connect (*NM*), the architecture shown in Fig. 12.4 can be adopted, allowing the highest flexibility to be obtained. In this architecture, channel selection is accomplished by a combination of passive power splitters and tunable filters. The large switch is thus substituted by several low-dimensional switch matrices.

There is a trade-off between costs and flexibility of the optical nodes. Consequently, the choice of the structure strongly depends on the different network applications.

The importance of the OXC, and the closely related WDM optical add/drop multiplexer (OADM), which simply adds channels into the network and extracts channels from the network [4–6], is that they allow the optical network to be reconfigured on a wavelength basis to optimise traffic, congestion, network growth and survivability. They also allow special circuits to be configured for signals in alternative format. As a result, both

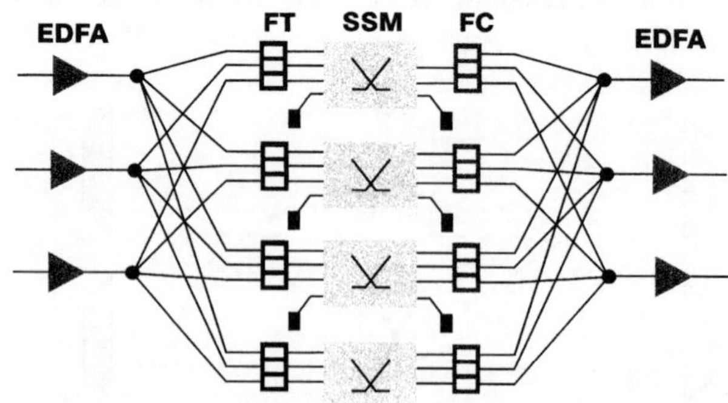

Fig. 12.4 A flexible and strictly non-blocking cross-connect: FT = tunable filter, SSM = 4 × 4 space switching matrix, FC = frequency (wavelength) converter.

OXCs and OADMs are the essential transparent elements upon which multiwavelength networks will be realised.

12.2.2 Scalability

Scalability means it is always being possible to add more nodes in the network; it is a critical requirement. By deploying more copies of the same equipment, a network can handle an increasing number of users located over a growing service region while offering higher aggregate capacity. Scalability dictates wavelength reuse [7]. As a matter of fact, if the number of wavelengths is eventually limited, the implication is that ultimately the number of nodes in the network must be completely independent of the number of wavelengths.

Wavelength routing alone is not sufficient to allow scalability in optical networks. A high level of scalability can be reached if any channel can be switched to a new wavelength path, which may be at a different wavelength, thereby requiring wavelength translation. This effectively requires both space and wavelength switching; not only are channels switched to different output space ports in each node, but they are also switched from one wavelength to another. This is done, in the OXC of Fig. 12.4. It can be performed on a circuit-switched basis or on a packet-switched basis [3]. If switching from one wavelength to another can be entirely optical, then a transparent all-optical network can be realised with 'true' scalability.

12.2.3 Wavelength translation

The necessity of wavelength translation within a network remains controversial. Two approaches have been reported in the literature dealing with the possibility of performing wavelength translation or not: wavelength path (WP) and virtual wavelength path (VWP) [3], depending on whether a channel is carried by the same wavelength along the complete transmission path throughout the network, or if it is translated onto other wavelengths at least once during transmission. Different results have been reported on this matter [8–11]. At present, it is not possible to draw general conclusions, butconsider this. Current telecommunications networks are divided up into regional administrative domains with simplified network interfaces in order to solve the problem of complexity. It is not considered essential to have current and complete network knowledge in a centralised location of a large network. Instead, each domain interacts with its neighbors to request call set up, for instance, or to isolate faults, without knowing the details of its neighbor networks' connectivities, etc. At the same time, it is not likely to be feasible to setup end-to-end transparent paths on a single wavelength across multiple administrative domains in an efficient and robust way. As a result, if not in each node and for each channel, wavelength translation will be introduced in optical networks at least partially [3].

12.2.4 Transparency

This term is used with different meanings, and with different weights. The absolute transparency can be intended as the property of a network for which any signal travels along the network independently of its transmission format, speed (bit rate in case of PCM signals), etc.; that is, only terminal equipment would determine the limitations on the signal format. However, due to the physical limitations of fiber propagation and the physical nature of optical devices crossed by the signal, the absolute transparency can never be reached. So it is more useful to specify a certain level of transparency. The simplest degree of transparency is to digital signals (independence of the bit rate, format and protocol). Then it is possible to define a transparency to intensity-modulated signals (analog and digital). Full transparency would require a network to be transparent to any optical signal with amplitude, phase or frequency modulation.

When considering transparency of an optical network, it is necessary to take into account the very real limitations of the physical medium and the physical nature of any optical devices used in the network, some of which are not easily avoided. What will work in a local network environment will not necessarily work in a national or international network. Hence it is necessary to comprehend the required level of transparency and to evaluate the costs of achieving this degree of transparency.

12.2.5 Network layering

The introduction of WDM technology opens the possibility of realising network layering (section 12.4). The optical network constitutes an infrastructure which can handle both synchronous transfer mode (STM), through the SDH/SONET hierarchies, and asynchronous transfer mode (ATM). This concept allows the extension of current networking directions and trends without having to displace what already exists. The only exception is related to the problem of OAM functions in transparent and rearrangeable systems.

12.2.6 OAM functions and control in transparent systems

There are two concerns about transparent networks. One relates to monitoring the state of the network, since in a transparent network the normal digital information about network performance is not available. The other one is that current transport systems, such as SDH and SONET, have well-defined internal means for dealing with faults and performance monitoring, and whatever is done at the optical layer must work together with the transport layers. The relevant topic is therefore the way in which the network control information should be carried. Furthermore, different transport systems may share the same medium, exploiting the transparency of the network. Each of these different systems has some internal means for dealing with faults, etc. Network management and control represent complex tasks to be investigated for future networks.

12.3 MULTIHOP NETWORKS

The multihop architecture represents a possible approach for realising scalable optical networks. Its basic structure [12, 13], reported by the ARPA consortium, is shown in Fig. 12.5. It consists of an all optical inner portion that contains passive wavelength-routing cross-connecting elements and a common network control to allow their dynamic rearrangeability utilising, for instance, the properties of acousto-optic wavelength switches [14]. This rearrangebility allows the dynamic allocation of wavelength and capacity through the network to meet changing traffic, service and performance requirements, and to provide a robust, fault-tolerant network. The transparency of the inner part accommodated multiple service formats, such as simultaneous digital and analog transport. Each network access node is able to transmit and to receive from several other nodes by selecting the appropriate wavelength. With an appropriate

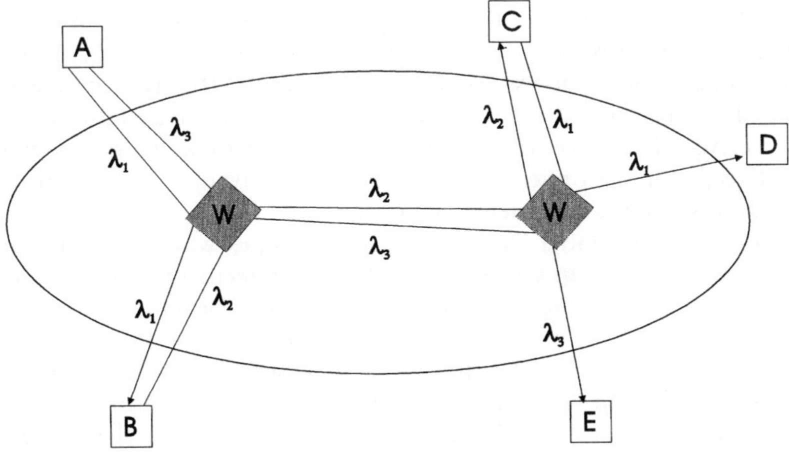

Fig. 12.5 Typical structure of a multihop architecture.

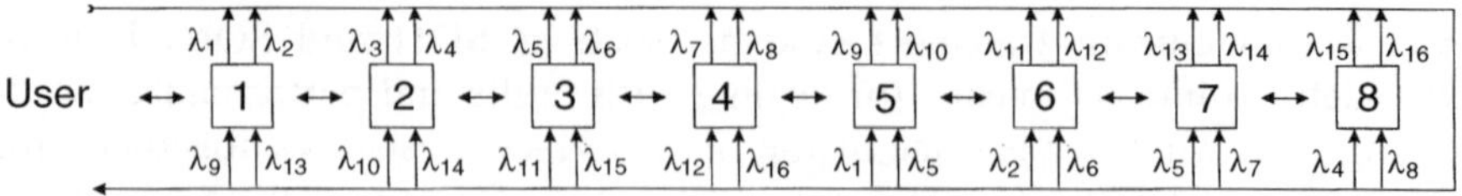

Fig. 12.6 Practical configuration of a multihop architecture with eight-nodes.

configuration of the wavelength-routing cross-connects, this transparent optical transmission may extend over large distances and through many WDM cross-connects.

Scalability, in this network, is achieved by incorporating the wavelength translation function, which consists of receiving a signal and retransmitting it on another wavelength. Because wavelength translation is performed in the access node, it can come under the same control as provided by the packet routing headers. Thus, the result is packet-by-packet wavelength translation. Besides providing the wavelength translation function, the access nodes also form the basis for the user interface by providing ATM switching and user access at the desired data rates and signal formats [12].

The basic approach behind the multihop architecture is shown in Fig. 12.6 [15], where an eight-node network is considered. The access nodes are positioned linearly along a bus, although the physical architecture can be a bus, a routed tree, a star etc. Each access node is provided with two optical transmitters and two optical receivers. Each transmitter operates at a wavelength unique to that transmitter. The signals produced by the different transmitters are combined on the shared medium by means of wavelength multiplexing. The signals launched by each access station are coupled onto the 'transmit' optical bus using passive couplers, and propagate along the upper bus toward the right. Then the combined signals are looped back and propagate to the left on the lower 'broadcast' bus. Using passive couplers, a portion of each signal is coupled to each station. Using optical filters, each access station then accepts the two wavelengths intended for local reception. Each receiver accepts a wavelength unique to that receiver. Each access station is equipped with a bidirectional user port. Except for the optical transmitters and receivers, all other functions of the access nodes are implemented in electronics.

Since each transmitter is assigned to a unique wavelength and each receiver accepts a unique wavelength, the assignment of a wavelength to the transmitter of one access node and the receiver of another node effectively creates a dedicated one-way channel existing at that wavelength between the two access nodes. In Fig. 12.6 the wavelengths are assigned to the access nodes in such a way as to provide the dedicated one-way connections shown in Fig. 12.7. In other words, Fig. 12.7 is the connection graph of the eight-node multihop network of Fig. 12.6. In fact, from a connectivity point of view, the eight nodes of Fig. 12.6 are arranged in two

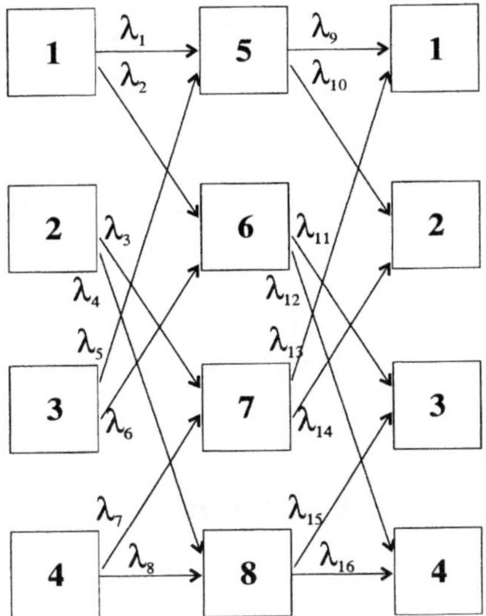

Fig. 12.7 Connection graph for the eight-node multihop network of Fig. 12.6.

columns of four nodes each. The four nodes in the first column are connected to the four nodes in the second column, via a perfect shuffle diagram using eight wavelengths; and the four nodes in the second column are reconnected back to those of the first one, via a perfect shuffle diagram using a second set of eight different wavelengths. This connection diagram is called recirculating shuffle [15].

The wavelength routing concept allows the network to be realised such that the number of wavelengths and the number of nodes are not related to each other. Aggregate network capacity would still suffer because the number of hops would become enormous, causing most of the capacity of each optical link to be consumed by relayed packets. However, it is practical to consider the use of up to a few tens of wavelengths. If each network access node can transmit up to m wavelengths, then an m-connected perfect shuffle can be generated, with the same form of wavelength routing shown in Fig. 12.7. It is easily shown [15] that the expected number of hops and the overall capacity, assuming uniformly distributed traffic, are given as

$$E(k,m) = \frac{km^k(m-1)(3k-1) - 2k(m^k-1)}{2(m-1)(km^k-1)},$$
(12.1)

$$C(k,m) = \frac{B \cdot N \cdot m}{E(k,m)}$$
(12.2)

where $N = k\,m^k$ is the number of access nodes and B is the bit rate being transmitted on each channel. Notice that for 8 wavelengths and $k = 8$, the network supports as many several hundred million nodes, with an average number of hops between users of about 12. The use of more wavelengths provides better performance by decreasing the expected number of hops. If each node were to transmit at about 1.25 Gb/s, such a network for 10^8 nodes would, in theory, be capable of carrying 10^7 Tb/s total traffic.

As a result, the principles of wavelength routing and translation can be used to produce truly scalable networks with a very modest number of wavelengths.

12.4 OPTICAL TRANSPORT NETWORKS

The advent of new telecommunication services and the eventual introduction of broadband communications will considerably increase communications traffic. Since a transport network is a large and complex network integrating different technologies and services, an appropriate network model, with well-defined functional entities, is necessary for its design and management. The transport network can be divided into independent transport network layers, as depicted in Fig. 12.8 [16]. This architecture consists of three layers: circuit layer, path layer and transmission media layer. The layering concept is being extensively discussed by ITU-T for SDH and ATM networks and leads to simplifications in the network design, development and management, and permits smooth network evolution in pace with user demand. The introduction of the layered structure even makes it easy for each network layer to evolve independently of the other ones. In particular, the path layer links the circuit layer with the transmission media layer by means of the path layer devices, the digital cross-connects (DXCs). Their introduction represents a significant element for the realisation of reliable and flexible networks.

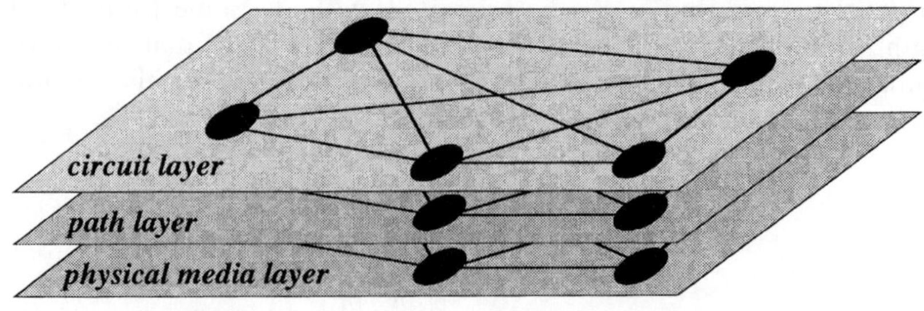

Fig. 12.8 The layered structure of the transport network.

The present evolution of the telecommunication network is characterised by the introduction of SDH and ATM transfer modes in the path layer, which require a basic support of electronic technologies for switching and processing. On the other hand, optical transmission is the only technology assuring the required transmission capacity. To date, optical technologies have only been introduced to the physical media layer: 2.5 Gb/s systems have already been deployed and systems with more than 10 Gb/s transmission capacity are under development, all employing time division multiplexing (TDM) techniques.

Technical advantages in WDM technology suggest that its practical application is now feasible. WDM techniques could also be suitable for the path layer, allowing enhancement of the network potential. In fact, WDM optical path technologies can be a viable approach to achieve very high transmission rates cost-effectively by exploiting the great optical fiber potential. Furthermore, WDM techniques allow the expansion of the cross-connect node processing capability, due to the fact that the electrical bottleneck is eliminated through the introduction of wavelength routing of paths at cross-connect nodes. In other words, wavelength routing will considerably extend cross-connect node throughput. As a consequence, optical technology can be reliably introduced in the path layer and optical cross-connects (OXCs) become important components when constructing optical path networks. Several examples of OXC architectures based on WDM multiplexing and wavelength routing have already been proposed in the literature [17–20]. Solutions based both on space and wavelength switching have been studied and the realisation of large-scale demonstrators is in progress in the frame of large research projects both in Europe [17] and in the United States [12].

The employment of wavelength translation in the optical layer of the transport network based on WDM is a very important issue that has to be taken into consideration [21]. Wavelength translation in a WDM transport network plays a role analogous to the exchange of time slots in a TDM frame in standard TDM networks: it can be used to design wavelength switches. Even in spac-switching architectures, wavelength conversion is useful as a means to avoid contentions inside the switch reducing the number of spatial cross-points. Moreover, by means of wavelength converters, different networks adopting the same set of carrier frequencies can be connected via an optical bridge.

12.4.1 Structure of the optical path layer

The layered structure of the transport network, as defined by ITU-T, is shown in Fig. 12.8, where three sublayers are shown with the main layer entities:

- The circuit layer manages the end-to-end connections (circuits) established dynamically or on the basis of short-term provision; each circuit is related to a particular service, such as telephone and data transmission.
- The transmission media layer provides point-to-point interconnections between network nodes.
- The path layer bridges these two layers with digital cross-connects (DXCs). In particular, different circuits, related to different services, are united to form a path and routed through the network. Network restoration is also realised in this layer.

A transport network, which has to be robust to future evolution, would be achieved by introducing an optical path layer between the transmission media layer and the electrical path layer [22]. Very high capacity data streams are routed through the optical layer by means of OXCs without optical-to-electrical conversion. To allow demultiplexing and routing at lower levels of the digital hierarchy, the OXC is interfaced with a DXC. However, if the high-speed optical signals has to be directly delivered to the access network, it can be done by including a ring of optical add/drop multiplexers (OADMs). A picture of a transport network including an optical path layer is shown in Fig. 12.9.

Functions such as optical path routing and network restoration can be directly realised in the optical path layer by optical technologies, allowing different electrical path technologies, e.g. plesiochronous digital hierarchy (PDH), synchronous digital hierarchy (SDH) and asynchronous transfer mode (ATM) to be integrated over the same optical platform, with a consequent advantage in network feasibility and flexibility [18, 22]. Even

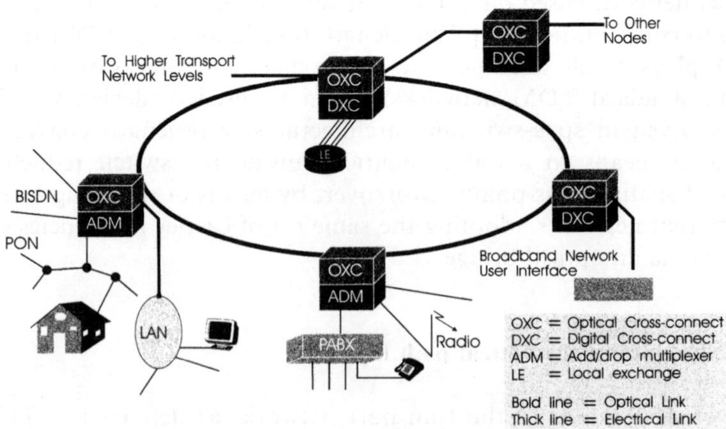

Fig. 12.9 A transport network including an optical path layer.

analog signals can be transported within optical paths, so that new application of the transport network can be included. For example, optical
cable television (CATV) signals, in a sub-carrier multiplexing format, can
be transported through the network, from a production centre to the local
exchange, where the CATV signals are sent to the access network to be
broadcast between different users.

Two main schemes can be achieved to route optical paths through the
optical path layer: the wavelength path (WP) and the virtual wavelength
path (VWP) [18]. In the WP scheme each optical path is established
between two nodes by assigning one wavelength to the complete path. In
the VWP scheme, the wavelengths are allocated on a link-by-link basis,
thus requiring a wavelength translation function inside the optical node.
The intermediate nodes along the transmission path (according to either
the WP or VWP scheme), perform the optical routing so that the electrical
processing bottleneck is removed. The VWP scheme requires wavelength
conversion at each node, so the node structure tends to be more complicated than with the WP scheme. On the other hand, the wavelength assignment problem has to be solved so that different wavelengths are assigned
to each WP existing in the network. Moreover, if fault restoration procedures are considered, the number of wavelengths for the WP scheme is
considerably larger than for the VWP scheme [23, 24].

A possible evolution of the state-of-the-art transport network toward a
network which contains an optical path layer is shown in Fig. 12.10.

The first block refers to a network in which optical technology is limited
to the transmission media layer, and the path layer is based on PDH and
STM technologies. The next step would be represented by the partial introduction of the optical technology in the path layer by WP or VWP. Such
an optical layer would interface the upper electrical layer based on PDH
or STM. Then the ATM technique, which is based on the virtual path

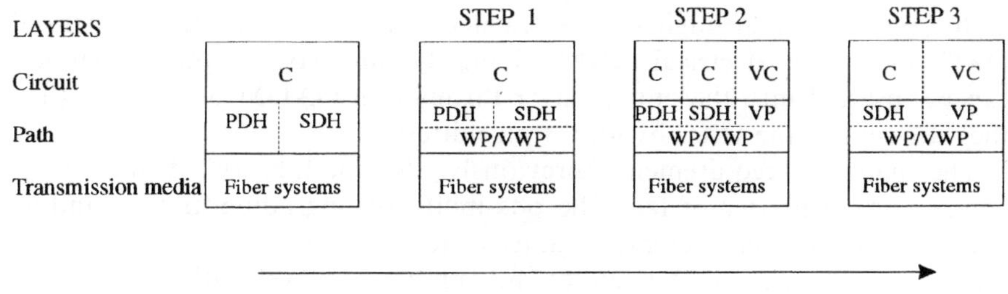

Fig. 12.10 A possible evolution of the state-of-the-art transport network toward a
network which contains an optical path layer.

(VP) principle, could be introduced in the path layer or even in the circuit layer by defining virtual circuits (VCs). Finally, both PDH in the electrical layer and WP in the optical layer would tend to disappear, and the network would have an optical path layer based on VWP, implementing STM or VP in the electrical path layer.

From these considerations, the requirements of the optical path layer can be easily stated:

(1) Transparency is the possibility of handling optical signal having any type of bit rate and transmission format. This means it is possible for the optical nodes to manage different types of signal. Moreover, an increase in the bit rate or the transmission format allows the structure of the OXCs to be manteined, but not the transmitters and the receivers.

(2) Scalability means it is always possible to add further nodes to the networks, allowing gradual upgrading without significantly changing the network structure.

(3) Modularity means the property of being able to add only the desired number of new nodes, and updating the logical connectivity diagram to include these new nodes.

These requirements are discussed in the following section.

12.4.2 Optical cross-connect functions

A generic node of a transport network path layer could be designed as an OXC and a DXC. The OXC performs the switching and routing functions of high-speed optical channels (digital or even analog channels). When the high-speed channels have to be directly sent to the optical access network, the OXC route them towards optical rings which contains OADMs. When the channels have to be demultiplexed to lower levels of the digital hierarchy and routed through the electrical network, they are sent to a local DXC and/or an electrical ADM. Similarly, high-speed optical channels can be inserted into the optical layer through an OADM or a DXC, then they are sent to the OXC by local transmitters.

The network requirements previously the modularity of the cross-connect. As a matter of fact, the possibility of upgrading and expanding the network in small increments allows the traffic to be increased (e.g. by augmenting the channel bit rate or the number of wavelength channels per fiber), or the number of nodes in the network to be increase. In this case the OXCs have to handle more input/output links, without an appreciable change in their structure except the addition of new components. In particular, it is useful to distinguish two types of modularity [18]:

- *Link modularity*: the structure of the OXC is modular if the number of input/output links is increased.
- *Wavelength modularity*: the OXC is modular if the number of wavelength channels per link is increased.

To fulfil network scalability and modularity, link modularity is the strongest feature. On the other hand, wavelength modularity is useful when it is necessary to increase the traffic volumes to be handled in the network, without augmenting the number of fiber links.

12.5 TRANSMISSION THROUGH OPTICAL TRANSPORT NETWORKS

The transmission performance of an optical signal throughout the optical layer of the transport network represents a key issue when considering flexible and transparent networks [25]. The ultimate vision would be a perfect transmission along the network whatever the distance covered by the signal and the number of network elements crossed by the signal itself. Unfortunately, the transmission is affected by several impairments. In particular, it is necessary to evaluate the transmission performance of a signal traveling to the network, for at least three reasons:

- to investigate the impact of a given technology, which allows the realisation of a particular device included in the network (e.g. laser transmitters, optical filters, switching matrices);
- to analyse the practicability of a given cross-connect architecture;
- to evaluate the feasibility of a given network structure from a physical viewpoint.

Henceforth, a tool is required to evaluate transmission performance. One simulation model [25] is based on the models discussed in the previous chapters. This model already been used in a number of investigations into different technologies, OXC architectures and geographical networks.

12.5.1 The transmission performance simulation model

To evaluate the network transmission performance, we can consider a generic signal path through the network that can be modeled as a chain of an originating OXC, containing the optical transmitter, N OXCs separated by fiber links and a final OXC containing the receiver. Without losing generality, we assume an OXC like in Fig. 12.11 (it is similar to Fig. 12.4 but with the adjunct of a set of optical transmitters and receivers for add/

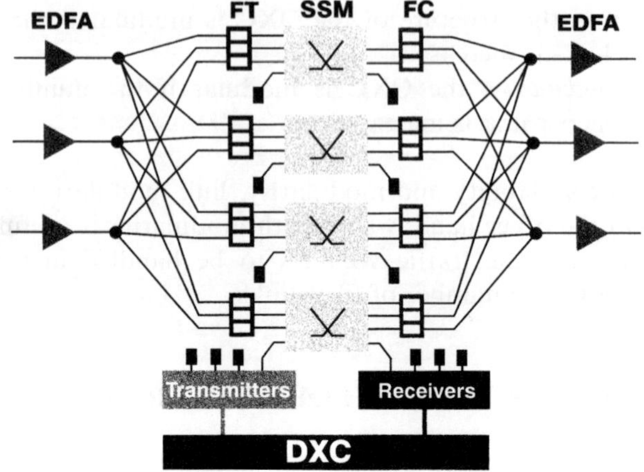

Fig. 12.11 Architecture of an OXC, employing space switching matrices and wavelength converters, interfacing an electrical cross-connect (DXC).

drop functions). Thus $N + 1$ wavelength conversions occur along the signal path and, in the case of uniformly spaced wavelength converters, the link is $(N + 1)$L kilometers long, where L is the OXC spacing.

The analysis is carried out through the semi-analytical method of Chapter 10, whose basic steps are summarised below:

- The evolution of the optical signal along the transmission path is numerically simulated, taking into account an accurate physical modeling of the optical devices. Phase noise is also simulated; in particular, when matching the transmitting laser and the FWM converters, a random phase following a Wiener process with a suitable variance is added to the signal phase.
- The noise power contributions in front of the receiver, due to amplified spontaneous emission (ASE) from both EDFAs and semiconductor optical amplifiers (SOAs) which are used for wavelength conversion, are analytically evaluated starting from the noise characteristics of the EDFAs and the SOAs. The electrical noise power at the receiver is also evaluated.
- The crosstalk is considered as an approximately additive Gaussian noise whose power is analytically evaluated and summed to the optical noise power affecting the receiver performances.
- The error probability at the receiver is analytically evaluated as a function of the average signal envelope and the optical and electrical noise power adopting the characteristic function method and the saddle point approximation [19]. This method is accurate since the non-linear

coupling between the signal and ASE noise along the fiber propagation can be neglected in all the cases considered [26, 27].

- The error probabilities, evaluated for each bit of a random signal pattern of a number of bits such that the optical link memory is taken into account,[1] are averaged to obtain the overall system error probability.

The approximation adopted in the crosstalk analysis is noteworthy. Crosstalk due to the tails of the channel spectra and to imperfect filtering allows the main crosstalk contribution to be simulated by the nearest neighbors of the selected channel. However, this ceases to be possible for non-linear crosstalk, induced by FWM during fiber propagation. In this case a single crosstalk component contains the contribution of all the transmitted channels, so the propagation of the whole transmitted optical spectrum should be simulated. This approach would lead to a tremendous computational complexity. In the case of a multicarrier optical network this approximation gives good results both for linear and non-linear crosstalk [28–30].

In the following, the modeling adopted to simulate the optical devices and the procedure to evaluate the error probability will be reported synthetically.

Transmitter

At high bit rates it is quite important to limit the spurious chirping introduced by the transmitter. Thus the transmitter is composed of a DFB semiconductor laser and an external intensity modulator. Either InP- or LiNbO3-based modulators can be simulated by the physical models reported in Chapter 3.

Optical fiber

In an optical transport network with very long links, a high transmission power has to be used to compensate the losses of fibers and OXCs. In this condition, fiber propagation cannot be assumed to be linear, and Raman and Kerr effects must be taken into account.

However, if the peak channel power is about 3 dB below the Raman threshold, this effect can be neglected; if it is above threshold, correct transmission is practically impossible. Thus the Raman effect can be avoided by limiting the peak optical power along the link 3 dB below threshold.

The Kerr effect introduces signal distortion that must be taken into

---

[1] In this way, phase noise and the pattern effects due to dispersion and chirping are taken into account.

account when simulating signal propagation. In the presence of the Kerr effect, neglecting the evolution of the signal polarisation, propagation of the optical field $E(z, t)$ along the fiber is described by the non-linear Schrödinger equation (Chapter 4). This equation is solved numerically by the split-step method, where the fiber is subdivided in small sections to separate the linear effects related to dispersion and attenuation from the instantaneous non-linear Kerr effect.

Fiber amplifiers and filters inside the OXC

The EDFAs can be considered as linear amplifiers of infinite bandwidth. A 3 dB bandwidth equal to 30 nm can easily be obtained by an EDFA amplifier, so this approximation is well satisfied in all the cases we will consider (e.g. the transmission of four 10 Gbit/s channels at 3 nm spacing).

The OXC filters can be modeled as in Chapter 6.

All-optical wavelength converter

Devices based on semiconductor optical amplifiers (SOAs) seem more promising for photonic switching applications because they provide a high operational speed and a high conversion bandwidth. Moreover, they are compact, they require a limited pump power (generally within a few milliwatts) and they can be integrated with other semiconductor devices (such as pump lasers and acousto-optic filters).

The modeling in Chapter 5 can be directly applied to the simulator, both to evaluate the expression of the optical signal at the output of the device itself, and the noise power contribution introduced by the converter.

Receiver

The optical receiver is modeled by a p-i-n photodetector followed by a wideband amplifier and a linear pulse-shaping filter. The quantum efficiency of the p-i-n diode has been assumed to be 85%. The electrical noise at the amplifier outputs is assumed to be thermal noise. This is a Gaussian white noise whose power spectral density is given by

$S_{rec} = F_{rec}k_B\theta / R_L$ where F_{rec} is the amplifier noise figure, k_B is the Boltzmann constant, θ is the absolute temperature and R_L is the load resistance.

The transfer function of the electrical filter can be chosen arbitrarily. Usually it is convenient to assume a Gaussian filter with a cutoff frequency equal to $R/2$ (R the bit rate).

ASE noise accumulation

ASE generated by EDFAs and SOAs accumulates along the optical link. The optical noise in front of the receiver can be assumed to be a complex white Gaussian noise. The power spectral density of the real and imaginary part of the optical noise can be written as

$$S_\eta(b) = 1/2[S_{orig}(b) + N_{amp}S_{EDFA}(b) + S_{dest}(b)] \qquad (12.3)$$

where $b = 0, 1$ represents the transmitted bit; the factor $\frac{1}{2}$ takes into account that the noise power is equally divided between the polarisation modes, $S_{orig}(b)$, $S_{EDFA}(b)$ and $S_{dest}(b)$, are respectively the power spectral densities of the originating OXC (derivided the active devices, from EDFA and semiconductor optical amplifiers, if present somewhere in that OXC), of the EDFAs crossed by the signal (inline amplifiers between two OXCs and EDFAs present inside any OXC), and of the last OXC. The various spectral densities present in (12.3) are evaluated starting from the structure of the OXC, the equation giving the power spectral density of the ASE noise introduced by an EDFA and by the equations relating to the SOA, in Chapter 5.

Crosstalk analysis

Three main crosstalk mechanisms affect system performance: hetero-wavelength linear crosstalk (HEC), homowavelength linear crosstalk (HOC), and nonlinear crosstalk (NLC).

Heterowavelength crosstalk

Heterowavelength crosstalk (HEC) arises from the spectral tails of the other channels entering the bandwidth of the selected channel and from the imperfect optical filtering. In a real network the spectrum of a signal is distorted by its transit through the OXC due to the presence of optical filters. In particular, the spectral tails are attenuated so that the amount of crosstalk depends on the path followed by the interfering channels through the network. Moreover, it depends on the relative polarisation of the channels, on the relative phase of the optical carriers at the beginning of the considered bit interval and on the relative phase of the signals transmitted on the different channels. Since it is not possible to account rigorously for the path followed inside the network by each interfering channel, a worst-case approach has been adopted. It is assumed that the spectra of the interfering channels are not altered in their transit through the OXCs before the crosstalk generation. Moreover, due to the random polarisations and phases of the interfering channels, the crosstalk term is a random process given by the sum of a generally large number of independent terms. Consequently, if the number of nodes N is large enough, the crosstalk term can be considered with a good approximation to a Gaussian additive noise. Under these hypotheses, it is possible to obtain the overall HEC power σ_{cr}^{2} as follows [25]:

$$\sigma_{HEC}^{2} = \int_{-\infty}^{\infty} |H_{s}(\omega)|^{2} \cdot \sum_{k=1}^{N} |H_{x}(\omega)|^{2(N+1-k)} \cdot \sum_{j=1,j\neq y}^{M} S(\omega + \Delta\omega)d\omega, \qquad (12.4)$$

In equation (12.2) the selected channel is indicated by y,
$H_{x}(\omega)$ is the transfer function of the OXC,
$H_{s}(\omega)$ is the transfer function of the selection filter at the receiver and
$S(\omega)$is the power spectral density of one of the transmitted channels.

Homowavelength crosstalk

Homowavelength crosstalk (HOC) happens when a channel interferes at the OXC output with crosstalk components at the same wavelength, as depicted in Fig. 12.12. These components can be originated either by the same channel that traversed the OXC along a spurious path, or by other channels. The former contribution can be easily eliminated through the use of protocols which guarantee wavelength translations inside the node [30].

HOC has been addressed for the case of OXCs based on space switching, not employing wavelength conversion [31].

It was shown that HOC is one of the main limitations to the network transmission performance. HOC generated by other channels at the same wavelength is taken into account using the same model as for HEC. In particular, equation (12.2) holds with $\Delta\omega = 0$. On the other hand, HOC

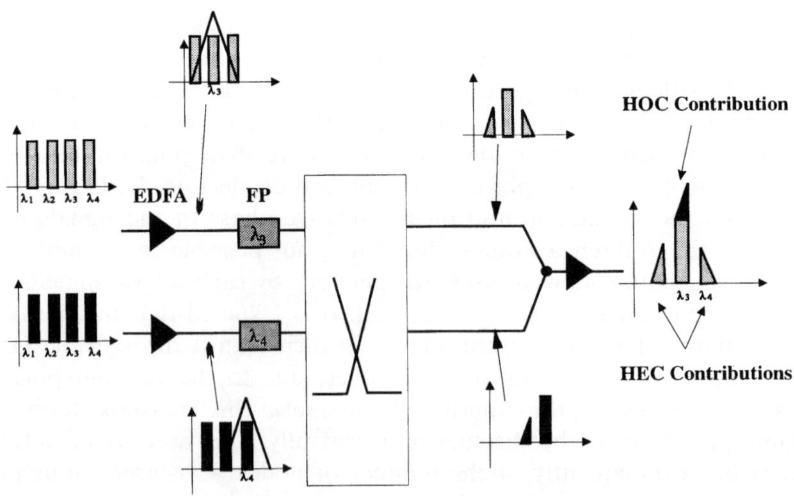

Fig. 12.12 Homowavelength crosstalk.

originating from self-interference can't be modeled as a noise, due to its coherent nature. Even in the presence of wavelength conversion, some routing configurations exists in which self-interference HOC occurs. However, these conditions are quite improbable and can be neglected.

As a result, the HOC noise power contribution can be expressed as

$$\sigma^2_{HOC} = \int_{-\infty}^{\infty} |H_s(\omega)|^2 \cdot \sum_{k=1}^{N} |H_x(\omega)|^{2(N+1-k)} \cdot S(\omega)d\omega, \qquad (12.5)$$

Non-linear crosstalk

Non-linear crosstalk (NLC) is due to four-wave mixing arising fiber propagation in the presence of Kerr non-linearity [29] (Chapter 11). NLC can be considered with a model similar to the model for HEC. The corresponding noise power contributions, relating to degenerate and non-degenerate FWMF arc given by the following relationships:

$$\sigma^2_{dg} = \frac{\Psi_{FWM}}{2} \int_{-\infty}^{\infty} |H_s(f)|^2 \cdot \sum_{r=1}^{N} |H_x(f)|^{2(N-r)} \cdot \sum_{j} \theta(j,j,k)|L_{j,j,k}|^2 D_{dg}(f)df$$

$$(12.6)$$

$$\sigma^2_{nd} = \frac{\Psi_{FWM}}{2} \int_{-\infty}^{\infty} |H_s(f)|^2 \cdot \sum_{r=1}^{N} |H_x(f)|^{2(N-r)} \sum_{j\neq k} \sum_{h\neq j\neq k} 4\theta(j,h,k)|L_{j,h,k}|^2 D_{nd}(f)df$$

$$(12.7)$$

where the spectrum functions are given by

$$D_{dg}(f) = \frac{1}{4}\delta(f) + \frac{1}{2R}\frac{\sin^2(\pi f/R)}{(\pi f/R)^2} + \frac{R}{4\pi^2 f^2}\left[1 - \frac{\sin(2\pi f/R)}{(2\pi f/R)}\right] \qquad (12.8)$$

$$D_{nd}(f) = \frac{1}{8}\delta(f) + \frac{3}{8R}\frac{\sin^2(\pi f/R)}{(\pi f/R)^2} + \frac{3R}{8\pi^2 f^2}\left[1 - \frac{\sin(2\pi f/R)}{(2\pi f/R)}\right]$$
$$+ \frac{3R}{16\pi^2 f^2}\left[1 - \frac{\sin(2\pi f/R)^2}{(2\pi f/R)^2}\right] \qquad (12.9)$$

being $\delta(f)$ the Dirac distribution. The function $\theta(j, h, k)$ is equal to 1 if $j \leq j + h - k \leq N$, to 0 otherwise. Furthermore, $L_{j,h,k}$ is the effective length in the presence of fiber dispersion [30], D_f is the fiber dispersion coefficient (ps/nm/km) and λ is the optical wavelength. Finally, the FWMF amplitude is given by

$$\Psi_{FWM} = \left(\frac{24\pi^3}{\lambda n_{rif}^2 c_0} \chi^{(3)} \frac{A_t^3}{A_{eff}} \right)^2 \qquad (12.10)$$

where n_{rif} is the fiber refractive index, $\chi^{(3)}$ is the third-order fiber suscept-ibility, A_{eff} is the effective mode area of the fiber and c_0 is the speed of light in a vacuum.

Evaluation of the transmission performances

The evaluation of error probability is by the method of Chapter 10. In order to take path memory into account, it is necessary to evaluate it before analysing the performance. This can easily be done by simulating the transmission of a single pulse and verifying, at the end of the simula-tion (i.e. before the decision circuit), when the transient effects are negli-gible.

12.5.2 Test of the simulation model

A comparison between simulation results and experimental data obtained from BER measurements on a real network is indicates the validity of the

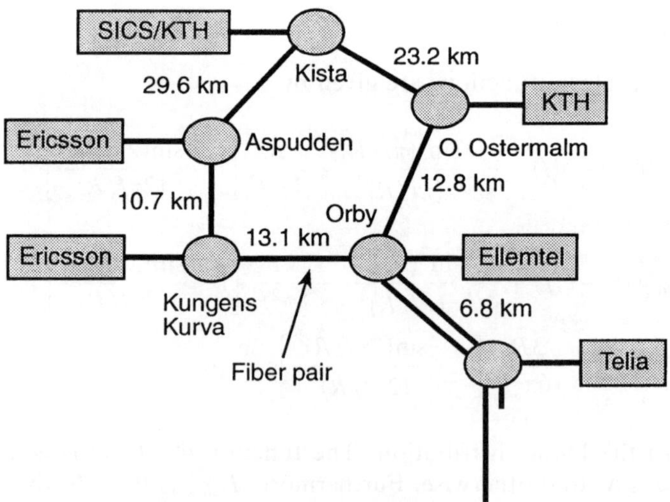

Fig. 12.13 The Stockholm Gigabit Network (SGN). This network was realised by the MWTN consortium in the framework of the RACE project 2028.

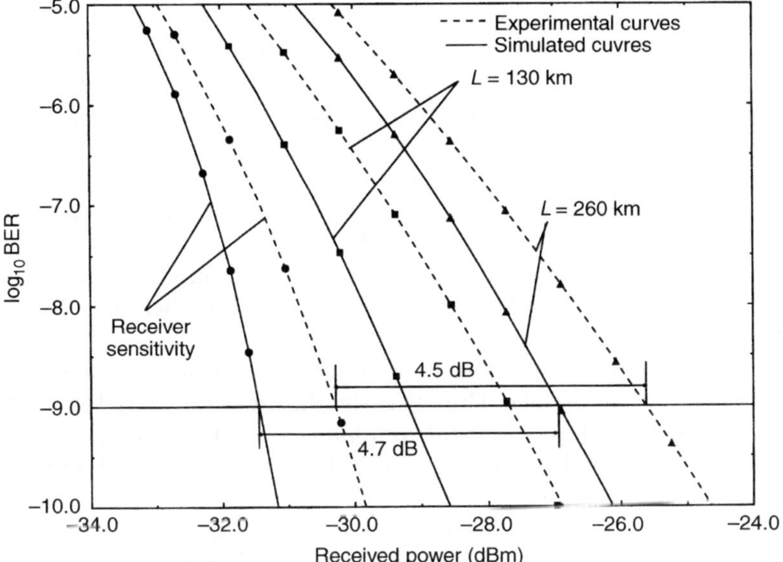

Fig. 12.14 Measures and simulated performance of the SGN. The difference between any simulated curve with the corresponding experimental curve, never exceeds 1 dB.

simulation model itself.

The network taken as a reference is the Stockholm Gigabit Network (SGN), realised by the MWTN consortium. Its structure of this is briefly illustrated in Fig. 12.13.

The SNG consists of an optical ring 130 km long. The experiments were performed using a bit stream at 2.5 Gb/s and standard single-mode fiber. The structure of the OXC is similar to Fig. 12.11, but without the wavelength converters. The BER is quoted for a back-to-back transmission, to assess the sensitivity curve, for a round trip and for two round trips, respectively. Figure 12.14 shows measured and simulated results. Notice that, if the received power required to obtain a BER of 10^{-9} is considered, the difference between any simulated curve with the corresponding measured curve, never exceeds 1 dB. Moreover, a penalty of 4.5 dB was found experimentally, in the case of a double round trip with respect to the sensitivity curve; whereas the same penalty, evaluated by simulation curves, is 4.7 dB. The good agreement between measured and simulated results confirms this approach can be used successfully to evaluate optical network performances.

12.6 OPTICAL CROSS-CONNECT ARCHITECTURES

Different OXC architectures have been reported in the literature [17–20], and some of them have been demonstrated in the framework of research projects. Here we report the more relevant architectures and compare them on the basis of some OXC functional characteristics and their transmission performance. Besides link and wavelength modularity, there is another characteristic which is worth considering: the *add/drop flexibility*, which ensures that the number of added/dropped channels can be varied from zero to a certain number MA depending on the traffic conditions. The higher M_A, the higher the degree of add/drop flexibility. The highest flexibility is obtained when M_A coincides with the number of OXC input channels.

Besides transparency, modularity and add/drop flexibility, the OXC cost must be as low as possible. Thus two significant parameters for the OXC are the number of adopted optical components and the complexity.

Finally, optical paths crossing several OXCs must provide satisfactory transmission performance.

Two basic switching schemes can be adopted in an OXC: space switching and wavelength switching. In the following we will consider seven different OXC architectures, four based on space switching and three based on wavelength switching. Among the four architectures based on space switching, two are based on the employment of optical space switching matrices, the others are based on deliver-and-coupling switches [18].

12.6.1 OXC architecture based on space switching

OXC architectures based on space switching matrixes

The principle block diagram for the first kind of OXC in this category is shown in Fig. 12.4. For completeness, each of the Nf fibers at the OXC input carries a comb of M WDM channels, and M channels are added/droped to the electrical level through local transmitters and receivers. The incoming channels are discriminated through the optical splitters and the tunable optical filters. Then they are routed, by optical space switch matrices, either to the proper output or to the set of optical receivers if they have to be dropped to the digital level. On the other hand, locally generated channels directly enter the optical switch to be routed to the proper output. The combiners blend the wavelength channels into the output fibers. At the node input and output, erbium-doped fiber amplifiers (EDFAs) allow the losses to be compensated. Up to M signals can be added/dropped at a node.

In the OXC architecture shown in Fig. 12.4, note that contentions could arise when channels entering the OXC at the same frequency from different fibers are sent to the same output. Contentions can be avoided by assigning a fixed wavelength to each optical path throughout the network (WP). In this case, channels with the same wavelength are always routed to different outputs, but the wavelength reuse and the network scalability are reduced.

Otherwise, contentions may be solved by detecting one of the channels using a local receiver and retransmitting it at a different wavelength. However, in this way, the OXC transparency is lost and some node resources are turned away from the add/drop function. As a result, this architecture cannot be consider strictly non-blocking, but rearrangeably non-blocking.

The OXC needs M crossbar switching matrices, each with $N = (N_f + 1)$ inputs/outputs. Therefore, provided the appropriate redundancy is built into the splitters, the OXC will be wavelength modular. The addition of one channel per input fiber only needs the addition of N tunable filters and of switch matrix. On the other hand, the OXC is not link modular since the addition of one input fiber requires the substitution of all the space switching matrices. The number of added/dropped channels can be changed from zero to M, providing an intermediate degree of add/drop flexibility.

To evaluate the complexity of the analysed OXC scheme, two basic elements can be individuated: spatial cross-points and tunable filters. Since an $N \times N$ crossbar matrix needs N^2 cross-points, each of which requires a single gate, MN^2 cross-points are present inside the OXC, besides MN tunable filters.

The blocking characteristics of the previous architecture can be overcome by introducing wavelength conversion inside the OXC. The block diagram of this second architecture, in which one wavelength converter is placed after any output of the switch matrices, is shown in Fig. 12.11. In this case, contentions are avoided by changing the wavelength of one of the channels and transparency can be preserved. This architecture is strictly non-blocking and allows the virtual wavelength path (VWP) routing scheme to be employed. This scheme is based on the possibility of changing the wavelength of the optical path inside the OXC and allows high wavelength reuse and network scalability to be achieved.

Three basic elements can be individuated to evaluate the complexity of the analysed OXC scheme: spatial cross-points, tunable filters and wavelength converters. Actually MN^2 cross-points are present inside the OXC, besides MN tuneable filters and NM wavelength converters. Moreover, using some kinds of wavelength converter, MN more tunable filters are needed after the wavelength conversion stage to suppress the optical pump needed in the wavelength conversion process. In this case the number of tunable filters is $2MN$.

Architectures based on delivering and coupling switches

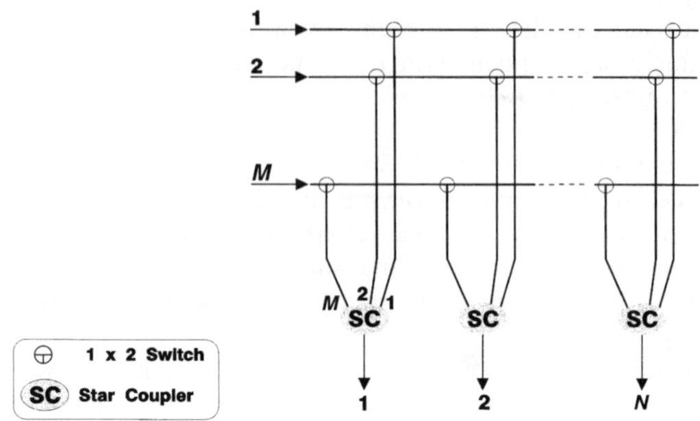

Fig. 12.15 Scheme of an $M \times N$ delivering and coupling switch (DCS).

Two architectures are considered in this section, both reported in the literature [18]. One is link modular the other is wavelength modular. The switch used in these schemes is a space division switch, and its structure is shown in Fig. 12.15 an $M \times N$ switch. The delivering and coupling switch (DCS) is based on star couplers and 1×2 optical switches (OS). Any OS has one input and two outputs and is characterised by four states: (i) the input signal is not transmitted, (ii) the input signal is transmitted to the first output, (iii) the input signal is transmitted to the second output, (iv) the input signal is transmitted to both the outputs.

The DCS is a very flexible switch and it allows WDM signal on its outputs, WDM signals obtained by multiplexing the input signals during the switch operation. Thus the DCS allows different OXC architectures to be designed. Only two schemes that seem to be promising for future application in the transport networks are presented here.

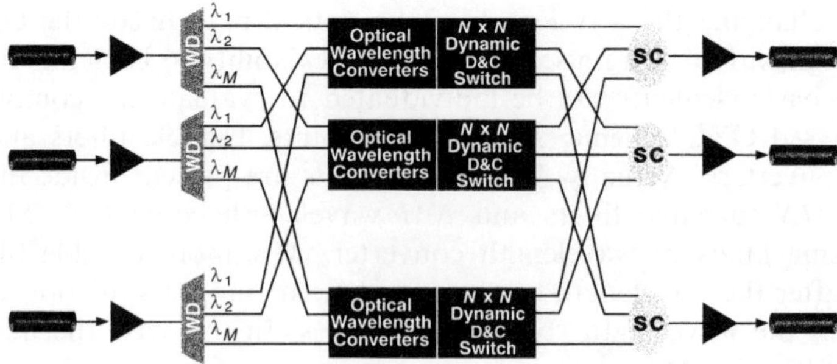

Fig. 12.16 Architecture of a wavelength modular optical cross-connect using DCS devices: DEM = demultiplexer; W – C = wavelength converter; DXC = digital cross-connect.

The block diagram of the wavelength modular is shown in Fig. 12.16. The OXC behaviour is somehow similar to that of Fig. 12.11, but with an important difference. Due to the combined effect of wavelength conversion and the adoption of DCSs, tunable filters at the OXC input are not needed and signal demultiplexing can be achieved by a static demultiplexer (e.g. a grating) thus simplifying the input stage structure. For the same reasons discussed in the previous section, the OXC is wavelength modular but not link modular. Moreover, it is strictly non-blocking and, due to the characteristics of the DCS, it has the maximum add/drop flexibility, provided that a sufficient number of transmitters and receivers are present in the TX and RX blocks.

Considering the OXC complexity, a single OS can be viewed as a single cross-point requiring two gates. Thus $MN2$ cross-points and $2\,MN\,2$ gates are needed inside the OXC, besides MN wavelength converters. Depending on the wavelength converters, NM tunable filters after the wavelength converters may be required.

The link modulator OXC architecture [18] is shown in Fig. 12.17. It is based on a set of N DCSs with M inputs and N outputs. At the OXC input, after amplification by a set of EDFAs, the incoming channels are demultiplexed, wavelength converted so to avoid contentions, and fed to the DCSs. Note that, even in this case, static demultiplexers are required, as in Fig. 12.16. Each input fiber feeds a DCS and a DCS is reserved for the local channels. Each DCS output is connected to a different output fiber or to a local receiver so that, inside the DCS, all the channels directed towards the same fiber are delivered to the same DCS output. The channels from different DCSs are multiplexed onto the output fibers by a set of couplers and amplified by a set of EDFAs.

Link modularity is ensured by the fact that each input fiber has its own DCS so that, adding an input/output fiber, requires adding two star

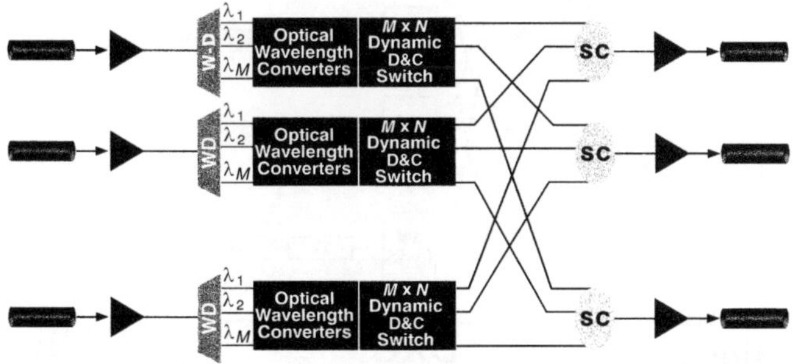

Fig. 12.17 Architecture of a link modular optical cross-connect using DCS devices. DEM = demultiplexer, W – C = wavelength converter, DXC = digital cross-connect.

couplers, M tunable filters and wavelength converters, a DCS and two EDFAs. On the other hand, the OXC is not wavelength modular, since adding a new channel needs to change all the DCSs. Maximum add/drop flexibility is achieved.

Inside the OXC MN^2 cross-points are needed, besides MN wavelength converters. Thus the overall OXC complexity is the same as for the wavelength modular OXC previously considered.

12.6.2 OXC architectures based on wavelength switching

Three OXC schemes, in which switching is performed directly in the frequency domain, are considered in this section.

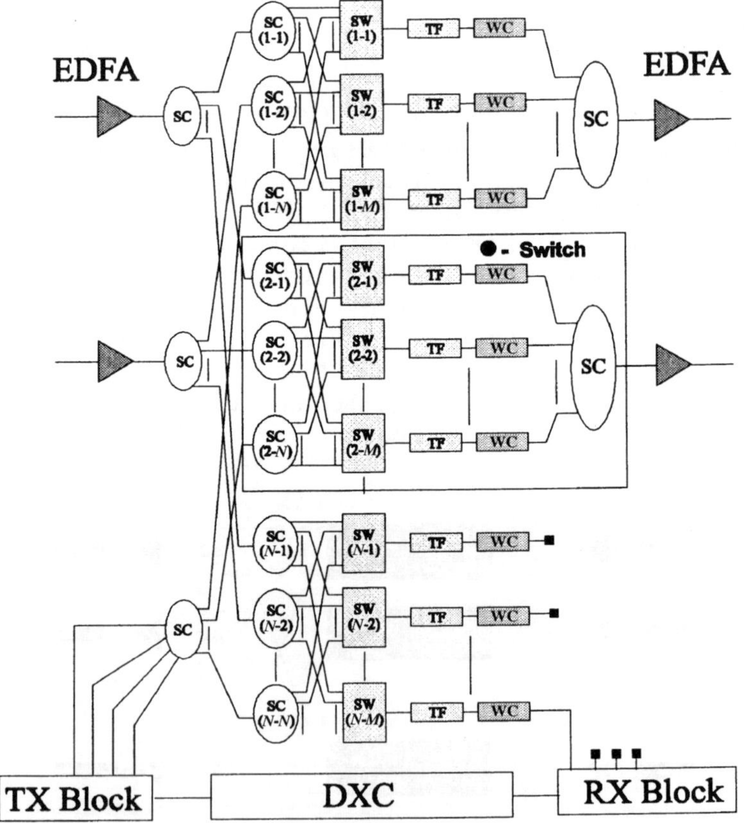

Fig. 12.18 Optical cross-connect architecture using parallel λ-switches: SC = star coupler; SW: (N:1) switch, TF = tuneable filter, W – C = wavelength converter, DXC = digital cross-connect.

The first one [18] shown in Fig. 12.18. There are N switching unit, called λ-switch, which are related to N-1 input/output fiber links and to one link for local traffic (one input port from the local transmitters and one output port for the local receivers). Each λ-switch contains the following elements:

- N star coupler of (1 × M) dimension,
- M switch of dimension (N × 1),
- M pairs of tuneable filters and wavelength converters,
- one optical combiner, i.e. an (M × 1) star coupler.

The WDM combs from any of the N links are delivered to all the λ-switches. Each λ-switch receives all the incoming WDM combs, one per input star coupler. The switch (N × 1) inhibits all the input combs but one, any tunable filter/wavelength converter pair selects the respective channel from the comb and shifts it onto the proper optical wavelength. Then the M channels are combined before reaching the output link.

This architecture is strictly non-blocking. Furthermore, it is link modular. In fact, the addition of one link requires the addition of one λ-switch and one star coupler connected to the incoming link. The number of additional components matches the number of additional links. On the other hand it is not wavelength modular, because an increase in the number of WDM channels leads to a change in all the λ-switches. Complete add/drop flexibility can be achieved if enough transmitters and receivers are available.

This scheme requires NM tuneable filters, NM spatial cross-points and NM wavelength converters. Depending on the type of the wavelength converters, NM tunable filters may be required after the wavelength

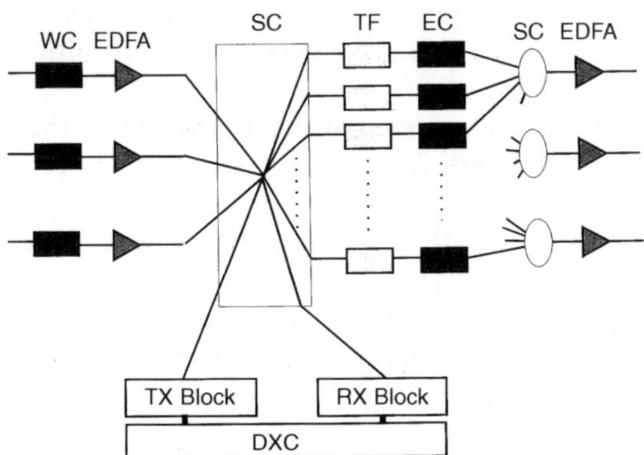

Fig. 12.19 Scheme of an optical cross-connect based on wavelength converters as switching elements: SC = star coupler, TF = tunable filter, W – C = wavelength converter, DXC = digital cross-connect.

converters.

The second architecture [32] is shown in Fig. 12.19. In some sense, this architecture is the equivalent of a T-switching stage for a WDM network. As a matter of fact, all the incoming channels are multiplexed together inside the switch fabric.

At the OXC input, the WDM combs carried by the input fibers are wavelength translated to occupy adjacent parts of the optical spectrum. This operation can be carried out by $(N\text{-}1)$ wavelength converters since there exist optical wavelength converters that are able to translate an entire WDM comb [33]. After translation, the signals are amplified by a set of EDFAs and feed the inputs of an $N \times NM$ star coupler. At the coupler center, all the incoming optical channels are wavelength multiplexed onto a single comb. At the coupler output, any tunable filter selects one channel, and the succeeding wavelength converter sets its wavelength to a suitable value for multiplexing onto the selected output fiber. Multiplexing occurs by optical couplers and, at the OXC output, the signal is amplified by a set of EDFAs. At the OXC input, wavelength conversion is set before amplification since wavelength converters based on FWM works more efficiently with a low-power input signal [33].

The OXC depicted in Fig. 12.19 is strictly non-blocking and is both fiber and wavelength modular, provided that the central star is suitably over-dimensioned. The addition of a new input/output fiber needs an extra M new filters, $M + 1$ new wavelength converters, a new output coupler and two new EDFAs, but the rest of the OXC remains unchanged. The addition of a new channel per fiber needs an extra N new filters and wavelength converters, besides a new setting of the wavelength converters in front of the OXC to avoid contentions at the star center. This setting can be made by tuning the semiconductor lasers that provide the optical pumps to the converters without substituting any optical device. Complete add/drop flexibility can be achieved if enough transmitters and receivers are available.

Since no space switching is realised inside the OXC, there are no crosspoints. On the other hand $N(M + 1)$ wavelength converters and N tuneable filters are required after the first wavelength converter stage. Depending on the wavelength converters type, $N(M + 1)$ more tunable filters may be required after the second wavelength converter stage. Moreover, the star coupler needed to multiplex all the input channels is more complex than the star couplers that are present in the OXCs based on space switching; they have many more ports.

This kind of OXC based on wavelength switching provides better modularity and a lower complexity compared with the previous scheme based on wavelength switching. On the other hand, it requires very high performance wavelength converters. As a matter of fact, the maximum conversion interval in this case is $NM\Delta f$, where Δf is the channel spacing, but it

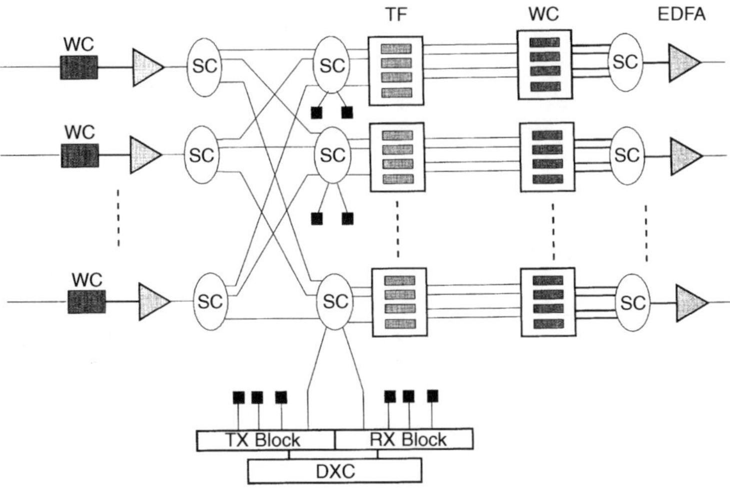

Fig. 12.20 Optical cross-connect based on wavelength converters as switching elements and using different input star couplers: SC = star coupler, TF = tuneable filter, W – C = wavelength converter, DXC = digital cross-connect.

is only $M\Delta f$ for an OXC based on space switching. Moreover, converters able to shift an entire WDM comb are needed.

The last architecture considered in this section is shown in Fig. 12.20. This scheme is quite similar to the previous one. At the OXC input, the incoming WDM combs are properly set in adjacent portions of the optical spectrum by means of the input wavelength converters. Then they are delivered to N star coupler, of dimension ($N \times N$). At any output of a star coupler there is the entire comb of the channels entering the node. Then the system works as in the previous scheme.

This architecture is strictly non-blocking and presents characteristics of modularity very similar to the previous architecture's. However, the single ($N \times NM$) star coupler is substituted by N smaller star couplers that are easier to realise. The main advantage of this architecture compared with the previous one, is that any upgrading or maintenance of the system can be accomplished without traffic disruption. Furthermore, a few small star couplers are needed instead of one of large dimensions, which is more complex to realise. The number of wavelength converters and tunable filters is the same as for the previous architecture.

12.6.3 Comparative analysis

In this section the different OXC architectures will be compared on the basis of their modularity, complexity and transmission performance. The

Table 12.1 Comparison of the routing ability, add/drop flexibility, modularity and complexity of the different OXC architectures

OXC architecture	Blocking	Routing strategy	Link modularity	Wavelength modularity	Add/Drop flexibility	Cross-points	Gates	Tuneable filters	Wavelength converters
space switch-1	RNB	WP	NO	YES	M	MN^2	MN^2	NM	—
space switch-2	SNB	VWP	NO	YES	M	MN^2	MN^2	NM $(2\,NM)$	NM
DCS switch-1	SNB	VWP	NO	YES	MAX (MN)	MN^2	$\frac{2\,MN}{2}$	—— (NM)	NM
DCS switch-2	SNB	VWP	YES	NO	MAX (MN)	MN^2	$\frac{2\,MN}{2}$	—— (NM)	NM
Wavelength switch-1	SNB	VWP	YES	NO	MAX (MN)	MN	MN	NM $(2\,NM)$	NM
Wavelength switch-2	SNB	VWP	YES	YES	MAX (MN)	—	—	$N(1+M)$ N $(1+2\,M)$	$N(1+M)$
Wavelength switch-3	SNB	VWP	YES	YES	MAX (MN)	—	—	$N(1+M)$ $N(1+$ $2\,M)$	$N(1+M)$

RNB = rearrangeably non-blocking, SNB = strictly non-blocking, WP = wavelength path, VWP = virtual wavelength path.

characteristics of the different OXC architectures are summarised in Table 12.1.

Apart from the first architecture, all the others are strictly non-blocking (SNB in the table) and can implement the VWP routing algorithm simultaneously saving the network transparency if suitable optical devices are used. The first architecture is rearrangeably non-blocking and can realise transparency only if VP routing is implemented.

Considering modularity, the architectures based on space division switching allow the achievement of only one of the two types of modularity. On the other hand-two architectures employing wavelength switching fulfil both link and wavelength modularity. Maximum add/drop flexibility can be achieved by all the considered architectures, but those based on space switching matrices which can add/drop up to M wavelength channels.

Concerning architectural complexity, notice that the number of gates for architectures based on space switching matrices is lower than for architectures based on DCSs. This figure is counterbalanced by the fact that the architectures based on DCSs can perform optical demultiplexing by static demultiplexers (e.g. gratings) that are simpler and cost less than the arrays of tunable filters needed in the OXCs based on space switching matrixes.

The table quotes two figures for the number of tunable filters. The first one refers to the case in which no filter is needed after the wavelength converters; the numbers in brackets refers to the case in which a tuneable filter is needed after the wavelength converters. In the last two architectures based on wavelength switching, filters are certainly present after the first stage of wavelength conversion, since a filter is needed for the wavelength converters that can shift an entire WDM comb. On the other hand, filters need not be present for the second stage of wavelength conversion.

In all architectures based on space division switching, the number of cross-points increases quadratically with the number of input/output fibers. But in the last two architectures, based on wavelength switching, the number of key optical devices (wavelength converters and tunable filters) increases linearly with respect to the number of input/output fibers and the number of WDM channels per fiber.

A significant indication of transmission performance can be given in terms of the OXC noise factor. The noise factor of a given OXC architecture depends on the adopted devices and on the OXC dimension, that is the number of input fibers and channels per fiber.

It is not possible to evaluate the transmission performance for all the described OXCs, considering all the possible technologies, so we fix some reference combinations of bit rate, number of input/output fibers, channels per fiber and technological options. In the selected configurations, the transmission performance of the different OXCs is evaluated. In particular, only two types of space switch are considered: space switches based on silica waveguides using the thermo-optic effect (TO) and space switches based on InP active device technology. Two types of wavelength converter are also considered: wavelength converters based on XPM and wavelength converters based on FWM in SOAs. As far as the system parameters are concerned, bit rates of 2.5 and 10 Gb/s are considered.

The EDFA gain at the OXC input is selected to ensure the right optical power at the wavelength converter input when using XPM converters. In the other cases, the amplifier gain is roughly optimised to achieve the lowest error probability. In all cases the maximum power at the output and the maximum amplifier gain are both limited by the amplifier physical characteristics of the amplifier. Typical optimum gain values are between 18 and 35 dB when using FWM converters.

The EDFA gain at the OXC output is chosen in order to have the same power level at the output of all the OXCs in the chain. The same power level is also present at the output of each inline EDFA.

No attempt has been made to optimise the pump and the reference signal power for the FWM and XPM wavelength converters, respectively.

A summary of the considered reference configurations is given in Table 12.2. The data are divided into two groups: data about the path into the

Table 12.2 Description of the considered system configurations

	C1	C2	C3	C4	C5	C6	C7
Total length (km)	2100	2100	2100	420	420	420	420
Number of OXCs	4	4	4	7	7	7	7
OXC spacing (km)	700	700	700	70	70	70	70
EDFA spacing (km)	50	50	50	35	35	35	35
OXC P_{out} (peak, dBm)	−3	−3	0	−3	−3	−6	−6
Input/output ports	8	8	4	8	8	4	4
Channels per fiber	8	8	4	8	8	4	4
Bit rate (Gb/s)	2.5	2.5	10	2.5	2.5	10	10
Channel spacing (GHz)	100	100	400	100	100	400	400
Wavelength converters	XPM	FWM	XPM	XPM	FWM	XPM	FWM
Wav. conv. (first stage)	FWM	FWM	FWM	FWM	FWM	FWM	FWM
Space switches	TO	TO	InP	TO	InP	TO	InP
Tunable filters	DS AO	DS AO	DS AO	DS AO	DS AO	DS AO	DS AO

TO = thermo-optic switch, DS AO = double stage acousto-optic filters.

Table 12.3 References for the physical models and parameters to reproduce the behavior of the various OXC optical devices

Device	Physical Model	Parameters	Values of the main parameters
Single-mode, dispersion-shifted fiber	[31]	[28]	Attenuation : 0.25 dB/km Dispersion : 4 ps/nm/km
Dispersion-compensating fiber	[32]	[28]	Attenuation : 0.25 dB/km Dispersion : −50 ps/nm/km
EDFA inside the OXC	[33]	[34]	Maximum gain : 35 dB Maximum output power : 18 dBm Noise factor : 3.8 dB
In-line EDFA	[33]	[33]	Works in linear regime Noise factor : 3.8 dB
Acousto-optic filter	[12]	[12]	Double-stage standard filters minimum bandwidth 0.1 nm
Switch matrixes (thermo-optical)	[35]	[35]	Losses (8 × 8 matrix) : 10 dB Losses (4 × 4 matrix) : 3 dB
Switch matrixes (InP)	[18]	[36]	Internal gain compensates for losses Internal gain (8 × 8) : 10 dB Internal gain (4 × 4) : 3 dB Noise factor : 6 dB
DCS (Thermo-optical)	[17]	[17]	Losses (8 × 8 DCS) : 10 dB Losses (4 × 4 DCS) : 3 dB
DCS (InP)		estimated from [36]	Internal gain compensates for losses Internal gain (8 × 8) : 15 dB Internal gain (4 × 4) : 5 dB Noise factor : 6 dB
Converters based on FWM in SOAs	[10]	[30]	SOA length : 1 mm SOA linear gain : 48 dB SOA saturation power : 7 mW
Converters based on XPM in SOAs	[24]	[24]	SOA length : 500 μm SOA linear gain : 28 dB SOA saturation power : 10 mW
IM-DD receivers (2.5 and 10 Gb/s)	[28]	[28]	Noise factor: 3 dB Load resistance : 50 ω Temperature : 300 K

network and data about the single OXC. Notice that the transmission performance has been evaluated by considering two different network paths: a chain of four OXCs, 700 km spaced, and a chain of seven OXCs 70 km spaced. In the first case inline EDFAs are 50 km spaced, in the

Table 12.4 Noise factors (in dB) of the seven considered OXC architectures for seven relevant network configurations

OXC Architecture	C1	C2	C3	C4	C5	C6	C7
Space Switch-1	6.15		4.4		8.7		4.8
Space Switch-2	8.81	8	10.5	11.9	11.5	9	9.9
DCS Switch-1	8	8.3	10.5	12.7	11.9	9	10.2
DCS Switch-2	8	8.3	10.5	12.7	11.9	9	10.2
Wavelength Switch-1	12.1	12.4	13.5	16.4	14.8	13	15.2
Wavelength Switch-2	10.4	31.5	13.5	14.6	33	11.7	28
Wavelength Switch-3	10.5	31.5	13.5	14.7	33	11.8	28

The shaded boxes indicate an error probability higher than 10–12.

second case one EDFA is present in the middle of the link between adjacent OXCs. The first case depicts a signal route in a wide area network, whereasthe second case represents a signal route in a regional network.

For the sake of brevity, the complete set of parameters used in the calculations is not reported here. They have been taken from the literature, and the relevant references are summarised in Table 12.3. The numerical values of some of the most important parameters are included in the table.

The noise factor F (dB) is reported in Table 12.4 for the seven considered OXC architectures. If the receiver error probability is not lower than 10^{-12}, the corresponding cell in Table 12.4 is shaded. Since the configurations C1, C2 are identical in the absence of wavelength converters (that is for the first OXC based on space switching), and similarly for C4 and C5, C6 and C7, the corresponding cells are merged.

The noise factor is higher for $R = 2.5$ Gb/s than for $R = 10$ Gb/s. This is because, for $R = 2.5$ Gb/s, the OXC dimension is higher, thus the OXC internal losses are greater with respect to the case $R = 10$ Gb/s. While losses are one of the key factors determining F, the optical bandwidth has no impact, since F is defined starting from the noise power spectral density.

The error probability of 10^{-12} is not reached in the considered configurations, for two OXC architectures adopting wavelength switching, if all the wavelength converters are based on FWM in SOAs. The performance of these OXC architectures can be improved by optimising the pump power and the channel spacing, especially in the case $R = 10$ Gb/s. This is shown in Table 12.5, where the noise factor is reported for the last two OXC architectures, in the cases C2 and C7, assuming a channel spacing of 75 GHz at 2.5 Gb/s and 200 GHz at 10 Gb/s and with a peak optical power

Table 12.5 Noise factor (in dB) for two OXC architectures based on wavelength switching, adopting wavelength converters based on FWM in SOAs

	C2	C7
Wavelength switch 1	24.4	18.1
Wavelength switch 2	24.8	18.4

The network parameters are reported in the text.

at the OXC output of 1dBm at 2.5 Gb/s and 0 dBm at 10 Gb/s. In all the cases reported in Table 12.5, the error probability is lower than 10^{-12}.

These OXC architectures are quite sensitive to the system optimisation and the power needed to reach a fixed error probability is higher than for OXCs adopting space switching or converters based on XPM. On the other hand, OXCs adopting wavelength switching and FWM in SOAs are strictly non-blocking and can provide full modularity and transparency to the transmission format.

The best transmission performance is exhibited by the OXC without wavelength conversion and which is transparent to the modulation format. However, it is not link modular and it is only rearrangeably non-blocking.

The other OXC architectures occupy an intermediate position: they have acceptable transmission performances in all the considered cases.

Using the data in Table 12.4, can be compared with the second architecture based on space switching the architectures based on DCSs. The difference in transmission performance for these architectures are quite small. In particular, in some configurations the architecture based on space switching matrices exhibits better performance, in other configurations it is outperformed by the architectures based on DCSs. This depends on the different positions in which the wavelength converters are placed in the different OXCs.

12.7 LIMITS AND FEATURES

Optical networks have already been investigated and demonstrated within the framework of several research and development projects. The use of wavelength conversion, with regeneration in each optical node, has not yet been demonstrated. Some relevant results obtained by network simulation including different types of wavelength conversion, are presented below.

The analysis has been conducted considering different technologies for the wavelength converters, different OXC architectures (one based on space switching and one based on wavelength switching), and different practical applications: low-density and high-density WDM systems and large geographical area networks. In the following subsections, we first consider near-term applications using optoelectronic wavelength converters, then we consider medium-long-term applications using all-optical wavelength converters (with the currently available data). In this case we distinguish low-density, high-density and geographical optical networks. The analysis uses data reported in the literature [25].

12.7.1 WDM optical networks using optoelectronic wavelength converters

Two types of wavelength converter can be considered: non-regenerative and regenerative. Non-regenerative converters have already been discussed in Chapter 7.

The use of regenerative converters saves us from the analysis of the performance, since at each conversion the original transmission quality of the signal is restored. However, if the transmission issues are alleviated by the use of this type of converter, transparency is strongly limited, since the regeneration occurs for a fixed bit rate and signal format. The trade-off between high transmission performance and high transparency drives the choice between the two types of device.

12.7.2 Low-density WDM systems

Here we consider low density systems based on space switching obtained by the use of InP switch matrices [34]. The WDM comb carried by each fiber at the OXC input is composed of four channels with a spacing of 3 nm. In this situation, with a selection filter bandwidth of twice the bit rate and a link length within 1200 km, the crosstalk is not the main factor limiting the system performance, and the maximum wavelength conversion interval is 6 nm (about 720 GHz at $\lambda = 1.55\ \mu$m).

The fiber losses are assumed to be compensated by a single inline EDFA between adjacent OXCs, and the two EDFAs inside the OXC are used to compensate the losses of the node. In particular, when FWM converters are used, the input EDFA compensates the loss of the optical devices which precede the converter inside the node (power splitters, optical filters and space switch matrix) and the EDFA at the OXC output compensates the other losses of the node. This is a suboptimum solution and an accurate optimisation of the EDFA gains would enhance the transmission performance. For XPM converters the input EDFA must have a gain which allows a fixed value of the signal power to be obtained at the input of the converter (–5 dBm in our case), to ensure the converter works properly. In this situation the OXC losses are compensated by suitably

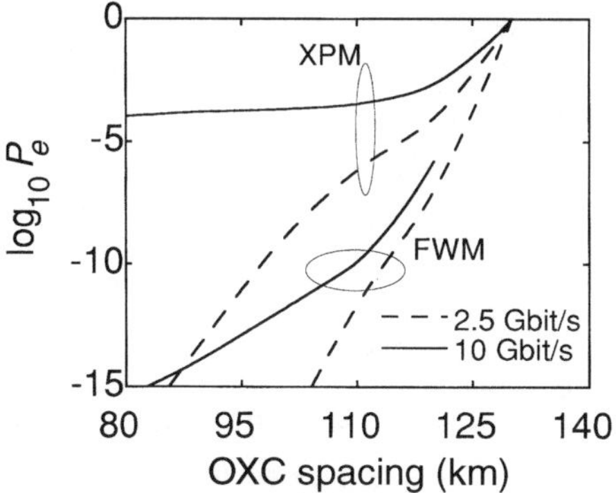

Fig. 12.21 Transmission performance of a network employing space switching. The BER is plotted versus the length between any pair of adjacent nodes, using FWM and XPM converters at 2.5 and 10 Gb/s.

setting the gain of the EDFA at the OXC output. In the simulations, the values of the gain never exceed 30 dB, so the EDFA can be assumed to working linearly [35, 36]. The transmission performance of networks employing 10 nodes, using either XPM or FWM converters, at 2.5 and 10 Gb/s are depicted in Fig. 12.21, where the error probability is reported versus the OXC spacing for a peak power at the OXC output of 0 dBm (in the following when we mention the peak power we always mean the power associated with a mark at the node output). The worst conversion bandwidth is chosen in the case of the FWM converter. By increasing the OXC spacing, the gain of the inline EDFA increases, enhancing the ASE power at the receiver. Moreover, the effect of chromatic dispersion is evident with XPM converters, whereas with equally spaced OXCs and FWM converters, dispersion is completely compensated by spectral inversion. For the same conditions Fig. 12.21, Fig. 12.22 shows the error probability for an OXC spacing of 100 km as a function of the peak power. At a bit rate of 10 Gb/s, the system adopting XPM converters does not work since it is above the dispersion limit; in the other cases that are illustrated, the system performance is satisfactory.

Figures 12.21 and 12.22 show that at 2.5 Gb/s both types of converter can be favorably used, allowing a reasonable spacing between the nodes with a moderate power at the node output. At a bit rate of 10 Gb/s the penalty induced by fiber dispersion severely compromise the employment of XPM converters [25, 37]. On the other hand, the dispersion is compensated by FWM converters due to the spectral inversion. If the nodes were placed with unequal spacing, the compensation of dispersion accomplished by FWM devices would be partial. The good performance of FWM is also due to the modest conversion interval. When the conversion interval

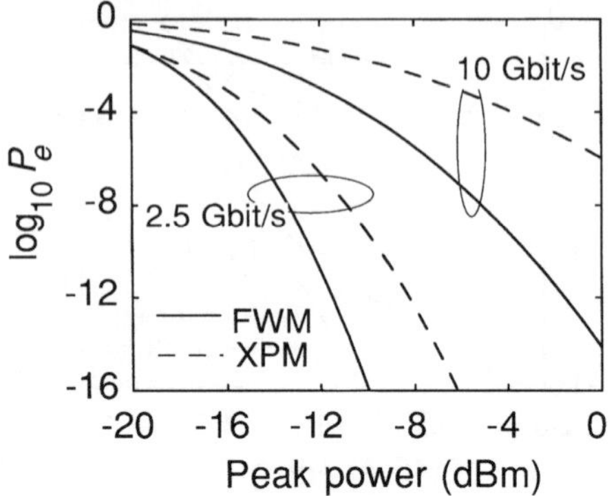

Fig. 12.22 Transmission performance of a network employing space switching. The BER is plotted versus peak power at the node output, using FWM and XPM converters at 2.5 and 10 Gb/s.

increases, the efficiency of an FWM converter decreases and the performance degrade; accordingly but as long as the signal is in the constant gain bandwidth of the SOA, the efficiency of the XPM converter weakly depends on the conversion interval.

Since fiber dispersion is the main impairment for a network employing XPM converters, from now on when XPM devices are considered, dispersion compensation is always assumed in the fiber link between adjacent OXCs. In particular, compensation by a passive fiber compensator is assumed [38, 39].

12.7.3 High-density WDM systems

The use of EDFAs limits the available optical bandwidth to about 30 nm; moreover, increasing the optical bandwidth occupied by the WDM comb decreases the efficiency of FWM converters. To increase the transmission capacity, for a given bit rate, it is necessary to increase the number of channels inside the available optical bandwidth, with a consequent amount of channel crosstalk. Linear crosstalk can be reduced by using proper filtering [25]. On the other hand, non-linear crosstalk is almost completely confined within the signal bandwidth if the channel spacing is uniform, thus it is practically independent of the shape of the optical selection filter. The impact of crosstalk is shown in Fig. 12.23 for two OXC architectures. For XPM converters the dispersion of the fiber link between adjacent OXCs is compensated by a passive fibre compensator with an efficiency 90% (residual medium dispersion of 0.4 ps/nm/km). Figure 12.23 shows the peak power required to obtain an error probability of 10^{-10} versus the

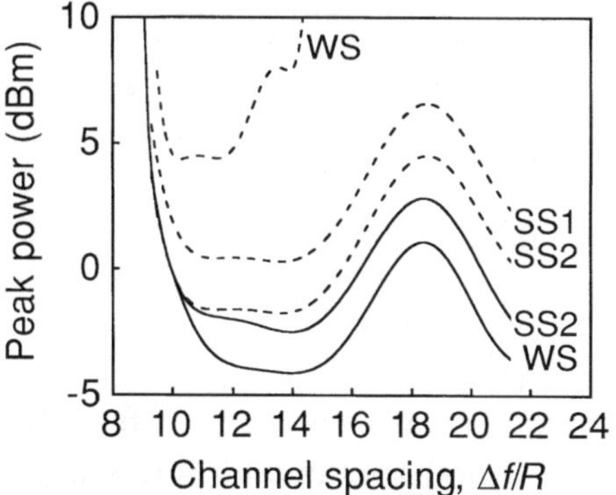

Fig. 12.23 Peak power required to reach a BER of 10^{-10} versus the channel spacing (normalised to the bit rate R). WS = wavelength switching. SS1 = architecture based on space switching using LiNbO3 devices [40]. SS2 = architecture based on space switching using InP devices [34]. The dotted curves relate to the FWMM; the continuous curves relate to the XPM.

channel spacing Δf normalised to the bit rate R. The bit rate is 10 Gb/s, a link composed by 5 nodes is considered (one source, 3 inline OXCs and one destination), with an OXC spacing of 90 km without inline EDFAs. The gain of the EDFA at the OXC input is set to 33 dB when using FWM converters and the gain of the EDFA at the OXC output is set to obtain an overall gain for the OXC + fiber system. The OXC has four input ports, each carrying four WDM channels. An EDFA preamplifier with a gain of 15 dB and a double-stage acousto-optic filter [25], with a bandwidth of twice the bit rate, are placed in front of the receiver. For the OXC architecture based on space switching two different matrices are considered: the first realised in InP technology (SS2) [34] and the second based on LiNbO3 technology (SS1) [40]. The oscillatory behavior of the curves in Fig. 12.23 arises because the main crosstalk contribution is HOC, and its level is determined by the acousto-optic filters used inside the OXC. Thus the presence of non-negligible sidelobes in the transfer function of these filters immediately affects the curves of Fig. 12.23. This behavior would be much less evident if apodied filters were used (Chapter 6).

The best transmission performances are attained by wavelength-switched OXCs adopting XPM converters (continuous curve marked WS). This is because wavelength-switched OXCs have low internal losses and the XPM converter has high efficiency, independent of the conversion interval. As a consequence, the performance gets better and better with increasing channel spacing, asymptotically tending to a situation where there is no crosstalk. On the other hand, when FWM devices are used, increasing the channel spacing decreases the efficiency of the wavelength converters. This

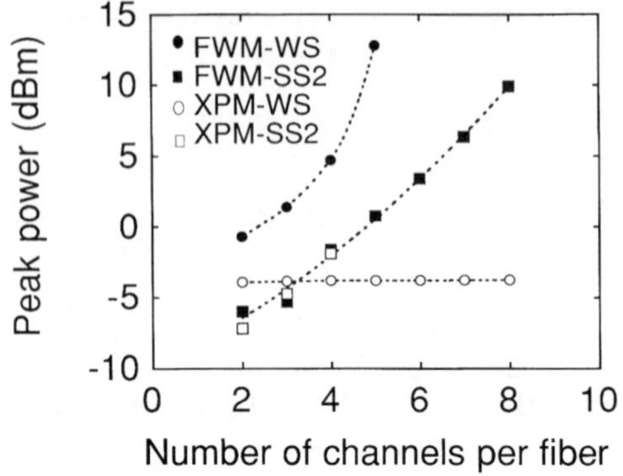

Fig. 12.24 Peak power required to reach a BER of 10^{-10} versus the number of channels per fiber at $R = 10$ Gb/s. WS = wavelength switching. SS2 = architecture based on space switching using InP devices [34].

is particularly evident in the case of OXCs based on wavelength switching, where an optimum channel spacing exists. This optimum value is caused by the trade-off between the decrease in the crosstalk level and the efficiency of the converters as the channel spacing increases.

Another key parameter to evaluate the network capacity is the number of channels carried by each fiber. The losses in space switches are assumed proportional to the number of cross-points in the matrix, that is to the square of the number of channels at the matrix input. This assumption is not rigorous, but the technological issues related to optical switch matrices and the aspects related to the blocking characteristics of complex switching stages are beyond the scope of this book. Figure 12.24 reports the peak power required to obtain an error probability $P_e = 10^{-10}$ versus the number of channels per fiber, at the same conditions as for Fig. 12.23, considering three different cases: space switch with FWM converters; wavelength switch adopting either FWM or XPM devices. The channel spacing in Fig. 12.24 is $\Delta\phi = 12R$. In evaluating the results shown in Fig. 12.24 remember that HOC is largely independent of the number of channels per input fiber. If HEC is considered, it is mainly due to adjacent channels and it depends on the channel spacing, so it grows very slowly with an increasing the number of channels. On the contrary, non-linear crosstalk strongly depends on the number of channels, but it is not crucial here since the overall distance is limited to 360 km.

If FWM converters are adopted, the transmission performance degrades as the number of channels increases. In particular, the use of wavelength switching gives poorer results due to the large penalties introduced by FWM converters when a large conversion interval is required. For example, if the bit rate is 10 Gb/s and the channel spacing is 10 times the bit rate, the maximum conversion interval is 2 THz (about 16.4 nm) if five channels are carried by each fiber at the OXC input. Conversely, the wavelength switching architecture using XPM devices provides better performance. In fact, at the increasing of the number of channels, the performance does not appreciably degrade. This behavior can be explained as follows: when the number of channels per fibre Nc increases by ΔN_c the star coupler loss increases of a factor $N_c + \Delta N_c / N_c$, much more slowly that the space switch matrix loss.

12.7.4 An optical network covering a national geographical area

The analysis of the network performance in a geographic area is a very important issue in view of future applications for optical networks. The distances covered by a geographical network can be quite long: this means that non-linear effects which arise during fiber propagation are quite relevant. As an example, we consider a possible path through a national optical transport network [25], whose length is 1673 km. Assume a node architecture based on space switching by an InP switching matrix and 30 km spaced inline EDFAs.

Figure 12.25 shows the network performance when FWM converters are used. Channels at 2.5 or 10 Gb/s are considered with a channel spacing of 3 nm. Continuous lines refer to the case in which standard SOAs (Param. I in Table 3 of Sabella *et al.* [25]) and dispersion-compensating fibers are adopted. Dashed lines indicate standard SOAs without dispersion compensation whereas dotted lines indicate optimised SOAs (Param. II in Table 3

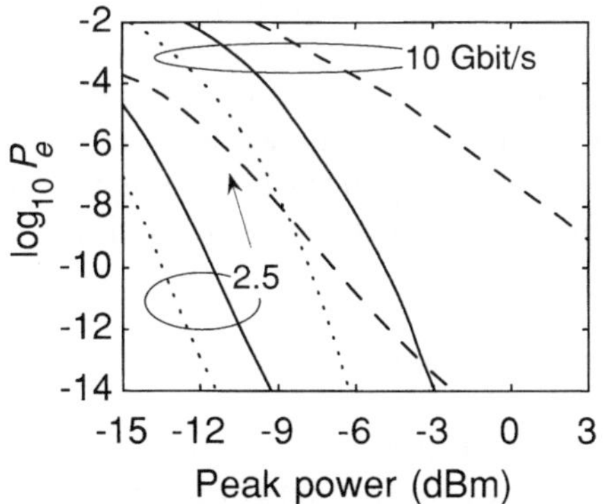

Fig. 12.25 Performance of a geographical network employing four-wave mixing (FWM) converters. The BER is reported versus the node peak output power at R = 2.5 and 10 Gb/s. Continuous lines indicate a standard SOA and a dispersion-compensating fiber, dotted lines to indicate an optimised SOA and a dispersion-compensating fiber, dashed lines indicate a standard SOA without dispersion compensation.

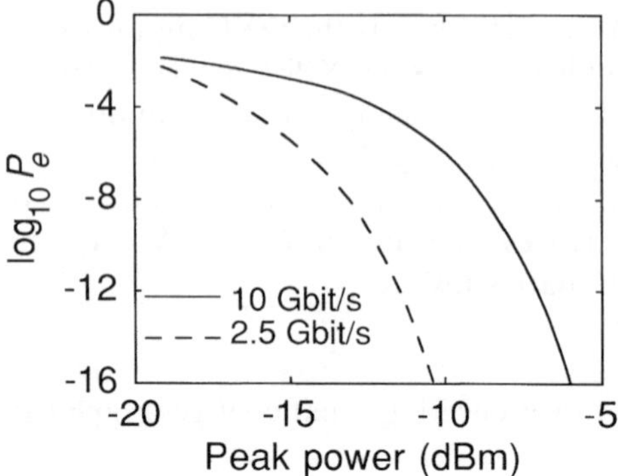

Fig. 12.26 Performance of a geographical network employing XPM converters. The BER is reported versus the node peak output power at R = 2.5 and 10 Gbit/s.

of Sabella *et al.* [25]) and dispersion-compensating fibers are adopted. The performance of the network employing FWM devices is quite good. Where no compensating fibers are used, the performances can still be acceptable at R = 2.5 Gb/s. Compensation techniques are required at 10 Gb/s since the FWM converters in the nodes do not sufficiently compensate dispersion and the Kerr effect because the nodes are not equally spaced along the link. And spectral inversion by FWM converters only partially compensates the dispersion and the Kerr effect. Moreover, notice that the adoption of optimised SOAs allows a saving of 3 dB in the peak power. In fact, an increase of about 10 dB in the converter efficiency allows the gain of the EDFAs inside the OXC to be decreaesd, hence the ASE power can be decreased. However, the various phenomena that limit the network performances restrict the peak power improvement to just 3 dB.

Figure 12.26 shows the network performance with XPM converters and dispersion compensation. The performance is quite good, at both 2.5 and 10 Gb/s, even if the Kerr effect induced penalty is not negligible. Without any dispersion compensation the performance would be unacceptable.

12.7.5 Some general considerations

From the reported results, it is possible to derive several considerations.

Physical layer design of the optical network involves taking into account many technological issues, concerning the choise of devices, and different effects influencing transmission, such as power budget, crosstalk and optical amplifier placement.

Different technological solutions are possible for realising cross-connecting nodes, with or without wavelength converters. Different technologies are also available for realising novel devices like space switching matrices, optical filters and wavelength converters. In general, it is not possible to state which is the best technology or the best architecture. In any case we would say that the realisation of optical networks employing wavelength conversion is now feasible.

REFERENCES

1. P.E. Green, Jr., Optical networking update, *IEEE J. Select. Areas in Commun.* **14**, 764–779 (1996).
2. L.C. Blank, A.D. Ellis, and D.M. Spirit, Optical time division multiplexing, in *High capacity optical transmissions explained,* Eds. D.M. Spirit and M.J. O'Mahony, John Wiley, New York (1995).
3. C.A. Bracket, Foreword: is there an emerging consensus on WDM networking?, *IEEE J. of Lightwave Technol.* **14**, 936–941 (1995).
4. T.H. Wu, D.J. Kolar, and R.H. Cardwell, Survivable network architectures for broadband fiber optic networks: model and performance comparison, *IEEE J. Lightwave Technol.* **6**, 1698–1709 (1988).
5. S.S. Wagner and T.E. Chapuran, Multiwavelength ring networks for switch consolidation and interconnection, in *Proceding of an International Conference on Communication,* paper 340–5 (1992).
6. A.F. Elrefaie, Self-healing ring network architecture using WDM for growth, in *Proc. ECOC '92*, paper Tu P1–16 (1992).
7. C. A. Brackett *et al.*, A scalable multiwavelength multi-hop optical network: a proposal for research on all-optical networks, *IEEE J. Lightwave Technol.* **11**, 736–752 (1993).
8. R. Ramaswami and K.N. Sivarajan, Routing and wavelength assignment in all-optical networks, *IEEE/ACM Trans. Networking* **3**, 489–500 (1995).
9. S. Baroni and P. Bayvel, Analysis of restoration requirements in wavelength-routed optical networks, in *Proc. NOC '96*, pp. 56–63 (1996).
10. N. Nagatsu *et al.*, Optical path cross-connect system scale evaluation using path accommodation design for restricted wavelength multiplexing, *IEEE J. Sel. Areas on Commun.* **14**, 893–902 (1996).
11. M. Listanti, M. Berdusco, and R. Sabella, A new strategy for employing wavelength conversion in WDM optical networks, in *Proc. IEEE/LEOS '97*, San Francisco CA, (1997).
12. S.B. Alexander *et al.*, A precompetitive consortium on wide-band all optical networks, *IEEE J. Lightwave Technol.* **11**, 714–735 (1993).
13. A.S. Acampora, A multi-hop local lightwave network, *in Proc. IEEE GLOBECOM '87*, Tokyo, Japan, pp. 1459–1467 (1987).
14. D.A. Smith *et al.*, Integrated-optic acoustically tunable filters for WDM networks, *IEEE J. Select. Areas in Commun.* **8**, 1151–1159 (1990).
15. A.S. Acampora, *An Introduction to Broadband Networks*, Plenum Press, New York (1994).
16. ITU-T Recommendation G. 803, Architectures of transport networks based on the synchronous digital hierarchy (SDH), 03/93 (1993).

17. G.R. Hill *et al.*, A transport network layer based on optical network elements, *IEEE J. Lightwave Technol.* **11**, 667–679 (1993).
18. A. Watanabe, S. Okamoto, and K. Sato, Optical path cross-connect node architecture with high modularity for photonic transport networks, IEICE Trans. on Commun. **E77B**, 1220–1229 (1994).
19. E. Iannone and R. Sabella, Performance evaluation of an optical multi-carrier network using wavelength converters based on FWM in semiconductor optical amplifiers, *IEEE J. Ligthwave Technol.* **13**, 312–324 (1995).
20. E. Iannone and R. Sabella, Optical path technologies: a comparison among different cross-connect architectures, *IEEE J. Lightwave Technol.* **14**, 2184–2196 (1996).
21. R. Sabella and E. Iannone, Wavelength conversion in optical transport networks', *Fiber and Integrated Optics* **15**, 167–192 (1996).
22. K. Sato, S. Okamoto, and H.Hadama, Network performance and integrity enhancement with optical path layer technologies, *IEEE J. Select Areas in Commun.* **12**, 159–170, (1994).
23. Y. Hamazumi, N. Nagatsu, and K. Sato, Number of wavelengths required for optical networks with failure restoration, *Tech. Dig. OFC '94*, pp. 67–68 (1994).
24. N. Nagatsu, Y. Hamazumi, and K. Sato, Optical path accommodation design applicable to large scale networks, *IEICE Trans. on Commun.* **E78**, 597–607 (1995).
25. R. Sabella, E. Iannone, and E. Pagano, Optical Transport Networks Employing All-Optical Wavelength Conversion: Limits and Features, *IEEE Journal of Selected Areas in Commun.* **14**, (1996).
26. A. Mecozzi, Long distance transmission at zero dispersion: the combined effect of Kerr nonlinearity and noise of the in-line amplifiers, J. Opt. Soc. of Amer. B. **11**, 462–469 (1994).
27. F. Matera and M. Settembre, Nonlinear evolution of amplitude and phase modulated signals and performance evaluation of single channel systems in long haul optical fiber links, J. Opt. Commun. in press.
28. L.G. Kazowsky and J.L. Gimlett, Sensitivity penalty in multichannel coherent optical communications, *IEEE J. Lightwave Technol.* **6**, 1353–1365 (1988).
29. E. Lichtman, Performance degradation due to four-wave mixing in multi-channel coherent optical communications systems, J. Opt. Commun. **12**, 53–58 (1991).
30. E. Iannone and R. Sabella, Analysis of wavelength-switched high-density WDM networks employing wavelength conversion by four-wave mixing in semiconductor optical amplifiers, *IEEE J. Lightwave Technol.* **13**, 1579–1592 (1995).
31. J. Zhou, M. J. O'Mahony, and S. D. Walker, Analysis of optical crosstalk effects in multi-wavelength switched networks, *IEEE Photon. Technol. Lett.* **6**, 302–305 (1994).
32. R. Sabella and E. Iannone, A new modular optical path cross-connect, *Electron. Lett.* **32**, (1996).
33. A. D'Ottavi *et al*, Efficiency and noise performances of wavelength converters based on FWM in semiconductor optical amplifiers, *IEEE Photon. Technol. Lett.* **31**, (1995).

34. M. Gustavsson *et al.*, Monolithically integrated 4×4 InGaAsP/InP laser amplifier gate switch arrays, Electron. Lett. **28**, 2223–2225 (1992).
35. Y. Kinura, K. Suzuki, and M. Nakazawa, 46.5 dB gain in Er3+ -doped fibre amplifier pumped by 1.48 μm GaInAsP laser diodes, Electron. Lett. **25**, 1656–1657 (1989).
36. Data sheet of the Ampliphos OP-980-F-15 EDFA produced by Pirelli Cavi.
37. E. Iannone, R. Sabella, L. de Stefano, and F. Valeri, All-optical wavelength conversion in multi-carrier networks, IEEE Trans. on Commun. **44**, 716–724 (1996).
38. A. J. Antos and D. K. Smith, Design and characterization of dispersion compensating fiber based on the LP01 mode, IEEE J. Lightwave **12**, 1739–1745 (1994).
39. C.D. Poole, J.M. Wiesenfeld, D.J. DiGiovanni, and A.M. Vengsarkar, Optical fiber-based dispersion compensation using higher order modes near cutoff, IEEE J. Lightwave Technol. **12**, 1746–1758 (1994).
40. P. Granestrand *et al.*, Pigtailed tree-structured 8×8 LiNbO3 switch matrix with 112 digital optical switches, *IEEE Photon. Technol. Lett.* **6**, 71–73 (1994).

13
Long-Haul Optical Communications

13.1 INTRODUCTION

The advent of EDFAs has had a profound impact on the design, operation
and performance of optical communications. In optical networks they give
a big contribution to transparency. In long-haul optical transmission their
impact is even more evident, since costs and reliability can be considerably
improved by adopting EDFAs instead repeaters.

The first long-haul EDFA systems were installed in 1995 with a single
5 Gb/s optical channel [1], twice the capacity of the most advanced digital
regenerator undesea fiber-optic system. Laboratory experiments have
demonstrated 100 Gb/s over transoceanic distances using WDM techniques [2–4]. Important progress has been made in dispersion management
techniques, gain equalisation and modulation formatsprogress which
hasallowed a large data transmission capacity. In practice it is possible to
divide the area of long-haul optical communications in two basic areas:
non-soliton methods and soliton transmission. In this chapter we first
review some experimental techniques that have been developed to improve
the performance of long-haul WDM transmission systems based on the
common NRZ format and other non-soliton methods, then we briefly
examine soliton transmission.

13.2 NON-SOLITON LONG-HAUL OPTICAL TRANSMISSIONS

Non-linear interactions between WDM channels can be managed by
cleverly tailoring the chromatic dispersion of the fiber. This is usually
called dispersion management [2]. Passive gain equalisation can appreciably increase the usable bandwidth in long amplifier chains. Bit-synchro-
nous polarisation and phase modulation in NRZ transmission systems can
remove the excess noise accumulation cause by polarisation hole-burning
and they simultaneously improve the transmission performance by
decreasing FWMF between channels and increasing the received eye
opening.

13.2.1 Dispersion management

Chapter 10 shows that several phenomena affect the transmission performance of high-speed systems. In particular, for long-haul systems, the non-

linear refractive index can couple different signal channels, as well as coupling signal with noise. Until recently the best solution was thought to be operation around the zero-dispersion wavelength of the fiber. However, in this case both signal and noise travel at similar velocities. So the signal and noise waves have long interaction lengths and can mix together. As a result, chromatic dispersion can reduce phase matching at the propagation distance over which closely spaced wavelengths overlap, and can reduce the number of non-linear interactions in the fiber. Consequently, in a long undersea system, the non-linear behavior can be managed by tailoring the accumulated dispersion so that the phase-matching lengths are short, and the end-to-end dispersion is small.

The accumulated dispersion returns to zero for just one wavelength near the average zero-dispersion wavelength for the transmission line. The consequent divergence in the accumulated dispersion for the WDM channels results fom the non-zero slope of the dispersion curve. The detrimental effects of higher-order dispersion are minimised by compensating any non-zero accumulated dispersion at the receiver. Another solution is to use a dispersion-compensating fiber with an opposite sign for the dispersion slope [5].

13.2.2 Gain equalisation

According to section 11.3.3, the usable bandwidth is reduced due to non-uniform gain of EDFAs. However, the usable bandwidth of a long chain can be significantly increased by using passive gain-equalising filters. These filters can be designed to approximate the inverse characteristic of the combination of EDFA and fiber span. In practice it is often simpler to design gain equalisers that correct for several amplifier spans; they do add insertion losses.

Another concern is gain saturation of the amplifiers by accumulated ASE noise. Since the gain of an amplifier chain with gain is equalised for many WDM channels, optical noise that occupies the same wavelength range can also accumulate. If this significantly augments accumulated noise the combined signal power, the power in the optical data channels will decrease at the expense of the noise accumulation [6].

13.2.3 Polarisation scrambling

Long-haul optical links also suffer from polarisation hole burning (PHB) in the amplifiers [7–9]. It arises from anisotropic saturation created when a polarised saturating signal is launched into the EDFA. In a chain of saturated EDFAs, PHB can cause ASE noise to accumulate in the polarisation orthogonal to the signal faster than it accumulatesalong the parallel axis. This impairment can be avoided by polarisation scrambling of the signal at a rate faster than the EDFA can respond. The consequence of PHB is the depression of EDFA gain for light with the same polarisation as a saturating signal. Besides the PHB effect, which follows the saturating signal, there is an additional polarisation-dependent gain that follows the pump polarisation, which has a similar system effect as polarisation-dependent loss.

In amplified transmission systems, the natural gain compression of the EDFA provides an automatic gain control function. The automatic gain control is accomplished by designing each amplifier in the chain to have a small-signal gain a few decibels larger than the span attenuation. The EDFA's output power achieves an equilibrium where the gain for the total signal is equal to the span losses. Thus, near the beginning of the system (where the accumulated ASE is still small) the EDFA's gain equals the span loss for the signal. However, the ASE noise in the polarisation orthogonal to the signals sees an excess gain ΔG in each amplifier and therefore accumulates faster than the standard noise accumulation theory would suggest [6]. Fortunately, in long-haul EDFA transmission systems, scrambling the transmitter's polarisation reduces PHB-induced transmission impairment, since the dynamics of the EDFA gain are relatively slow [10]. As a result, when the signal polarisation is scrambled faster than the amplifier can respond, there is no preferred polarisation axis for the gain to be depleted, and the transmission performance returns to the expected value. Since the measured PHB characteristic time constant is about 130 μs [11], polarisation scrambling should change the optical signal between orthogonal polarisations at a frequency higher than about 7 kHz.

Bit synchronous polarisation scrambling represents the optimal trade-off between the two regimes of low-speed and high-speed scrambling. In fact, scrambling at speeds lower than the bit rate reduces the effects of PHB but introduces unwanted amplitude modulation (AM) through polarisation-dependent loss (PDL) in the optical elements of the systems [12]. On the other hand, scrambling at frequencies higher than the bit rate reduces the polarisation-dependent loss (PDL) [13] but causes an increase in the transmitted bandwidth, which can limit the number of channels packed into a fixed optical bandwidth. Furthermore, superimposed phase modulation (PM) can significantly increase the eye opening of the received bit pattern, originating from the conversion of PM into bit synchronous AM through chromatic dispersion and non-linear effects.

13.3 SOLITON OPTICAL TRANSMISSIONS

A soliton is a waveform which propagates without changing its shape [14, 15], remaining unchanged even after a collision with another soliton. The invariance of pulse shape is of great value since it avoids the impairments of intersymbol interference (ISI), which heavily affect the transmissions of dispersion-limited systems.

Solitons have been proposed and demonstrated for allowing ultrahigh-speed long-haul optical communications via an exact balance of dispersion and non-linearity [14, 15]. From a mathematical point of view, solitons are obtained as stable solutions of the non-linear Schrödinger equation (4.44).

The non-linearity in this equation can compensate for dispersion: a pulse need not to disperse since it can dig its own potential well, which provides confinement. This happens when $\beta 2 < 0$, when the group velocity dispersion (GVD) is negative (anomalous). Indeed, a solitary-wave solution of equation (4.44) is

$$
\begin{aligned}
A_s(z, t) =& A_0 \mathrm{sech}\left(\frac{t - \beta_2 \Delta\omega_0 z}{\tau}\right) \\
& \exp(i\omega_0 t) \cdot \exp\left[i\left(\frac{\gamma|A_0|^2}{2} + \frac{\beta_2}{2}\Delta\omega_0^2\right)z\right] \cdot \exp)i\phi)
\end{aligned}
\tag{13.1}
$$

with the constraint that the parameters τ and A_0 obey the following relationship:

$$
\frac{1}{\tau^2} = -\frac{\gamma|A_0|^2}{\beta_2}.
\tag{13.2}
$$

This constraint shows that the GVD has to be negative. The physical interpretation of (13.1) is relatively straightforward. The soliton carrier frequency is detuned from the nominal frequency by $\Delta\omega_0$. Setting $\Delta\omega_0$ to zero, we see that the solution is a simple hyperbolic secant of amplitude A_0. As it propagates, it accumulates a phase delay $\gamma|A_0|^2 z / 2$ which is due to the Kerr effect produced by the average intensity. The area of the soliton amplitude is fixed at

$$
Area = \int_{-\infty}^{\infty} |A_s(z, t)dt = \pi\sqrt{\frac{|\beta_2|}{\gamma}}
\tag{13.3}
$$

independent of its amplitude. The energy of a soliton is thus proportional to the inverse of its pulsewidth τ. Detuning by $\Delta\omega_0$ has simple consequences. The propagation constant changes by $\beta_2 \Delta\omega_0^2 / 2$ and modifies the phase factor accordingly. The inverse group velocity changes by $\beta_2 \Delta\omega_0$, thus producing the timing shift in the argument of the hyperbolic secant function. The area in (13.3) is unaffected by detuning.

It is conventional to refer to a normalised form of the Schrödinger equation. In fact, a normalising distance is chosen so that the optical Kerr effect, acting alone, would produce one radian of phase shift per unit distance. The normalisation of the time variable (choice of normalised bandwidth) is chosen so as to produce equal and opposite effects due to GVD and the optical Kerr effect on a standard pulse of unit width. In normalised form becomes the solution to (13.1)

$$E(t) = A \sec h\left(\frac{t}{\tau}\right) \exp(i\beta z), \tag{13.4}$$

where τ represents the pulsewidth. To have a suitable solution, certain relationships must be verified between the parameters. In particular, the peak power P_{sol}, pulsewidth () and fiber dispersion D are related by

$$P_{sol} = \frac{0.766\lambda^3 D A_{eff}}{\pi^2 c n_2 \tau_{FW}}, \tag{13.5}$$

in which case the soliton is indeed a balance between non-linearity ($P_{sol}\, n_2$) and dispersion (D/τ^2) and propagates without change in shape.

This exact mathematical balance between dispersion and non-linearity can be investigated numerically, revealing that pulses with powers above the soliton period initially compress, before the excessive spectral broadening leads to enhanced temporal dispersion [16].

Besides the exact solution, the inverse scattering also reveals several other 'higher-order' solitons, related to the fundamental soliton [16]. Such higher-order solitons have complex pulse shapes, which evolve with propagation along a fiber. However, the pulse shape repeats periodically, and the length scale of this repetition is known as the soliton period. The required power P_N for these higher-order solitons is given by

$$P_N = N^2 P_{sol}, \tag{13.6}$$

Together with the CW solution, they represent all the eigenvalues of the propagation equation. Intrinsically, any optical field can be expressed as a linear superposition of these eigenvalues. However, due to the complexity of the higher eigenvalues, this approach is rarely adopted.

Finally, the constraint on the exponential factor of (13.4) also indicate a characteristic length scale for such waveform. This is directly related to the soliton period for higher-order solitons, and by convention this soliton period is normally used. For a fundamental soliton (i.e. $N = 1$), the period z_0 is given by

$$z_0 = \frac{0.322\pi^2 c \tau_{FW}^2}{\lambda^2 D}. \tag{13.7}$$

13.3.1 Basic limitations on soliton transmission

The general expression of the soliton waveform (13.4) represent an exact solution of the propagation equation. However, in practical cases, different aspects cause it to depart from from ideality: soliton-to-soliton interactions, periodic behavior of the waveform amplitude due to localised amplification, the Gordon-Haus effect relating to the addition of amplifier noise.

Soliton-to-soliton interactions: the effect of jitter

A modulated pulse stream used in communications, having the form of a series of pulses (13.4), like

$$S(t) = \sum_n A_n e^{i\beta z} \sec h\left(\frac{t - nt_{spacing}}{\tau}\right), \tag{13.8}$$

does not represent a solution of the propagation equation. In equation (13.8) A_n $(= 0, 1)$ represents the data stream, and $t_{spacing}$ represents the pulse spacing with time. It has been shown [17] that neighboring pulses may be considered in terms of interaction forces that decrease exponentially with separation, and vary sinusoidally with phase. These interaction forces cause soliton pulses to attract each other, creating a pattern-dependent arrival time jitter.

This effect, analysed both theoretically [18] and experimentally [19], leads to a significant result: pulses initially in phase will periodically collapse and separate with a period given by

$$L_{collapse} = \frac{2\pi z_0}{4\sec h(1.76T/2\tau_{FW}}, \tag{13.9}$$

where T represents the pulse spacing.

Since the pulse sequence is formed by random data, interaction forces must simply be avoided by spacing the pulses sufficiently to ensure that the resultant jitter is minimised. Usually it is satisfactory to impose the condition that T is large enough to ensure the collapse period is at least four times the system length.

Soliton power requirements: the effect of localised amplification

Since fiber loss has to be compensated, optical amplification has to be used. In principle, two types of amplification could be adopted: distributed amplification using the stimulated Raman process, and localised amplification EDFAs. If a truly distributed amplification were used, fiber losses would be compensated exactly, and a soliton could be transmitted in the fiber for practically infinite distance. However, if the Raman pump is periodically injected into the fiber, the Raman gain changes periodically along the transmission distance of solitons, creating periodic variations of the soliton width and inducing interactions with adjacent solitons [16].

Anyway, in practical cases, localised amplification is mostly used. This leads to a periodic evolution of the pulse amplitude and also to the accumulation of spontaneous emission along the transmission system. Besides noise, which will be considered later, if the amplitude deviates from the soliton amplitude, the pulse adjust itself to form a new soliton. For instance, if the power is increased from P_0 to $(1 + \Delta)P_0$ by local amplification, a new soliton is generaetd with the asymptotic power amplitude given by $(1 + 2\Delta)P_0$, and the rest of the power, $\Delta^2 P_0$, is lost to a linear dispersive wave.

One method for transmitting solitons in an EDFA amplified system has been reported [20]. In fact, if a pulse is launched with an initial amplitude up to 50% greater than the soliton power (for a lossless fiber), then soliton transmission is possible in a real fiber, provided the power never falls below 150% of the required soliton power. This technique, named pre-emphasis or dynamic soliton control, places a limit on the amplifier spacing for soliton transmission due to the large but finite soliton dynamic range (i.e. the range in which soliton transmission is possible).

Another practical technique is based on the assumptions that, provided

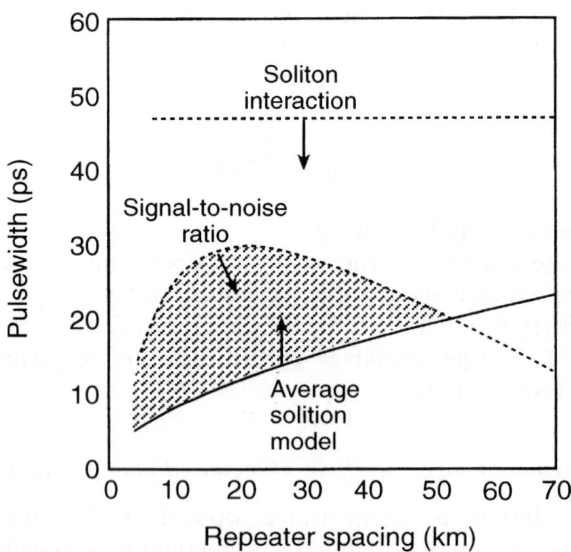

Fig. 13.1 Soliton system design. The system conditions are: bit rate = 5 Gb/s, length = 10 000 km, amplifier noise figure = 6 dB, D = 0.5 ps/nm/km. The shaded portion of the diagram indicates the operating area.

any perturbation to the soliton power, fiber dispersion or non-linear coefficient occurs periodically with a length scale substantially less than the soliton period, the soliton will behave according to the average values of these parameters. This principle, known as the average soliton model, has been verified in many experiments e.g. in [21, 22]. Consequently, it is asserted that, for stable soliton transmission, periodic disturbances to the transmission path are allowed, provided the length scale is less than one-sixth of the soliton period z_0.

Therefore, by considering the pulsewidth and the extinction ratio of a data sequence, the mean output power of the amplifiers and the S/N ratio can be evaluated. Three major constraints are placed on the system: low pulsewidth to avoid interactions hence ISI; high launch powers to allow a good S/N ratio; and short amplifiers spacings to allow average solitons. Under these constraints, it is possible to compose a design diagram (Fig. 13.1) where the shaded region depicts the allowed values of the pulsewidth and amplifier spacing that fulfil all three criteria, for a given bit rate and system configuration.

Impact of noise in soliton transmission: the Gordon-Haus effect

So far, amplifier noise has been considered as a linear addition to the signal. Unfortunately, non-linear interactions between signal and noise affect the transmission performance of systems [23].

Amplifier noise may be added to both the amplitude and phase of the soliton. In the first case, the soliton peak power is modified, and provided the soliton remains within its dynamic range, a new soliton corrects with slightly different energy. However, this process can be schematised as the simple addition of noise. On the other hand, phase noise has a significant effect on these systems. The intermingling of phase noise and chromatic dispersion causes changes in the pulse arrival times. The random nature of this effect is something like a timing jitter and is known as Gordon-Haus jitter. The magnitude of this jitter is given as follows [22]:

$$< \delta t >^2_{rms} = \frac{4.14 \cdot 10^{-6} L^3_{sys} D(G-1)k_{out}n_{sp}(1 - e^{-\alpha L_\alpha})}{\alpha L^2_\alpha A_{eff}\tau} \tag{13.10}$$

where L_{sys} is the overall system length (km), D is the fiber dispersion (ps/nm/km), G is the amplifier gain, kout the output coupling loss, n_{sp} is the inversion factor, α is the fiber loss (km), L_a is the amplifier spacing (km), A_{eff} is the effective area of the transmission fiber (μm^2) and τ is the FWHM pulsewidth (ps). This gives rise to the famous maximum bit rate-distance product limit for soliton systems.

A new design diagram can be built, as shown in Fig. 13.2 [22]. Soliton-soliton interactions become more important at higher capacities, since

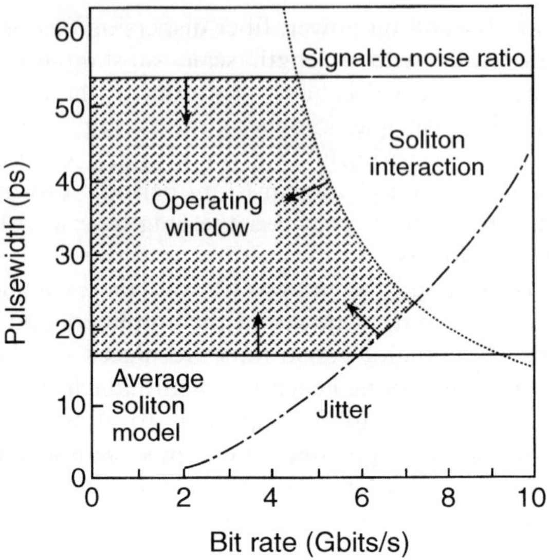

Fig. 13.2 Taking into account the Gordon-Haus effect. The system conditions are length = 7000 km, amplifier noise figure = 5.5 dB, D = 0.5 ps/nm/km, amplifier spacing = 33 km. The shaded portion of the diagram indicates the operating area.

reducing the soliton spacing will increase the interaction strength and significantly reduce the collapse period. The S/N ratio remains unperturbed as the capacity is increased, since both the noise power and the signal power per bit remain constant. However, the reduced bit period increases the severity of the Gordon-Haus jitter, as the width of the allowed time slot is reduced.

13.3.2 Soliton control: overcoming the Gordon-Haus limit

The limitation imposed by the Gordon-Haus jitter can be reduced by the introduction of filters periodically in a fiber link [24, 25], and the dynamic range can also be increased. Let's consider a system in which an optical filter is introduced after any EDFA, whose bandwidth is approximately 10 times larger than the soliton bandwidth. The filters push solitons that may have drifted away in frequency back towards the filter center frequency. The corrective action of the filter is shown in Fig. 13.3. As the carrier frequency deviates from the center frequency of the filter, parts of the spectrum farther away from the center frequency experience greater attenuation than parts nearer the center. So the spectrum is pushed towards the center of the filter response. The carrier frequency exposed to the noise driving sources does not experience a random walk since the

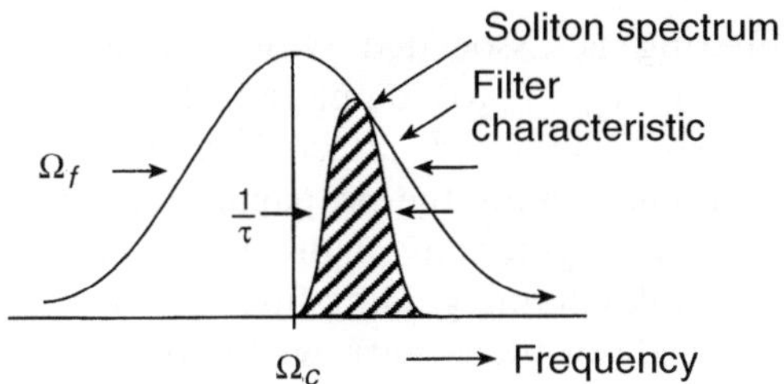

Fig. 13.3 Corrective action of the filter.

filter limits the deviation. This reestablishing force in the spectral domain therefore reduces the Gordon-Haus effect. The filters also reduce the effect of the S/N deterioration due to spontaneous emission noise by filtering out a good portion of the noise itself. As an example, Fig. 13.4 pots the range of distances reachable using filters [14]. The range of permissible powers for transatlantic and transpacific distances is now greatly increased.

Another benefit of the filters is that they provide stabilisation against excessive energy changes of the solitons as they propagate along the fiber link. An increase in the soliton energy above the design average shortens the soliton and broadens its spectrum. Pulses with a broader spectrum experience excess loss, therefore energy increases are reduced by filtering. Energy decreases are similarly contrasted. This effect is particularly advantageous when solitons are wavelength division multiplexed [15].

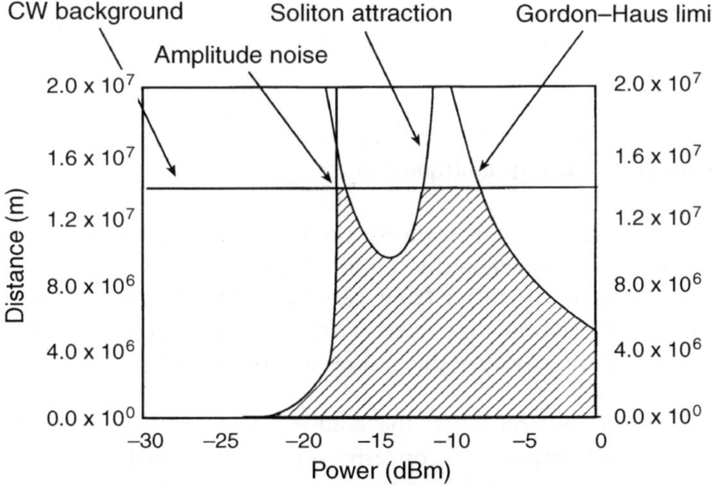

Fig. 13.4 Transmission distance of a transatlantic soliton link with filters for a bit error rate of 10^{-9}. The filter bandwidth is assumed to be 0.72 ps^{-1}.

Nevertheless, filtering is associated with a noise penalty, since the solitons require increased gain to compensate for the loss of the filters. Noise at the center frequency is not affected by the filters and sees excess gain. This noise eventually limits the propagation distance. However, this effect can be eliminated by gradually changing the center frequency of the filters along the fiber link (sliding guiding filters) [26].

Notice that the cascade of filters narrows the overall optical bandwidth, which limits the maximum bit rate.

13.3.3 The effect of polarisation

An important issue relating to soliton propagation is the effect of the fiber birefringence on soliton propagation. If the birefringence were fixed and did not vary randomly along the fiber, the effect would be severe. When two polarisations are coupled by birefringence, they are described by two non-linear Schrödinger equations that in general are not integrable. Thus, it could be expected that soliton propagation would be possible only in polarisation-maintaining fibers, which are much more expensive and lead to higher losses with respect to normal fibers. Fortunately, it is possible to demonstrate [27] that the coupled non-linear Schrödinger equations are integrable, and give rise to solitons of arbitrary polarisation [28].

Polarisation hole burning is another important effect that is related to the saturation properties of EDFAs [29, 30]. Nominally, the amplifier gain is polarisation insensitive. However, if one polarisation saturates the amplifier, a slight excess gain is left over in the other polarisation. Noise can grow in this polarisation and affect the BER. This effect can be overcome [31] by varying the input polarisation at a rate faster than the relaxation rate of the EDFA.

13.3.4 Soliton wavelength multiplexing

The stability of individual solitons permits the passage of streams of solitons at different wavelengths. Unlike their linear counterparts, soliton systems are predicted to suffer from very little interchannel crosstalk. This is because the interaction time is reduced (narrow pulses) and when inter-actions (or collisions) occur, the pulses are predicted to suffer negligible variation in shape.

After each pair of collisions, the solitons recover their original shape. Unfortunately, an important concern is associated with soliton interactions. As two solitons at different frequencies approach, they create poten-tial wells for each other, and the potential wells increase the relative velocity between the two solitons during their advance. After passage through each other, the velocity is decreased, again due to mutual attraction. The net change of velocity is zero, but the integral of the velocity is not. This effect would be disastrous, were it not for the fact that in the process of many collisions, the solitons in a particular wavelength channel experience this net displacement collectively and are therefore displaced collectivel; collective displacement is not a cause of error. But because the random code in each of the channels leaves net displacements uncompensated, some errors can be introduced.

The introduction of the filters decreases the frequency change during collision and thus decreases the errors introduced by this mechanism. A periodic filter, such as a Fabry-Perot filter, may be used to provide a guiding filter function, besides restricting the wavelength shifts due to deviations from the average soliton model over each collision. Control of the source wavelength is required to align the pulse sequences to the passband of the filters.

Partial regeneration may also be employed in multichannel soliton systems, where full control of all sources of timing error (jitter from ideal interactions and breakdown of the average soliton model during collision) will be achieved. However, in order to avoid the use of a large number of wavelength multiplexers and modulators at each retiming stage, the wavelength spacing should be arranged such that the propagation velocity difference produces a difference in arrival time equal to an integral number of bit periods. This must be true for all retiming stages, so besides careful source wavelength control, the total fiber dispersion must also be controlled for each fiber span.

13.3.5 Ultrahigh-speed soliton systems

Figure 13.2 shows that as the data rate is increased, soliton interactions and the soliton period impose the major limitations for soliton systems. In fact, an increase in the bit rate demands a reduction in the pulse spacing, with an enhancement of interactions and a decrease in the pulsewidth, hence a decrease in the soliton period. As expressed in (13.7), the soliton period is proportional to the square of the pulsewidth divided by the dispersion; whereas from (13.9) the interaction length, depend on the relative spacing between pulses.

Consequently, the following two conditions have to be fulfilled for the design of a soliton system:

$$z_0 > 6L_{amp}, \tag{13.11}$$

$$l_{Collapse} > 4L_{sys} \tag{13.12}$$

where L_{amp} is the amplifier spacing and L_{sys} is the system length.

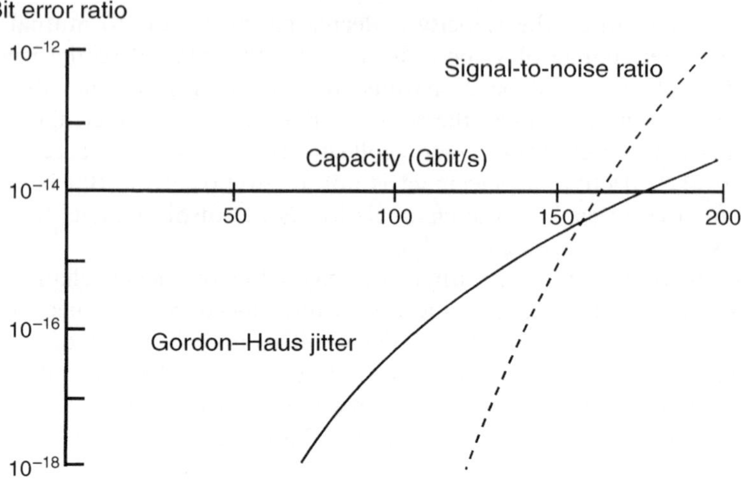

Fig. 13.5 Ultimate performance limits for a single-channel soliton system: the assumed length is 1000 km.

The second condition leads to a maximum allowed pulsewidth as a function of the bit rate, whereas the first one leads to a maximum dispersion coefficient.

These relationships allow the performance degradation to be plotted as a function of the bit rate. In fact, Fig. 13.5 shows the ultimate performance limits for a 1000 km single-channel soliton system [16]. Notice that an ultimate maximum capacity of 140 Gb/s for an error ratio of 10^{-15} is due to Gordon-Haus jitter at this distance.

And notice that in ultrahigh-speed transmissions near at zero dispersion, other effects impair the performance, such as higher-order dispersion and Raman self-frequency shift. As a matter of fact, different parts of the pulse experience different dispersions (higher-order dispersion) and energy transferring from one spectral component to another, via the Raman effect, leads to a wavelength change (Raman self-frequency shift).

REFERENCES

1. P. Trischitta *et al.*, The TAT-12/13 cable network, *IEEE Commun. Mag.* **34**, 24–28, (199&).
2. N.S. Bergano and C.R. Davidson, Wavelength division multiplexing in long-haul transmission systems, **14**, 1299–1308 (1996).
3. A.R. Chraplyvy, A.H. Gnauck, R.W. Tkach, and R.M. Derosier, 8 × 10 Gb/s transmission through 280 km of dispersion-managed fiber, *IEEE Photon. Technol. Lett.* **5**, (1993).

4. N.S. Bergano *et al.*, 40 Gb/s WDM transmission of eight 5 Gb/s data channels over transoceanic distances using the conventional NRZ modulation format, in *Proc. OFC '95* San Diego CA, paper PD19 (1995).

5. A.H. Gnauck *et al.*, Transmission of 820 Gb/s channels over 232 km of conventional fiber, in *Proc. OFC '95* San Diego CA, paper PD23 (1995).

6. C.R. Giles and E. Desurvire, Propagation of signal and noise in concatenated erbium-doped fiber amplifiers, *IEEE J. Lightwave Technol.* **9**, (1991).

7. E. Lichtmann, Performance degradation due to polarization dependent gain and loss in lightwave systems with optical amplifiers, *Electron. Lett.* **29**, (1993).

8. F. Bruyere and O. Audouin, Penalties in long-haul optical amplifier systems due to polarization dependent loss and gain, *IEEE Photon. Technol. Lett.* **6**, (1994).

9. N.S. Bergano, V.J. Mazurczyk, and C.R. Davidson, Polarization hole-burning in erbium-doped fiber-amplifier transmission systems, in *Proc. ECOC '94*, Florence, Italy (1994).

10. C.R. Giles, E. Desurvire, and Simpson, Transient gain and crosstalk in erbium-doped fiber amplifier, *Opt. Lett.* **14**, (1989).

11. N.S. Bergano, The time dynamics of polarization hole-burning in an erbium-doped fiber amplifier, in *Tech. Dig. OFC '94* San Jose CA, paper FF4 (1994).

12. E. Lichtmann, Limitations imposed by polarization-dependent gain and loss on all-optical ultra-long communication systems, *IEEE J. Lightwave Technol.* **13**, (1995).

13. M.G. Taylor and S.J. Penticost, Improvement in performance of long haul EDFA link using high frequency polarization modulation, *Electron. Lett.* **30**, (1994).

14. H.A. Haus, Optical fiber solitons, their properties and uses, *Proc. IEEE* **81**, 970–983 (1993).

15. H.A. Haus and W.S. Wong, Solitons in optical communications, Rev. Mod. Phys. **68**, 423–444 (1996).

16. A.D. Ellis *et al.*, Nonlinear Propagation Effects, in *High capacity optical transmissions explained,* edited by D.M. Spirit and M.J. O'Mahony, John Wiley, New York (1995).

17. J.P. Gordon, Interaction forces mong solitons in optical fibers, *Opt. Lett.* **8**, 596–598 (1983).

18. V.I. Karpman and V.V. Solov'ev, A perturbational approach to the two soliton system, Physica D **3**, 487–502 (1981).

19. F.M. Mitschke and L.F. Mollenauer, Experimental observation of interaction forces between solitons in optical fibers, *Opt. Lett.* **12**, 355–357 (1987).

20. H. Kubota and M. Nakazawa, Long distance optical soliton transmission with lumped amplifiers, *IEEE J. Quantum Electron.* **26**, 692–700 (1990).

21. K.J. Blow and N.J. Doran, Average soliton dynamics and the operation of soliton systems with lumped amplifiers, *IEEE Photon. Technol. Lett.* **3**, 369–371 (1991).

22. A.D. Ellis *et al.*, 5 Gb/s soliton propagation over 350 km with large periodic dispersion coefficient perturbations using erbium doped fibre amplifier repeaters, *Electron. Lett.* **27**, 878–879 (1991).

23. J.P. Gordon and H.A. Haus, Random walk of coherently amplified solitons in optical fiber transmission, *Opt. Lett.* **11**, 665–667 (1986).

24. A. Mecozzi, J.D. Moores, H.A. Haus, and Y Lai, Soliton transmission control, *Opt. Lett.* **16**, 1841–1843 (1991).

25. Y. Kodama and A. Hasegawa, Generation of asymptotically stable optical solitons and suppression of the Gordon-Haus effect, *Opt. Lett.* **17**, 31–33 (1992).

26. L.F. Mollenauer, J.P. Gordon, and S.G. Evangelides, The sliding-frequency guiding filter: an improved form of soliton jitter control, *Opt. Lett.* **17**, 1575–1577 (1992).

27. S.V. Manakov, *Zh. Eksp. Teor. Fiz.* **65**, 1394 Sov. Phys. JETP **38**, 248–253 (1974).

28. P.K. Way, C.R. Menyuk, and H.H. Chen, Stability of solitons in randomly varying birefringent fibers, *Opt. Lett.* **16**, 1231–1233 (1991).

29. M.G. Taylor, *Optical Solitons: Theory and Experiment*, Cambridge University Press, Cambridge (1992).

30. M.G. Taylor, Observation of new polarization dependence effect in long haul optically amplified system, *IEEE Photon. Technol. Lett.* **5**, 1244–1246 (1993).

31. F. Heismann *et al.*, Electrooptic polarization scramblers for optically amplified long-haul transmission systems, *IEEE Photon. Technol. Lett.* **6**, 1156–1158 (1994).

Index